PROCEEDINGS OF THE SYMPOSIA ON

ELECTROCHEMICAL PROCESSING IN ULSI FABRICATION I and INTERCONNECT AND CONTACT METALLIZATION: MATERIALS, PROCESSES, AND RELIABILITY

Editors

P. C. Andricacos
IBM T. J. Watson Research Center
Yorktown Heights, New York

J. O. Dukovic
IBM T. J. Watson Research Center
Yorktown Heights, New York

G. S. Mathad
Siemens Microelectronics
Hopewell Junction, New York

G. M. Oleszek
University of Colorado
Colorado Springs, Colorado

H. S. Rathore
IBM Microelectronics
Hopewell Junction, New York

C. Reidsema Simpson
Motorola, Inc.
Austin, Texas

DIELECTRIC SCIENCE AND TECHNOLOGY, ELECTRONICS, AND ELECTRODEPOSITION DIVISIONS

Proceedings Volume 98-6

THE ELECTROCHEMICAL SOCIETY, INC.,
10 South Main St., Pennington, NJ 08534-2896, USA

Published by:

The Electrochemical Society, Inc.
10 South Main Street
Pennington, New Jersey 08534-2896, USA

Telephone (609) 737-1902
Fax (609) 737-2743
e-mail: ecs@electrochem.org
Web site: http://www.electrochem.org

Library of Congress Catalogue Number: 99-62262

ISBN 1-56677-200-1

Printed in the United States of America

PREFACE

The Symposium on **Electrochemical Processing in ULSI Fabrication I** was held on May 4 and 5, 1998 in San Diego, California in the context of the 193th meeting of the Electrochemical Society. It was the first of a series of symposia to be held annually during subsequent Spring meetings of the Society.

The goal of this symposium was to capture from the beginning the explosive growth that electrochemical processing is experiencing as a result of the immense opportunities that semiconductor fabrication offers, as witnessed by the recent emergence of electroplating as the process of choice for copper deposition in on-chip interconnections. Another goal of the symposium was to bring together practitioners from all research aspects of electrochemical processes, from the most fundamental to the most applied. We thought that this kind of integrated approach to information exchange would help dispel myths about the predominantly empirical nature of electrochemical processes.

We are grateful to all the authors for their in-depth treatment of their subjects and for contributing to the success of the symposium. We also thank the Society staff for their assistance in the publication of this volume.

March 1999.

Symposium Organizers:

Panos C. Andricacos
Cindy Reidsema Simpson
Gerald M. Oleszek

This symposium on **Interconnect and Contact Metallization: Materials, Processes, and Reliability** was held in Boston during the 194th meeting of The Electrochemical Society, November 1-6, 1998. The symposium was sponsored jointly by the Dielectric Science & Technology, Electronics, and Electrodeposition divisions of the Society. The symposium papers are grouped as follows:

- Low-k Materials
- Copper Interconnects
- Interconnect Processes & Reliability

The aim of the symposium was to provide a forum for scientists and engineers active in the area of metal interconnections to present their recent work, discuss concerns and identify areas for future activities. It has been known for sometime that interconnections ultimately decide the performance of the final fully integrated circuit. From various low-k material formulations to copper interconnect systems to new contact metallurgies, the interconnect field is teaming with research, development, and manufacturing activities worldwide. With the anouncement of copper wiring by IBM, Motorola, AMD in their logic products, a new dimension has been added, giving credence and confidence to the viability of copper as an interconnection metallurgy at the device level. The papers presented in this symposium reflect this attitude. It is certain that future interconnect symposia will report many significant results in this area of copper and low-k dielectric materials.

The organizers would like to thank, first of all, the authors for their presentations, the attendees for their participation, and the staff of the Society for their assistance in organizing the meeting and the publication of this proceedings volume.

March 1999.

Symposium Organizers:

G.S. Mathad
H.S. Rathore
J.O. Dukovic
C. Reidsema-Simpson
D.D. Snyder
R.K. Ulrich

TABLE OF CONTENTS

SYMPOSIUM ON ELECTROCHEMICAL PROCESSING IN ULSI FABRICATION

SYMPOSIUM ON INTERCONNECT & CONTACT METALLIZATION

Low-k Materials

Copper Interconnects

Interconnect Processes & Reliability

FACTS ABOUT THE ELECTROCHEMICAL SOCIETY, INC.

The Electrochemical Society, Inc., is an international, nonprofit, scientific, educational organization founded for the advancement of the theory and practice of electrochemistry, electrothermics, electronics, and allied subjects. The Society was founded in Philadelphia in 1902 and incorporated in 1930. There are currently over 7,000 scientists and engineers from more than 70 countries who hold individual membership; the Society is also supported by more than 100 corporations through Contributing Memberships.

The Technical activities of the Society are carried on by Divisions and Groups. Local Sections of the Society have been organized in a number of cities and regions. Major international meetings of the Society are held in the Spring and Fall of each year. At these meetings, the Divisions and Groups hold general sessions and sponsor symposia on specialized subjects.

The Society has an active publications program which includes the following:

Journal of The Electrochemical Society - The *Journal* is a monthly publication containing technical papers covering basic research and technology of interest in the areas of concern to the Society. Papers submitted for publication are subjected to careful evaluation and review by authorities in the field before acceptance, and high standards are maintained for the technical content of the *Journal*.

Electrochemical and Solid-State Letters - *Letters* is the Society's rapid-publication, electronic journal. Papers are published as available at http://www3.electrochem.org/letters.html. This peer-reviewed journal covers the leading edge in research and development in all fields of interest to ECS. It is a joint publication of the ECS and the IEEE Electron Devices Society.

Interface - *Interface* is a quarterly publication containing news, reviews, advertisements, and articles on technical matters of interest to Society Members in a lively, casual format. Also featured in each issue are special pages dedicated to serving the interests of the Society and allowing better communication among Divisions, Groups, and Local Sections.

Meeting Abstracts *(formerly Extended Abstracts)* - Meeting Abstracts of the technical papers presented at the Spring and Fall Meetings of the Society are published in serialized softbound volumes.

Proceedings Series - Papers presented in symposia at Society and Topical Meetings are published as serialized Proceedings Volumes. These provide up-to-date views of specialized topics and frequently offer comprehensive treatment of rapidly developing areas.

Monograph Volumes - The Society sponsors the publication of hardbound Monograph Volumes, which provide authoritative accounts of specific topics in electrochemistry, solid-state science, and related disciplines.

For more information on these and other Society activities, visit the ECS Web site:

http://www.electrochem.org

CHALLENGES IN THE EXTENSION OF HIERARCHICAL WIRING SYSTEMS

Thomas N. Theis
IBM T.J. Watson Research Center
P.O. Box 218, Yorktown Heights, NY 10598

Copper wiring and strongly hierarchical wiring schemes are capable, for now, of meeting performance needs for on-chip interconnections in CMOS logic. Based on a performance scaling scenario, future wiring demands can be estimated. Meeting these demands for 0.1 micrometer ground rules and beyond will pose many challenges for further improvement of the electrolytic-plating process which is used to fabricate the wiring.

INTRODUCTION: COPPER INTERCONNECTIONS

IBM recently announced a six-level copper wiring process as a feature of its CMOS 7S logic technology,(1) and the technology is now in full-scale manufacturing. Along with the complete replacement of aluminum by copper for on-chip wiring, the technology represents a number of additional firsts in microelectronics. The industry-standard subtractive metal etch process has been replaced by a damascene (metal inlay) process -- wire patterns (trenches and via holes) are etched in an insulator, then filled with copper, and the excess copper is then removed by chemical-mechanical planarization. For each wire level, both the via and trench structures are filled in a single step -- the first use of a dual-damascene process for chip wiring. Furthermore, the copper is electrolytically deposited. With a minimum wiring pitch of 0.63 micrometers and aspect (height-to-width) ratios of the combined via and trench structures on the order of 3 to 4, CMOS 7S marks the first industrial fabrication of deep sub-micrometer-scale features by an electrochemical process.

Copper interconnections fabricated by a dual-damascene process offer advantages of performance, cost, and reliability over existing aluminum wiring processes.(1) Performance is gained because the resistivity of copper is approximately 40% lower than that of aluminum, so that copper wires exhibit 40% lower RC delay than aluminum wires of the same cross-section. Cost reduction comes from the elimination of some process steps and the simplification of other process steps in the dual-damascene process. Reliability is improved because the electrolytically-deposited copper, when compared to aluminum, exhibits far less electromigration and far less stress-migration. A detailed discussion of these advantages can be found in reference 1.

A representative cross-section of CMOS 7S wiring is shown in the scanning electron micrograph in figure 1. The use of dual-damascene metal fill is evidenced by the absence of a seam between each copper wire layer and the via layer immediately

below. The liner isolating the copper from the surrounding insulator is difficult to discern in the figure because it is very thin compared to the wire widths. Perhaps the most striking feature is the dramatic difference in size between the bottom and top levels of the wiring stack. Minimum contacted pitch is 0.63 micrometers at the high-density first copper level, 0.81 micrometers at succeeding levels, and the pitch and thickness of the fifth and sixth levels can be optionally scaled by a factor of 2x, as shown in the figure, for low RC delay.

HIERARCHICAL WIRING

CMOS 7S is thus an example of a hierarchical wiring system in which successive wire levels at increasing thickness and width enable long wire runs with low RC delay. Increasing the height and width of a wire and thickness of surrounding insulators, all by a factor, λ, leaves capacitance per unit length unchanged while resistance per unit length is reduced by a factor of $1/\lambda^2$. In principle, RC delay can be reduced to arbitrarily low values by implementation of such "fat wires". This is sometimes referred to as reverse scaling.

A more complete model of RC delay in a logic circuit includes the effective internal resistance, R_t, of a driver transistor and the input capacitance, C_t, of the transistors that form the load, as well as the lumped resistance, R_w, and lumped capacitance, C_w, of the connecting wire. The total delay of the circuit is approximated by a weighted sum of delay terms of the form $R_w C_w, R_w C_t, R_t C_w$, and $R_t C_t$, with coefficients that depend on the circuit that is modeled.(2) Thus, scaling wire cross-sectional dimensions by λ while leaving wire length and transistor dimensions fixed causes $R_w C_w$ and $R_w C_t$ to decrease as $1/\lambda^2$while the third wire-related delay term, $R_t C_w$ is unchanged. This last term, if significant, can be reduced by use of a "wide" driver transistor which costs little in terms of additional chip area. So in this simple model circuit, all wire-related RC delays can easily be reduced. In general, the combination of reverse scaling and appropriate circuit design techniques prevents interconnect delays from overwhelming transistor delays in current microprocessor designs. Since a relatively small fraction of the wires represent long or critical paths, the cost of the additional metal area required to implement fat wires has so far been acceptable.

Will this situation continue? Can an existing hierarchical wiring system be scaled to future transistor densities, with acceptable wire-related RC delay at acceptable cost? First, the minimum wiring pitch must continue to scale with transistor dimensions. The scaling rate can be estimated from historical data as shown in figure 2. At the same time, an ever-increasing fraction of the total wires on chip will have to be implemented with larger pitches and thicknesses in order to meet RC constraints. To see that this is true, consider a circuit in which all transistor and wire dimensions are shrunk by a factor, $1/s$. Since the length of every wire in the circuit shrinks as $1/s$, R_w for any wire

increases as s, while C_w decreases as $1/s$, and $R_w C_w$ remains constant. However, in the simplest transistor scaling scenario, R_t is constant, C_t decreases as $1/s$ and hence $R_t C_t$ decreases as $1/s$. Hence, some fraction of the wire runs which are acceptable in terms of RC delay before scaling will become unacceptable (compared to transistor switching delays) after scaling. To maintain parity with the improved transistor performance, such wires must be moved up the wiring hierarchy to the next larger pitch. This is awkward because it means that an existing wiring layout cannot be simply shrunk, but must instead be redesigned. Furthermore, the area required for the new wiring layout does not scale as s^2, so either the circuit area also fails to scale as s^2, or additional wiring levels must be added to contain those wires with pitches that do not scale. Since these fat wires require more area than traditional wires implemented at, or close to, minimum lithographic width, aggressive implementation of hierarchical wiring could drive the addition of wiring levels at a rate above the historical norm and the current industry projections which are shown in figure 3. The problem is compounded if, instead of shrinking an existing design (constant number of transistors), a new design is developed with more transistors and consequently additional long wires that must be implemented at large pitches.

Until recently the introduction of fat wires and truly hierarchical wiring structures has been avoided by the use of modified scaling approaches in which pitch is reduced while aspect ratio is increased (3). However, with current aspect ratios as high as 2,(4) there is little incentive for further increases, at least from the point of view of RC reduction. This is because the well-known contribution of "sidewall" capacitance makes negligible the net reduction in RC (5), and in addition, greatly exacerbates the problem of cross-coupling between adjacent wires. Introduction of low-dielectric-constant (low-ε) insulators can compensate the increase in capacitance due to increasing aspect ratio. For instance, if the dielectric constant is reduced by $1/s$ while aspect ratio is increased by s and wire pitch and transistor dimensions are reduced by $1/s$, it can be readily shown for high-aspect-ratio wires that all wire-related RC components ($R_w C_w, R_w C_t, R_t C_w$) scale with transistor delay, $R_t C_t$. However, reducing the dielectric constant of the insulator does not reduce cross-coupling at all. This must be addressed by greater attention to circuit layout and improved design tools. Since the efficacy of future improvements in circuit layout methodology is unknown, it is difficult to predict the extent of future increases in wire aspect ratios. In the projection of future wiring needs which follows (figure 6 and associated discussion), some increase is allowed, roughly consistent with current Semiconductor Industry Association projections.

Of course, RC constraints are not the only factor driving the evolution of wiring systems. Wires for power distribution must be scaled to limit IR voltage drops, and must also avoid electromigration constraints. Long wires that operate as transmission lines must scale in width as the square root of the clock frequency for wires of constant length, while thickness of transmission lines need not increase at all with clock frequency once the thickness significantly exceeds the skin depth. Transmission line

considerations and power distribution requirements are additional reasons why, at this time, metal thickness and minimum pitch of last- or global- wiring levels are no longer decreasing and in fact are beginning to increase. The combined effects of shrinking the minimum pitch, adding intermediate metal levels and increasing the pitch of global wiring levels is schematically illustrated in figure 4. Wiring systems will become increasingly hierarchical, increasingly three-dimensional, with an increasing disparity between minimum and maximum wire dimensions.

THE WIRING BOTTLENECK

In principle, a hierarchical wiring system allows RC delay contributions of future wiring systems to be scaled to match improvements in device performance, but perhaps at the cost of introducing additional wire levels at a rate above the historical trend(6,7) of roughly 0.75 levels per lithography generation. The introduction of copper wiring (1,8) and, eventually, low-dielectric-constant insulators, is seen by some as a partial solution, containing the proliferation of additional wiring levels for a generation or two before chip-level performance becomes limited by wiring (6). The National Technology Roadmap for Semiconductors estimates that without radical material, design, or architectural innovations, this point will be reached at the 0.1 micrometer generation (9). We now demonstrate that any such estimate is extremely sensitive to some basic assumptions about chip design. In particular, wiring demands are very sensitive to the fraction of chip area that is devoted to random logic, as opposed to more regular arrays such as memory, registers, and so on.

For purposes of this demonstration, the total area occupied by wires on a chip is approximated by

$$A_{tot} = \int w(L) L f(L,N) dL \qquad (1)$$

where $w(L)$ is the wire pitch as a function of wire length, L, and $f(L,N)$ is the wire length distribution function which must depend on N, the total number of transistors. The usual starting point in determining $f(L,N)$ is Rent's Rule (10), an empirical relationship that specifies the number of wires which cross the boundary of a block of circuitry (input/output or I/O wires), K, in terms of the number of transistors or nodes within the block, N, and the number of wires, k, connecting each transistor to other transistors within the block.

$$K = kN^p \qquad (2)$$

The Rent exponent, p, is observed to vary from $0.55 < p < 0.85$, the lower values for highly regular circuits such as memory, and the higher values for random logic. Rent's rule is observed to hold for circuit blocks of widely varying size. If its validity is assumed at all length scales greater than the transistor-to-transistor spacing, then a power-law distribution is obtained for the number of wires as a function of wire length.

For wires on a chip, there is a natural cutoff in this distribution function at wire lengths longer than roughly the length of the chip. This cutoff has been treated with perhaps the greatest degree of rigor by Davis, *et al.* (11), and we use the wire length distribution function derived by these authors. The minimum wire pitch implemented in the first level of wiring, $w_{1,min}$, is always determined by lithography. Once $w_{1,min}$ and the aspect ratio are set, there will be maximum acceptable wire length at minimum pitch, $L_{1,max}$. The maximum run length, $L_{n,max}$, at minimum allowed pitch, $w_{n,min}$, for each higher lying wire level, n, in the wiring hierarchy is chosen to satisfy a constraint on minimum performance of the wiring system. The particular constraint considered here is,

$$L_{n,max} \propto (w_{n,min})^2, \qquad (3)$$

Assuming that aspect ratios and metal resistivity are unchanged between wiring levels, this constraint guarantees that maximum wire resistance is the same at each level in the wiring hierarchy. Thus, as successively longer lines are implemented at successively higher levels in the wiring hierarchy, the maximum $R_w C_w$ delay increases as L, and therefore increases no faster than the delay of a lossless transmission line. The maximum value for $R_t C_w$ also increases as L and the maximum value of $R_w C_t$ is independent of L. These favorable scaling properties ensure that the application of well-known design approaches (repeaters, cascaded drivers) can bring the delay of critical paths close to the physical limit set by transmission line delays. Note, however, that equation 3 is not a unique constraint. A less aggressive reverse-scaling scenario would require less wire area for a given circuit, and might still yield acceptable interconnect delay for a chip design. Thus, the use of equation 3 as a constraint on design of hierarchical wiring systems may be a conservative assumption, leading to an overestimate of the metal area required to implement future systems.

Equations 1 and 3 imply a simple and idealized wiring system. Equation 1 implies that $w(L)$ is a single-valued function of L, whereas in actual wiring systems, designers are generally free to run wires of a given length at various widths equal to or greater than the minimum specified width for each level. Such "wide" wires (as opposed to fat wires) are used to reduce wire resistance when this is a limiting factor (for example, IR drops in power distribution, $R_w C_t$ delays), or to increase wire cross-section where current density is constrained (for example by electromigration). Wide wires are also used to reduce $R_w C_w$ delays, but provide little benefit once the aspect ratio is reduced towards 1 or less. Equation 3 implies that all wires of length greater than $L_{n,max}$ are implemented in wire level $n+1$ or above, whereas in actual wiring systems, not all wires are critical in terms of delay, and the distribution of wire lengths in any particular wire level does not have a sharp cutoff. Thus the wire area calculated using equations 1 and 3 will differ from the wire area of a real wiring system even if the wire distribution function, $f(L, N)$, accurately models the actual wire distribution function. However, we do not use equations 1 and 3 to generate values for the wire area, but rather to calculate the *relative increase* in wire area as a hierarchical wiring system

is implemented at successively smaller minimum lithographic dimensions while simultaneously adding wire levels and pitches to maintain the RC scaling scenario discussed above. Our results will be most accurate if wiring practices, such as the relative use of wide versus fat wires, do not change from lithography generation to lithography generation. Our underlying assumption is that if we underestimate the wiring area for present generations, we underestimate it for future generations as well by the same factor, and the relative increase in wire area from generation to generation is accurately estimated. Actually, this underlying assumption is probably unduly pessimistic. With more levels available in future wiring systems, and with better circuit design tools able to optimize the use of the hierarchical levels, we may expect *less* use of wide wires, more area-efficient designs, and thus *less* relative increase in wiring area than is projected by this simple model.

It would then be straightforward to choose values of $L_{n,max}$, $n > 1$, so as to distribute the wire area (equation 1) evenly among a discrete set of higher lying metal levels, and to repeat this procedure iteratively, varying the total number of metal levels so as to optimize the metal fill factor at each level. However, here we avoid this tedious iterative procedure by further idealizing our model wiring system -- we assume that for wires longer than $L_{1,max}$, wire pitch varies smoothly as a function of wire length,

$$L(w) \propto w^2 \qquad \textbf{(4)}$$

so that each wire just satisfies the reverse scaling constraint on resistance as a function of length. Although such an ideal minimum-area wiring system cannot be realized in practice, a hierarchical wiring system becomes a better approximation of the ideal system as more wire levels are added. Again, future wiring will therefore better approximate the ideal system of our model. Again, since we compare future lithography generations to a current generation, the *relative* increase in wire area projected from our idealized model is expected to be pessimistically large.

To establish a baseline for our projections, we pick $L_{1,max}$ roughly appropriate for an existing CMOS generation. The pitch of all runs of length $L \leq L_{1,max}$ is $w_{1,min}$, while the pitch of all runs of length $L \geq L_{1,max}$ is $w = w_{1,min}(L/L_{1,min})^{1/2}$], consistent with equation 4. As $w_{1,min}$ is scaled proportional to the minimum lithographic dimension, $L_{1,max}$ is scaled according to equation 3, so that the maximum wire resistance in the down-scaled wiring system remains constant, and wire RC delay contributions scale with transistor delay. The relative increase in metal area required to wire the system can then be calculated from equation 1 as a function of $w_{1,min}$.

In figure 5 we show the relative increase in the number of wire levels required at each lithography generation for a hierarchical wiring system in which wire RC delay is scaled with transistor performance. In this and the other projections that follow, the value for the Rent exponent is chosen as $p = 0.85$, valid only for random logic, and a

pessimistic estimate even in that case. Each lithography generation corresponds to a full $1/\sqrt{2}$ reduction in all minimum lithographic dimensions or a doubling of transistor density. At lithography generation "zero", the maximum run at the minimum lithographically allowed wire width, is chosen as $L_{1,max} = 0.4N^{1/2}$ (that is, 0.4 chip lengths), a value roughly appropriate for the 0.25 micrometer lithography generation and aluminum wiring. Thus the projection of wiring needs extends about 6 lithographic generations beyond 0.25 microns. (The general results and trends obtained here are insensitive to the precise choice of $L_{1,max}$.) Three scenarios are shown: 1) the number of transistors doubles with each lithographic generation ($N \propto s^2$) to fill a chip of constant size, 2) the number of transistors doubles every two lithographic generations ($N \propto s$) to fill a diminishing fraction, $1/s$,of a chip of constant size, and 3) the number of transistors is constant over lithography generations, filling a rapidly diminishing fraction, $1/s^2$, of a chip of constant size. The author believes that scenarios 2 and 3 are closer to reality than scenario 1, since the largest chips for the most demanding logic applications are being increasingly filled with memory and other regular arrays, while random logic appears to be shrinking in absolute area.

As can be seen, filling a chip of constant size with an exponentially increasing number of transistors soon leads to an explosion in the number of wire levels required to avoid an RC delay bottleneck. If the number of wiring levels at generation "zero" is taken to be 6, then scenario 1 indicates about 9 times as many levels (54 levels!) would be required 4 lithography generations later. On the other hand, the intermediate scenario and the constant transistor count scenarios appear much more manageable.

What is the value of introducing new materials? In figure 6 we show the same three scenarios as in figure 5, but we introduce copper and a low-dielectric constant insulator ($\varepsilon = 2.7$) and keep the aspect ratio constant at lithography generation 2, and introduce a lower-dielectric constant insulator ($\varepsilon = 1.3$) and increase the wire aspect ratio by a factor of $\sqrt{2}$ at lithography generation 4. In this case, the number of wire levels still quickly becomes unmanageable in scenario 1, but remains plausible at least to generation 4 or 5 in scenario 2, and does not increase at all in scenario 3. Note that scenario 3 implies that we could eventually wire 64 present-day microprocessors on a single chip, each surrounded by memory and each core running at a multi-gigahertz clock rate, all with about 6 levels of metal. Perhaps 1 or 2 additional levels of metal could be devoted to a high-speed bus connecting the individual processors to one another? The point is, while a wiring bottleneck must eventually limit progress in the implementation of ever-more-complex, ever-faster, random logic circuits, wiring issues need not halt the exciting progress in microelectronics technology for the foreseeable future.

AN OPTIMISTIC VIEW

There are many reasons to take a guardedly optimistic view of future interconnect technology development. We are no where near the end of the road in terms of materials, processes, and structures. We must learn to integrate low dielectric constant insulators with copper. We can continue, to some extent, to increase aspect ratios as we simultaneously improve our design tools and methodology to avoid the penalties imposed by increased cross-coupling. We can add wire levels in increasingly hierarchical wiring schemes as illustrated in figure 4. We can also expect "straight-ahead" progress in circuit design layout, especially the inclusion of signal delay at all levels of the wiring hierarchy and in all stages of the design process. Beyond such incremental advances, we can foresee innovations such as stacked device structures and active devices in the interconnect levels (three-dimensional integration), and optical links on chip (no faster than electrical transmission lines, but with the potential for much higher bandwidth through wavelength division multiplexing). Such advances are more likely as increased resources are devoted to the "interconnect problem". The Microelectronics Advanced Research Corporation (MARCO),(12) a recently formed subsidiary of the Semiconductor Research Corporation, will greatly increase university funding for long-time-horizon research in architecture and design, as well as materials, processes and structures.

Also, some estimates of future wiring needs make the worst-case assumption that high-performance random logic completely fills chips of ever-increasing area. A more realistic assessment takes account of the fact that, in the most aggressive current microprocessor designs, random logic is actually shrinking while memory is growing as a fraction of total chip area. As discussed above, this significantly delays (but does not eliminate) the need for additional wiring levels.

Finally, we are just beginning to implement fully hierarchical wiring systems and copper/low-epsilon materials. Copper wiring patterned by the dual damascene method offers substantial process simplification, with an accompanying potential for cost reduction and improved manufacturing yields. Thus the addition of many more wire levels, perhaps more than the NTRS estimate of nine in the year 2012, may be economically viable. Certainly, an important and inescapable challenge for future interconnect technology is to continue to articulate the wiring hierarchy, with an ever increasing disparity between minimum pitches of first and last metal levels, and a steadily increasing number of intermediate levels at intermediate pitches -- an increasingly three-dimensional system.

CHALLENGES FOR ELECTROCHEMISTRY

The above-mentioned scenario raises some very interesting challenges for electrochemical technology. Can damascene plating continue to fill line and via structures at sub-0.1 micron dimensions and steadily increasing aspect ratios? Will electromigration and stress-migration improvements be maintained at such dimensions? Can the process

and tools be extended to 300 mm wafers and beyond? Can deposition rates be increased, deposition uniformity be increased, and process costs lowered to facilitate the further addition of thick metal levels? Can the process be modified to take maximum advantage of low-dielectric-constant insulators? The extension of the technology should keep many talented scientists and engineers busy for the foreseeable future.

REFERENCES

1. D. Edelstein, *et al.*, Proc. IEEE IEDM, 773 (1997).

2. H. B. Bakoglu, Circuits, Interconnections, and Packaging for VLSI, Addison - Wesley, 1990, pp. 202-204.

3. *Ibid*, p. 197.

4. S. Yang, *et al.*, Proc. IEEE IEDM, 197 (1998).

5. Bakoglu, p. 140.

6. M. Bohr, Proc. IEEE IEDM, 241 (1995).

7. J. G. Ryan, R. M. Geffken, N. R. Poulin, and J. R. Paraszczak, IBM Journal of Research and Development, 39, 371 (1995).

8. S. Venkatesan, *et al.*, Proc. IEEE IEDM, 769 (1997).

9. The National Technology Roadmap for Semiconductors: Technology Needs, pp. 101-102, Semiconductor Industry Association (1997).

10. W.E. Donath, IBM Journal of Research and Development, 25, 152 (1981).

11. J. A. Davis, V. K. De, and J. D. Meindl, IEEE Transactions on Electron Devices, 45, 580 (1998).

12. http://marco.fcrp.org

FIGURES

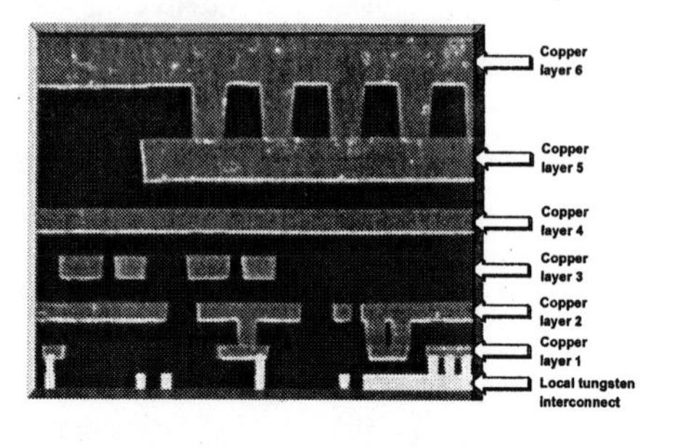

Figure 1. Cross-sectional scanning electron micrograph showing typical CMOS 7S interconnections with tungsten local interconnect and six levels of copper wiring.

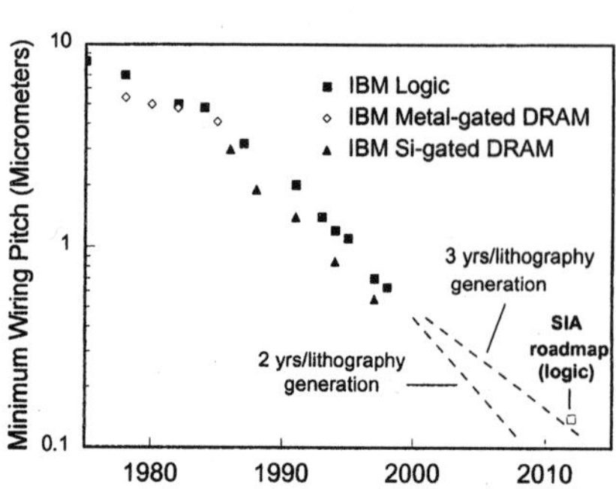

Figure 2. Minimum wire pitch used in IBM DRAM and CMOS logic technologies versus year of introduction, and extrapolation of the current scaling trend into the future. (After J.G.Ryan *et al.,* IBM J. Res. Develop. 39, 371 (1995).)

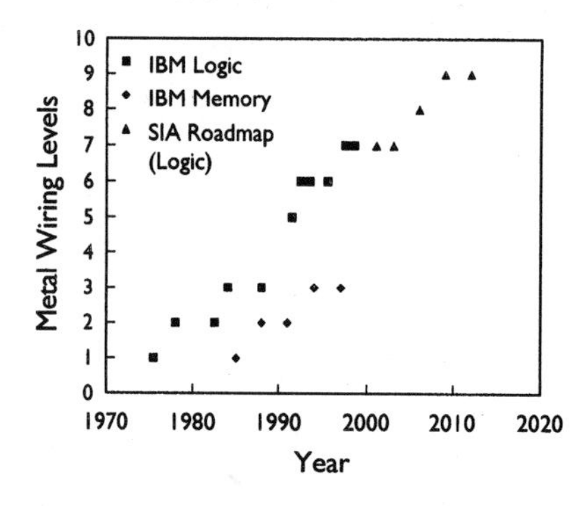

Figure 3. Number of wire levels used in IBM DRAM and CMOS logic technologies versus year of introduction (includes tungsten local interconnect), and SIA roadmap values for future years. (After J.G. Ryan *et al.*, IBM J. Res. Develop. 39, 371 (1995).)

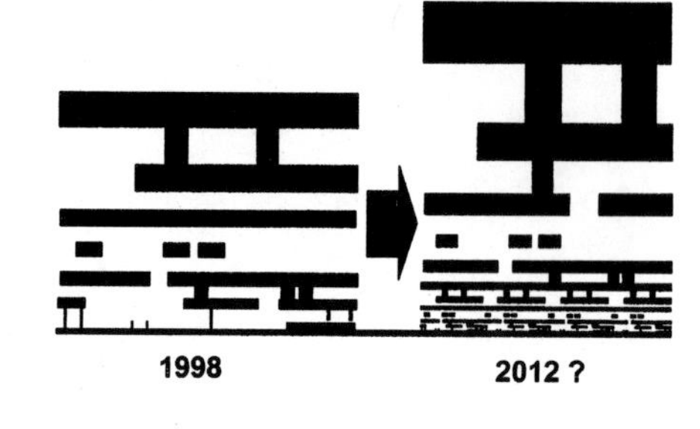

Figure 4. Schematic illustration of the likely evolution of interconnect architecture for high performance CMOS logic.

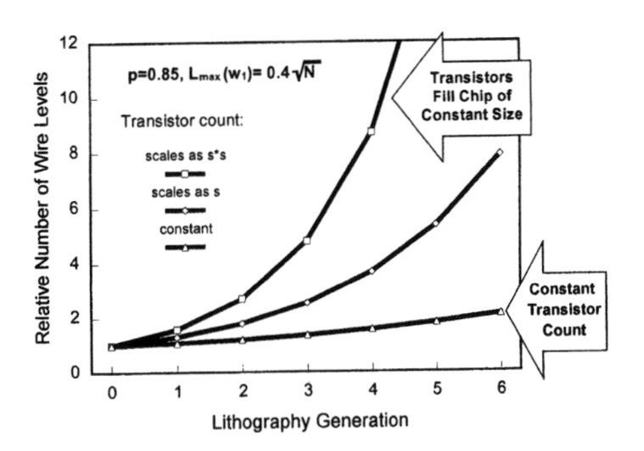

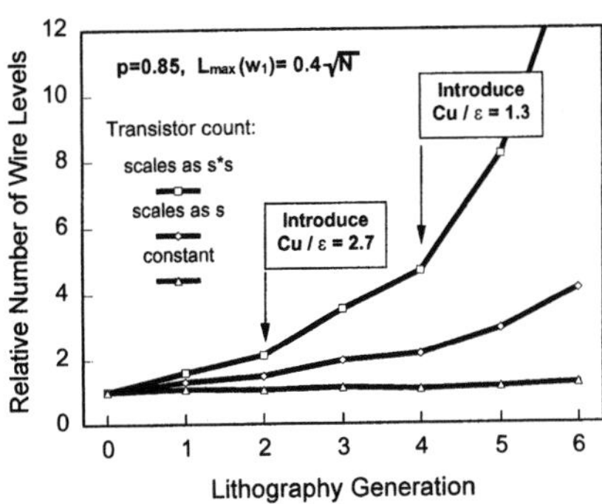

Figure 5. Relative number of wire levels required at each lithography generation for a hierarchical wiring system in which wire RC delay is scaled with transistor performance.

Figure 6. Relative number of wire levels required at each lithography generation as in Figure 5, but with the staged introduction of copper and insulators with successively lower dielectric constants.

COPPER ELECTROPLATING FOR ON-CHIP METALLIZATION

Valery Dubin*, Sergey Lopatin, Robin Cheung

Advanced Micro Devices
Sunnyvale, CA 94088-3453

ABSTRACT

Copper electroplating processes for filling sub-0.5 μm trenches and vias with very high aspect ratios (> 4:1) were developed. Copper was electroplated on sputtered Cu seed layer with Ta (or TaN) diffusion barrier. An enhanced Cu deposition at the bottom of trenches/vias and defect-free filling sub-0.5 μm trenches (down to 0.25 μm width) of high aspect ratio (up to 4:1) were achieved. Large grains occupying the entire trench were observed. Bottom step coverage of electroplated copper in sub-0.5 μm trenches was estimated to be about 140%, while sidewalls step coverage was about 120%. Via resistance for sub-0.5 μm vias was measured to be below 0.55 Ω. Strong <111> texture, large grains, and low tensile stress were observed in electroplated Cu films and in-laid Cu lines after low temperature anneal.

INTRODUCTION

Copper is going to replace aluminum in Ultra-Large-Scale Integrated Circuits (ULSI) metallization for better conductivity and reliability. Copper films can be obtained by several different deposition techniques including Physical Vapor Deposition (PVD), Chemical Vapor Deposition (CVD) and plating methods such as electroless plating and electroplating.

Plating techniques including electroplating and electroless plating are especially appealing because of its low cost, high throughput, high quality of deposited copper films, and excellent via/trench filling capability.

Electroless plating has advantages of using very thin seed layer, intrinsic selectivity and excellent uniformity. Electroless plating also provides conformal Cu deposition. Complete filling the sub-half micron trenches and vias of high Aspect Ratio (AR) (AR > 4:1) was demonstrated using electroless deposition [1-5].

Electroplating offers advantages of high deposition rate, stable plating solution, enhanced deposition at the trench/via bottom as well as high purity of electroplated copper [6,7].

There is a trade-off to use electroplating versus electroless plating for on-chip interconnects. Both plating techniques have low cost, good via/trench filling capability and low resistivity of plated copper (Table 1).

*Current address: Intel, Hillsboro, OR 97124-6497

Electroplating can provide high deposition rate and electroplating solutions for copper deposition are stable (Table 1).

Table 1. Characterization of electroless and electroplating methods for copper interconnect formations.

Dep. method	Equip. cost	Seed layer thick., nm	Dep. rate, nm/min	Solution stability	Cu growth in trench	Impurities, ppm	Electrical uniformity	Surface roughness, nm	Trench filling	In-laid Cu line resist., Ω·cm
Electroless	low	>10	35-75	limited	conformal	112	<5% 3 sigma	<20 1.5μm Cu	good	2.2-2.5
Electroplating	low	>50	500-1000	stable	Enhanc. at the bottom	56	<10% 3 sigma	<10 1.5μm Cu	good	2-2.2

We have developed copper electroplating processes to fill sub-0.5 μm trenches and vias with very high aspect ratios (AR > 4:1). Microstructure and properties of plated Cu films as well as electrical properties of Cu metallization such as effective resistivity of damascene Cu lines and via resistance (both Kelvin and String vias) are presented.

EXPERIMENT

Damascene Cu conductors were fabricated by burying the conductor into the etched SiO_2 dielectric pattern using a gap filling Electro-Chemical Deposition (ECD) process followed by Cu Chemical Mechanical Polishing (CMP). Sputtered 20-30 nm thick Ta (or TaN) was used as a barrier and adhesion promoter layer. A Cu seed layer was sputtered with the thickness in the range of 50-100 nm to perform electroplating. Continuity of seed layer was found to be crucial for filling by electroplating the sub-0.5 μm vias and trenches with high aspect ratios.

The plating system mainly contains a copper plating solution, an inert or soluble anode, and the wafer, which serves as a cathode and a power supply. Electrons are supplied to the wafer-plating solution interface through a seed layer on the wafer surface. Electrical current in the plating solution is conducted by ions such as cations and anions. Copper plating solutions generally contains copper salt as source of copper ions, a support electrolyte to provide conductivity of the solution, and addition agents to modify kinetics of electroplating process and regulate the properties of deposits.

A simplified Cu plating system is presented in Fig. 1. It can be seen from Fig. 1 that near the surface of the cathode the concentration of the cations (Cu^{2+}) is increased while the concentration of anions is decreased. This layer is called the diffusion part of a double electrical layer. The thickness of this layer (δ) is in the range of 1-10 nm. This layer has positive charge while the surface of the cathode is negatively charged. The thickness of the diffusion layer, δ_D, is where the concentration of ions is decreased due to electrochemical reactions on the electrode surface. The thickness of this layer (δ_D) is about 10-100 μm.

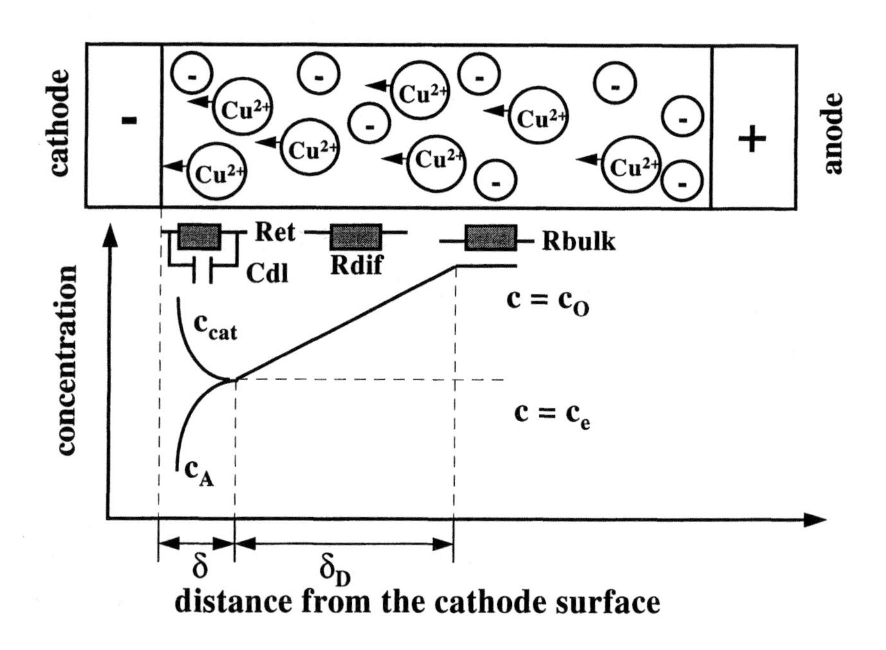

Figure 1. Schematic diagram of electrode/electrolyte interface for Cu electroplating system. Where **Ret** is the resistance of double-electrical layer, **Cdl** is the capacitance of double electrical layer, **Rdif** is the resistance of diffusion layer and **Rbulk** is the resistance of plating solution, $\mathbf{C_{cat}}$ is the concentration of cations in double electrical layer, $\mathbf{C_A}$ is the concentration of anions in double electrical layer, $\mathbf{C_e}$ is the concentration of reactants at the boundary of double electrical layer, $\mathbf{C_o}$ is the concentration of the reactants in bulk plating solution.

The following reactions may occur at the cathode (wafer) / plating solution interface:

$$Cu^{2+} + e = Cu^{+} \quad (E^{o} = 0.153 \text{ V}) \tag{1}$$

$$Cu^{+} + e = Cu^{o} \quad (E^{o} = 0.52 \text{ V}) \tag{2}$$

where E^o is the standard oxidation-reaction potential for the electrochemical reaction.

In case of an inert anode the following reactions will occur at the anode / plating solution interface:

$$4OH^{-} = O_2 + 2H_2O + 4e \quad (E^{o} = -0.4 \text{ V}) \tag{3}$$

$$2H_2O = O_2 + 2H^{+} + 2e \quad (E^{o} = -0.682 \text{ V}) \tag{4}$$

The sum of standard oxidation-reduction potentials for cathodic and anodic reactions (1) - (4) is negative and the change in free-energy is negative. Thus Cu electroplating reaction is not spontaneous and require external power supply.

Current distribution issues need to be addressed in Cu electroplating system to provide good thickness uniformity across the wafer and complete filling of submicron vias and trenches. Current from the external power supply is distributed on the wafer surface through the conductive seed layer. The local current density (j) on the seed layer and the plating rate depend on the electrical field strength (E) ($j = \sigma E$, where σ is the conductivity). The electrical field strength is higher at the top corners of trenches and vias as well as at the edge of the wafer. The conductivity or resistance of the cathode/electrolyte interface (Fig. 1) needs to be regulated to provide uniform current distribution across the wafer and in small features. Optimized pulse plating conditions for Cu electroplating can be also used to fill sub-half micron trenches/vias of very high aspect ratio (AR > 4:1).

Effective electrical resistivity of in-laid Cu lines of various widths has been determined by measuring the cross-section area and the resistance of in-laid Cu lines. Sheet resistance of electroplated Cu film was measured by four-point probe. Impurities in electroplated Cu film were analyzed by SIMS and TXRF. The chemical composition was determined by using XPS. Texture of electroplated Cu film was measured before and after low temperature annealing by using x-ray diffraction fiber pole plot technique. The grain size distribution of copper was obtained from the plan-view TEM micrographs (for blanket Cu films) and dual beam microscopy (for submicron Cu lines). Stress in electroplated Cu films was determined by using a FLEXUS™ stress measurements system. Device characteristics with Cu metallization have been measured before and after thermal stress at 400 °C for several hours.

RESULTS AND DISCUSSION

Defect-free filling of the 0.25 μm trenches and vias (Fig. 2) with high aspect ratios (>4:1) was observed due to the growth of electroplated deposits from the bottom of trenches. Moreover, single copper grains occupying the trenches were found by ion-beam images of Cu-filled trenches.

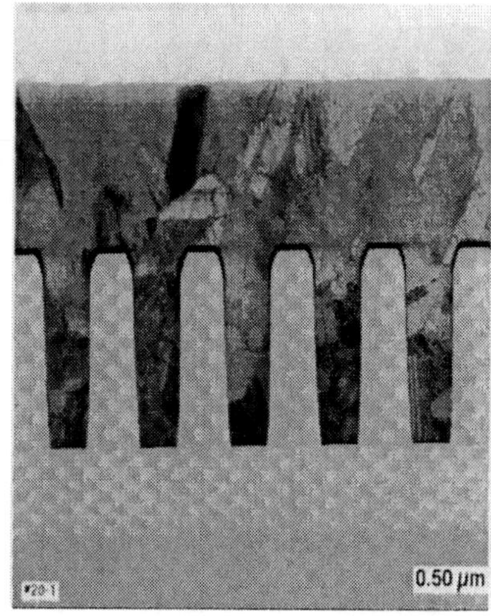

Figure 2. SEM cross-sectional view of 0.25 μm plated Cu-filled vias (AR > 4.5:1)

Pulse deposition was used to improve the via/trench filling capability with unipolar and forward-reverse pulse plating conditions. We have achieved enhanced copper deposition at the bottom of trenches that lead to complete planarization of metallization structures (Fig. 3). Bottom step coverage of electroplated copper in sub-0.5 μm trenches is estimated to be about 140%, while sidewalls step coverage is about 120%. Copper plating rate was in the range of 0.5-1 μm/min. The high copper deposition rate for electroplating allows single-wafer processing with a cluster tool.

Electroplated copper films of high purity were obtained. The contamination level of various elements such as K, Na, C, H, O, S, Cl etc. in electroplated copper films does not exceed a few ppm. We found a few nm thick Cu_2O layer on the surface of in-laid Cu lines by using XPS. This oxide is probably formed during copper CMP in H_2O_2-based slurry. This oxide can be simply removed in HCl solution for subsequent processing. TXRF and XPS data show that copper contamination on the field oxide and the backside of the wafers after copper CMP and post-CMP cleaning was the same as for standard AlCu metallization.

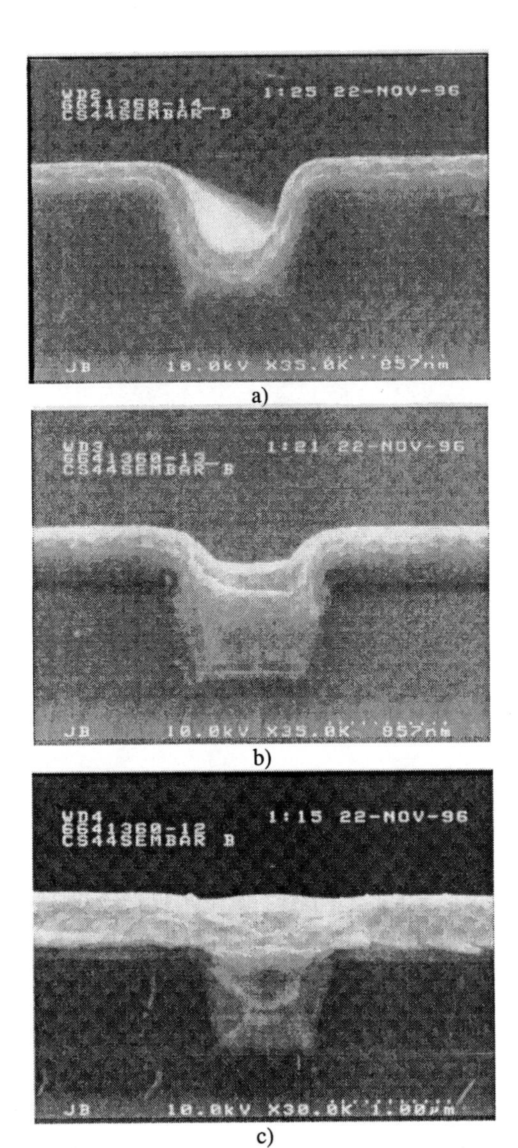

Figure 3. Kinetics of copper growth in trenches during electroplating;
a) 20 sec plating, b) 40 sec plating and c) 80 sec plating.

The grain size distributions of copper films were obtained from the plan-view TEM micrographs (for blanket Cu films) and dual beam microscopy (for submicron Cu lines). For 1.5 μm thick as-plated Cu film, the median grain size is 0.26 μm and the lognormal standard deviation, σ, is 1.05. We can also affect the grain structure and grain orientation by controlling the deposition conditions. Copper grains occupying the entire trenches were found by ion-beam images for Cu-filled trenches under suitable deposition conditions. Grain sizes of electroplated copper increase further after low temperature annealing. We observe grain growth and strong <111> texture for electroplated copper films and in-laid Cu lines after low-temperature annealing (<400 °C). After annealing, the median grain size increases to 1.04 μm, and σ is decreased to 0.45 (Fig. 4).

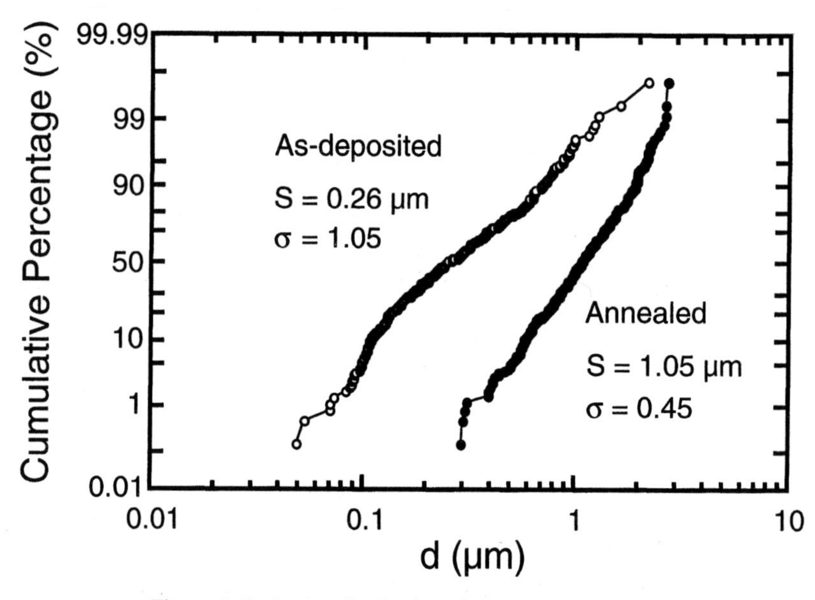

Figure 4. Grain size distribution of electroplated Cu films.

Elastic modulus was found to be about 150 GPa. Microhardness decreases from 1.5 GPa to 1 GPa when the grain size increases. If the grain size is large, a greater stress concentration is developed in the adjacent grain, and thus the applied stress needed to activate flow in this grain is relatively low, and vice versa. This is known as boundary strengthening and described by Hall-Petch relation. Adhesion of plated Cu films to substrate was measured by pull test in the range of 200 Kg/mm^2. No failure was observed on plated Cu/sputtered Cu seed interface.

Sputtered Cu seed layer on Ta barrier exhibits strong (111) texture (<3% random component and <4.5 degree tilting angle ω_{95}). Four percent of the Cu grains have the (111) direction oriented in a (511) pattern.

Electroplated Cu films also exhibit a strong (111) texture (<17 % random component and <5 degree tilting angle ω_{95}). After annealing, the (111) texture becomes stronger with narrower tilting distribution (<4% random component with less than 3° tilt range) (Table 2). Two secondary orientations were also characterized for electroplated Cu films. Five percent of the Cu grains have the (111) direction oriented in a (511) pattern and 4% oriented in a (311) pattern.

Table 2. Texture of electroplated copper films before and after anneal (400 °C in forming gas for 1 hour).

Sample	Texture direction	%Random	ω_{95}	ω_{90}	ω_{63}	ω_{50}
before anneal	111	17	4.29	3.4	1.49	1.07
after anneal	111	4	2.57	1.87	0.77	0.55

Electroplated Cu lines formed in sub-micron trenches also exhibit (111) texture (for 0.5 µm wide Cu lines, the random component is 48% with less than 10 degree tilting angle ω_{95}) which becomes stronger after annealing (<10% random component with less than 4° tilt range) (Table 3).

Table 3. Texture of in-laid electroplated Cu lines.

Line width	Texture direction	%Random	ω_{95}	ω_{90}	ω_{63}	ω_{50}
0.5 µm	111	9	2.18°	1.74°	1.41°	0.97°
1 µm	111	13	2.84°	2.01°	1.01°	0.73°
2 µm	111	5	3.23°	2.29°	0.99°	0.72°
4 µm	111	8	3.97°	2.75°	1.09°	0.77°

The driving force to improve texture of in-laid Cu lines and blanket Cu films after anneal is surface energy reduction because the (111) plane is the lowest surface energy plane for copper.

We also observe low tensile stress (in the range of 10^{-8} dyne/cm^2) in electroplated Cu films. Strong <111> texture, large grains and low stress in plated Cu will improve the reliability related attributes of Cu metallization such as electromigration and stress voiding. The electrical and thickness uniformity of electroplated Cu films was found to be in the range of 7-10%, 3 σ for 8" wafers. Average surface roughness of electroplated Cu was about 10 nm for 1.5 μm thick films. Resistivity of electroplated Cu films was about 1.8-2 μΩ cm when the thickness of electroplated copper films exceeded 0.5 μm. The resistivity increases with decreasing copper thickness. Low temperature annealing of electroplated Cu films resulted in further reduction in resistivity. The resistivity of as-plated in-laid 0.35 μm wide Cu lines with aspect ratio 3:1 was about 2 μΩ cm. The effective resistivity of in-laid Cu lines achieved was below 2 μΩ cm after low temperature annealing. Both Kelvin via and String (25000 vias) via resistance were measured. Very low via resistance was demonstrated. 100% of vias (down to 0.35 μm size) have via resistance below 0.55 Ohm (Fig. 5).

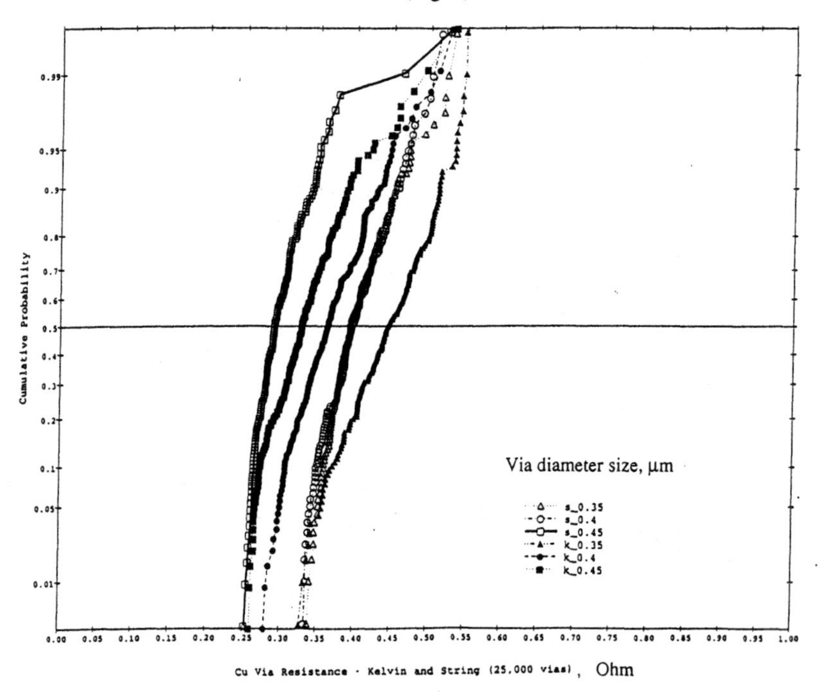

Figure 5. Electroplated Cu via resistance (Kelvin (k) and String (s) 25 000 vias) for different via sizes.

The newly developed copper metallization process was evaluated by using it to fabricate test vehicles containing different device and circuit structures. Normal device and circuit characteristics were obtained with this new copper metallization process. No negative impact of Cu metallization on device characteristics (drain current, threshold voltage etc.) or the quality of interlevel dielectric (leakage current, breakdown voltage etc.) was found after thermal stress of devices with Cu metallization at 400 °C for several hours.

CONCLUSIONS

Electroplating offers advantages of high deposition rate, stable plating solution, enhanced deposition at the trench/via bottom as well as high purity of electroplated copper. A novel electroplating method has been developed to provide enhanced Cu deposition at the bottom of trenches and vias. We measure step coverage for electroplated copper to be about 140% for bottom step coverage and about 120% for sidewalls step coverage. Step coverage of more than 100% for electroplated copper allows us defect-free filling the sub-0.5 μm features such as 0.25 μm wide trenches and 0.25 μm size vias of high aspect ratio (up to 4.5:1).

The median grain size of electroplated copper was measured to be about 1 μm and the lognormal standard deviation is about 0.4 μm. The electrical uniformity of electroplated Cu film was about 2-5%, 1 sigma. The total impurities level in electroplated Cu film was observed below 60 ppm. Strong <111> texture was observed in electroplated Cu film.

A very low effective resistivity of in-laid electroplated Cu lines was achieved (< 2 μΩ cm). The resistivity of in-laid Cu lines decreases (about 10-15%) after low temperature annealing. The via resistance (both for Kelvin and string vias) of sub-0.5 μm vias was measured to be below 0.55 Ohm. Normal device characteristics were obtained with electroplated Cu metallization.

ACKNOWLEDGMENTS

The authors thank all the staff members of Technology Development Group at Advanced Micro Devices who participate in Cu damascene process development. Specifically, J. Iacoponi, C. Woo, A. Preusse, T. Nogami, S. Chen, Y.C. Joo, R. Lee, D. Erb, K. Yang, M. Buynoski, S. Avanzino, D. Schonauer, E.H. Adem, J. Bernard, D. Brown, J. Bertrand, W. Stockwell, G. Morales and J. Barragan for all of their help with this work.

REFERENCES

1) V.M. Dubin, Y. Shacham-Diamand, B. Zhao, P.K. Vasudev and C.H. Ting. "Selective and Blanket Electroless Copper Deposition for Ultralarge Scale Integration" J. Electrochem. Soc. **Vol. 144**, 898-908 (1997)
2) V.M. Dubin, Y. Shacham-Diamand, B. Zhao, P.K. Vasudev and C.H. Ting. "Sub-Half Micron Electroless Cu Metallization", Mat. Res. Soc. Symp. Proc. **Vol. 427**, P. 179 San Francisco (1996)
3) Y. Shacham-Diamand, V.M. Dubin and M. Angyal. "Electroless copper deposition for ULSI", Thin Solid Films, **Vol. 262**, 93 (1995)
4) V.M. Dubin, Y. Shacham-Diamand, B. Zhao, P.K. Vasudev and C.H. Ting. "Electroless Cu deposition", MRS Conf. Proc. ULSI XI, P. 597 Portland (1995)
5) S. Lopatin, Y. Shacham-Diamand, V. Dubin, J. Pellerin, B. Zhao, P.K. Vasudev, in MRS Proc. ULSI XII, p. 169 (1996)
6) V. M. Dubin, S. Chen, R.Cheung, J. Iacoponi and C.H. Ting, "Copper electroplating for ULSI metallization" Semiconductor World, **Vol. 16**, 192 (1997)
7) V.M. Dubin, C.H. Ting and R. Cheung. "Electro-Chemical deposition of Copper for ULSI Metallization" Proc. of 14th VLSI Multilevel Interconnection Conference, P. 69 Santa Clara (1997)

Design Tools for Copper Deposition in the Presence Additives

James J. Kelly* and Alan C. West**
*Materials Science and Metallurgical Engineering
**Department of Chemical Engineering
Columbia University, New York, NY 10027

ABSTRACT

Considerations for the development of a plating tool that predicts the spatial uniformity of metallization rates are discussed. Models that attempt to account for the effect of additives on electrode kinetics, which may affect metal deposition rates, must be based upon significant experimental work. The mechanism of two of these additives, polyethylene glycol (PEG) and chloride ions, is investigated *via* an electrochemical quartz crystal microbalance (QCM) and electrochemical impedance spectroscopy (EIS) to understand how a plating tool should account for their effects. The addition of polyethylene glycol (PEG) and Cl^- to an acid copper electrolyte inhibits the deposition reaction for cathodic overpotentials of up to about 150 mV. Adding Cl^- only promotes the deposition reaction, while adding PEG alone has a relatively small effect on electrode kinetics. Frequency shifts of an electrochemical quartz crystal microbalance suggest the adsorption of a monolayer of PEG molecules that are collapsed into spheres provided chloride ions are present, with little adsorption occurring when Cl^- is absent. This behavior is the same for gold and copper surfaces. Simulated results from a model that assumes the adsorption of a nearly complete monolayer of PEG in the presence of chloride ions and no adsorption without Cl^- agree with experimental steady state and EIS results. The primary effect of PEG adsorption, which does not appear to vary with time during an EIS measurement, is a blocking of available surface sites for charge transfer.

INTRODUCTION

We are interested in developing mathematical models of metallization processes that can be used in the design of plating tools and for the determination of optimal operating conditions. Current focus of our efforts is on copper deposition of damascene structures. Such applications are of great technological importance as copper replaces aluminum and tungsten for on-chip interconnects and vias (1,2).

Simulation tools that describe transport and interfacial phenomena that dictate, for example, wafer-scale uniformity and the formation of voids in trenches may require detailed descriptions of fluid flow, homogeneous solution chemistry, and electrode kinetics. Frequently, one bottleneck in the development of an effective current-distribution solver is an adequate mathematical description of the electrode kinetics, especially for plating in the presence of additives.

Although we wish to emphasize that is desirable to develop design software that can treat a large number of transport phenomena and solution chemistries, we should also like to suggest the importance of relatively simple modeling approaches. For example, one can assume that the only source of spatial nonuniformity in deposition rate is an electrical field. This would indeed be realized when fluid flow is sufficiently large and deposition rate sufficiently low that concentration fields are uniform. Such a model is termed a secondary current distribution. Figure 1 shows the result of one such simulation. The average deposition rate is assumed to be 50 mA cm^{-2}. Furthermore, the assumed electrolyte conductivity is that of the acid-copper bath described below. The Tafel slope determined by Mattson and Bockris is also used (3).

The electrode used in the calculation is shown in Figure 1b. Deposition occurs on the entire surface, although the current distribution is only shown along the sidewall of the trench. When L = 250 microns, typical for a printed circuit board, the deposition rate is relatively nonuniform. However, when L = 0.25 microns, typical for an interconnect, the predicted distribution of plating rate is completely uniform. This implies that electric-field effects are not of importance in predicting the spatial uniformity on the feature-scale of an interconnect. The results of these simple calculations may indicate that the engineering design of the solution chemistry and the plating process for interconnect technologies may be radically different from those used for packaging applications. One must also bear in mind that these results only indicate that electrical-field effects are unimportant for determining trench-filling capability of a process. It is more likely that nonuniform concentration fields will be the major design consideration on the feature scale. Since convection is unlikely to significantly improve mixing in submicron features, a major design consideration may be the development of current waveforms other than direct current (4).

The above example uses well-established kinetic parameters for additive-free copper-deposition kinetics. However, a major uncertainty in model development is the incorporation of surface-active additives into descriptions of electrode kinetics. The uncertainty increases in the treatment of transients that would arise if a pulsed waveform were employed; an understanding of reaction mechanisms would appear essential. Because of the lack of a fundamental understanding of additives, the development of realistic design tools must include significant experimental work. The focus of the present communication is an experimental study of the influence of a commonly employed additive in acid-copper baths for packaging applications. Such baths and additive chemistries may serve as a starting point in the development of baths for interconnects.

EXPERIMENTAL

Modern electrolyte baths for electrochemical metallization processes contain several ingredients that are added in small quantities to effect various desirable material properties (5,6). Two additives often incorporated into industrial plating baths are polyethylene glycol (PEG) and chloride ions. Although PEG and chloride ions have been in use for some time, a fundamental understanding of these substances' mode of action is lacking (7). The polymer, sometimes referred to as a "carrier", along with chloride ions, is commonly used with substances known as "brighteners" and "levelers." The previous work on understanding the interactions between PEG and chloride is summarized elsewhere (8,9).

Potentiostatic polarization and electrochemical impedance spectroscopy (EIS) experiments were conducted at 25 ± 0.5°C. The working electrode (WE) was a copper rotating disk (0.60 cm diameter, 99.999% Johnson Matthey) with Plexiglas insulation, while a mercury/mercurous sulfate electrode (Metrohm) served as the reference electrode (MSE). The composition of the standard electrolyte was always 0.24 M $CuSO_4 \cdot 5H2O$ and 1.8 M H_2SO_4 (Fisher, Certified ACS), to which various quantities of PEG and chloride ions were added. PEG average molecular weights of 600, 3350, 6750, and 10^5 were used (Aldrich; Fisher; Scientific Polymer Products; and Matheson, Coleman, and Bell, respectively). EIS experiments were conducted potentiostatically after holding the system at the steady state applied voltage for 3 minutes; a perturbation ranging from 1 to 6 mV was employed. The response was verified to be linear. Experiments were conducted several times for each case presented to confirm the reproducibility of all EIS results. Other experimental details are given elsewhere (10).

A QCM experimental setup employing Maxtek 5 MHz AT-cut gold quartz crystals was used as described previously (11). QCM electrode surfaces were cleaned by a brief immersion in piranha reagent and subsequent rinsing with deionized water. Measurements on gold surfaces took place in a 1.8 M H_2SO_4 electrolyte. Copper surfaces for QCM measurements were prepared by electrodepositing 100 nm of copper from a 0.24 M $CuSO_4 \cdot 5H_2O$ and 1.8 M H_2SO_4 electrolyte containing no additives at 10 mA/cm^2 directly on the gold QCM electrode, rinsing the entire QCM probe in an aliquot of distilled water, and finally transferring the probe at the desired potential into a 1.0 mM $CuSO_4 \cdot 5H_2O$ and 1.8 M H_2SO_4 electrolyte. Inspection with an optical microscope revealed that these copper films were smooth and continuous. Additives were first dissolved in approximately 1 mL of deionized water and then gently poured into the electrolyte. The additives were not introduced into the electrolyte until the QCM frequency had been stable for at least two minutes. Data acquisition ceased two minutes after the additives entered the electrolyte. Measurements in which the introduction of the additive mixture caused a large disturbance in frequency were rejected. All QCM data points presented are the average of at least five measurements. The temperature for all QCM experiments was 23±2 °C.

RESULTS

Figure 2 shows the effect of the additives on the polarization behavior of the system. Chloride ions acting alone accelerate the deposition reaction, while the addition of PEG without Cl^- weakly inhibits the current for small cathodic overpotentials. The concomitant addition of PEG and chloride ions leads to a significant inhibition of current for cathodic overpotentials up to 150 mV. Other workers have observed similar effects (12). Using 300 ppm of the monomer (ethylene glycol, Aldrich) and 50 ppm Cl^- yields results nearly identical to those from an electrolyte having only 50 ppm Cl^-.

Figure 3 depicts a typical QCM frequency response to the addition of 300 ppm 3350 PEG and 50 ppm Cl^- to the electrolyte using both gold and copper QCM surfaces; the presence of 1.0 mM $CuSO_4 \cdot 5H_2O$ in the electrolyte for the case of the copper surface results in a steady decrease in frequency before and after the introduction of additives due to the electrodeposition of copper. The frequency shift due to metal deposition during the time over which polymer adsorption occurs (typically 2-3 sec) is subtracted from the total frequency shift to yield the frequency shift due to the polymer. For both cases, the current (resulting from oxygen reduction in the case of a gold surface and copper deposition in the case of a copper surface) substantially decreases after the introduction of the additive as well. Pouring in a "blank" aliquot of deionized water does not produce the effect shown in Figure 3. Moreover, the addition of 300 ppm of polymer to the electrolyte produces no significant changes in viscosity, as was verified by measurements with a capillary viscometer. In the simplest interpretation, the negative frequency shift resulting from the introduction of the additives is proportional to m, the adsorbed mass/area, and corresponds to the presence of an adsorbed layer, presumably the PEG. This would be in agreement with the findings of Healy *et al.* (13).

Figure 4 shows the frequency shift response Δf upon introduction of the additives as a function of PEG molecular weight. The degree of current inhibition mentioned above appears to saturate after a molecular weight of 3350 for both copper and gold surfaces, perhaps indicating the nearly complete blocking of the cathode by the PEG. A least squares regression on a log-log plot of Δf *vs.* polymer molecular weight yields a line having slope of 0.38; *i.e.*, assuming

$$\Delta f \propto m \propto n^{\alpha} \tag{1}$$

$\alpha = 0.38$, where m is the areal mass density of an adsorbed layer and n is the number of repeat units in the polymer and is proportional to its molecular weight. It can be shown that the experimentally observed frequency shifts are of the same order of magnitude as those expected from an adsorbed monolayer of PEG and that if the PEG molecules adsorbed as spheres, $\alpha = 1/3$ (11). Since other packing arrangements such as so-called polymer brushes, where a portion of the polymer dangles into the solution, would have $1/2 \lesssim \alpha \lesssim 1$, the data suggest that the adsorbed PEG molecules are more or less collapsed spheres (14,15).

Figure 5 demonstrates the effect of chloride ion concentration on the additive system. In agreement with the SERS observations of Healy *et al.*, no consistent frequency shift was obtainable without chloride ions (13). For measurements on gold surfaces, the scatter in the frequency shift is large when chloride ions are not present. It is possible that for these experiments, PEG adsorption is highly sensitive to the surface preparation of the gold substrate if chloride ions are absent. Nevertheless, the average frequency shift is zero.

DISCUSSION

Based upon the QCM experimental observations discussed previously, we developed a model for copper deposition in the presence of PEG for comparison to steady state polarization and EIS experimental results. The model assumes only that an adsorbed layer of PEG on the cathode competes with cupric ions for adsorption sites. As described below, good agreement is obtainable for both steady state and transient results despite making only a few simple assumptions.

Considering the previous adsorption study, it is proposed that chloride ions are necessary for the adsorption of PEG on metal surfaces. Cupric ions compete with adsorbed PEG and adsorbed cuprous ions for free adsorption sites; the adsorbed PEG molecules do not further affect electrode kinetics, and it is assumed that typical thin film growth processes involving the diffusion of cuprous ions to high energy incorporation sites remain largely unaffected (16,17). Furthermore, over the time scales of an impedance experiment, the adsorbed dimensionless concentration of PEG (Θ_{PEG}) is assumed to be invariant in time. Details of the model can be found in reference 10.

Experimental and simulated steady state polarization curves for an electrolyte with 300 ppm 3350 PEG and 50 ppm Cl^- are shown in Figure 5. Θ_{PEG} assumes a constant value of 0.97 so as to be in agreement with the steady state polarization results. This value for Θ_{PEG} is consistent with the idea that a full monolayer of PEG is adsorbed on the cathode. We would like to emphasize that we do not adjust other kinetic parameters in the model. A comparison of simulated and experimental EIS data at -0.750 V is shown in Figure 7 for an electrolyte having 300 ppm 3350 PEG and 50 ppm Cl^-. Θ_{PEG} was set to its steady-state fitted value of 0.97. The spectra are distinct in the appearance of a third capacitive loop for low rotation speeds. This feature was very reproducible. Reid and David obtained similar EIS results; they suggested that the PEG forms a dense adsorbed layer which inhibits deposition (18). The third capacitive loop, which does not appear in EIS spectra from additive-free or chloride-only electrolytes, is seen to become distinct for both experimental and simulated EIS spectra. This agreement, achievable by modifying only Θ_{PEG}, suggests that copper deposition kinetics are otherwise unaffected by the presence of PEG.

Since changes in electrode kinetics due to the presence of adsorbed PEG may be described by a simple model, accounting for these changes in a current distribution solver should be relatively straightforward. Further complications arise, however, when other additives typically used with PEG and Cl^- are introduced to the plating bath.

CONCLUSIONS

The addition of PEG and Cl^- to an acid copper electrolyte has a strong inhibitive effect on the cathodic deposition reaction. Adding chloride ions only increases the deposition current, while adding PEG alone has little effect on electrode kinetics. QCM frequency shifts, which suggest the adsorption of approximately a monolayer of spherically packed PEG molecules when chloride ions are present are consistent with theoretical predictions. A model that only assumes the adsorption of nearly a monolayer of PEG in the presence of chloride ions is sufficient to explain steady state and EIS experimental results. It is assumed that adsorbed PEG competes for adsorption sites with cupric ions on the cathode if chloride ions are present. The introduction of PEG and Cl^- does not necessitate modifying kinetic parameters from values observed in the presence of Cl^- alone. Assuming that the Cl^- concentration affects Θ_{PEG} yields reasonable agreement for experimental and simulated steady state polarization and EIS results.

ACKNOWLEDGMENTS

This work was partially supported by the National Science Foundation under Grant No. CTS-93-58380.

REFERENCES

1. A. B. Frazier, R. O. Warrington, and C. Friedrich, *IEEE Transactions on Industrial Electronics*, **42**, 423 (Oct. 1995)
2. D. Edelstein, J. Heidenreich, R. Goldblatt, W. Cote, C. Uzoh, N. Lustig, P. Roper, T. McDevitt, W. Motsiff, A. Simon, J. Dukovic, R. Wachnik, H. Rathore, R. Schultz, L. Su, S. Luce, and J. Slattery, *International Electronic Device Meeting Technical Digest*, IEEE, Washington, D. C., 773 (1997).
3. E. Mattson and J. O'M. Bockris, *Transactions of the Faraday Society*, **55**, 1586 (1959).
4. A. C. West, C. C. Cheng, and B. C. Baker, "Pulse Reverse Copper Electrodeposition in High Aspect Ratio Trenches and Vias", *J. Electrochem. Soc.*, submitted (1997).
5. D. Stoychev, L. Vitanova, R. Buyukliev, N. Petkova, I. Popova, and I. Pojarliev, *Journal of Applied Electrochemistry*, **22**, 987 (1992).
6. L. Mayer and S. Barbieri, *Plating and Surface Finishing*, **68**, 46, (Mar., 1981).
7. W. Plieth, *Electrochimica Acta*, **37**, 2115, (1992).
8. L. Mirkova, St. Rashkov, and Chr. Nanev, *Surface Technology*, **15**, 181 (1982).
9. M. Wuensche, W. Dahms, H. Meyer, and R. Schumacher, *Electrochimica Acta*, **39**, 1133 (1994).
10. J. J. Kelly and A. C. West, "Copper Deposition in the Presence of Polyethylene Glycol: II. Electrochemical Impedance Spectroscopy," *J. Electrochem. Soc.*, accepted (1998).
11. J. J. Kelly and A. C. West, "Copper Deposition in the Presence of Polyethylene Glycol: I. Quartz Crystal Microbalance Study," *J. Electrochem. Soc.*, accepted (1998).
12. M. Yokoi, S. Konishi, T. Hayashi, *Denki Kagaku*, **52**, 218 (1984).
13. J. P. Healy, D. Pletcher, and M. Goodenough, *J. Electroanal. Chem.*, **338**, 155 (1992).
14. de Gennes, P. G., *Macromolecules*, **13**, 1069 (1980).

15. A. Halperin, M. Tirrell, and T. P. Lodge, *Advances in Polymer Science*, **100**, 31 (1992).
16. M. Fleischmann and H. R. Thirsk, *Electrochimica Acta*, **2**, 22 (1960).
17. M. Fleischmann and H. R. Thirsk, in *Advances in Electrochemistry and Electrochemical Engineering*, Vol. 3, P. Delahay, Editor, p. 125, VCH Publishers, New York, 1963.
18. J. D. Reid and A. P. David, *Plating and Surface Finishing*, **74**, 66 (Jan. 1987).

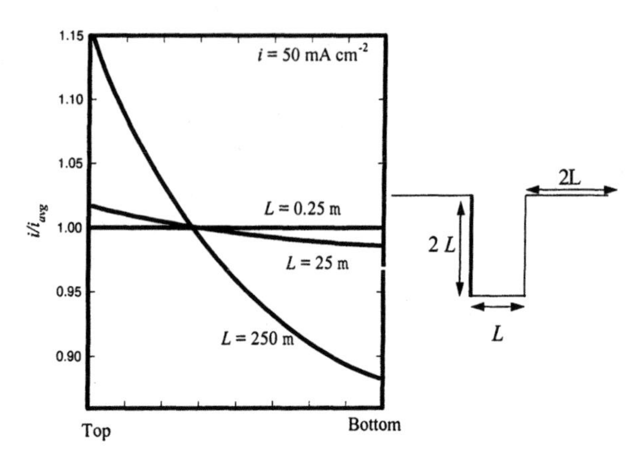

Figure 1. Normalized current distributions on the sidewall of the feature shown to the right for three characteristic lengths L. The entire surface to the right is an electrode. The counterelectrode is assumed to be far away from substrate, and insulating walls are assumed to be located at $2L$ from the feature sidewalls.

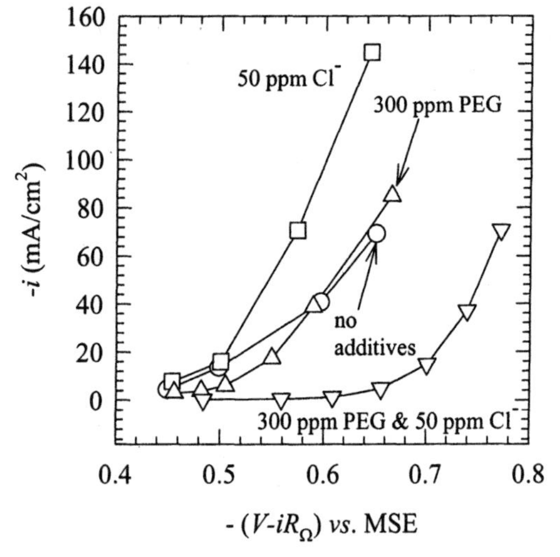

Figure 2. Effect of 3350 PEG and Cl^- on the polarization behavior of an acid copper sulfate electrolyte. Points represent steady state currents at 25 °C on a rotating disk electrode at 900 rpm.

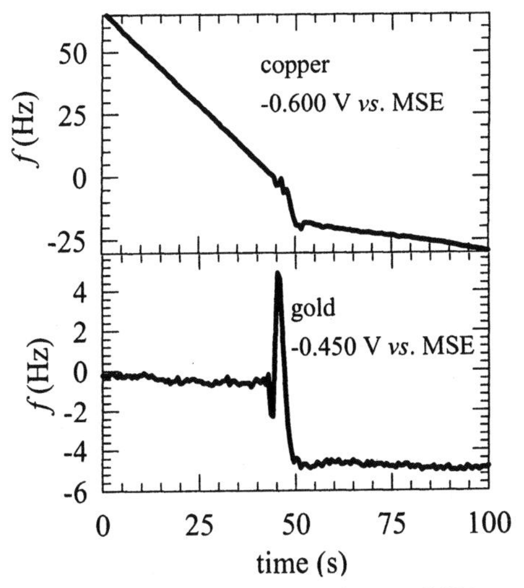

Figure 3. Typical QCM experiment. A mixture of 300 ppm 3350 PEG and 50 ppm Cl^- was introduced to the electrolyte at approximately $t = 45$ s.

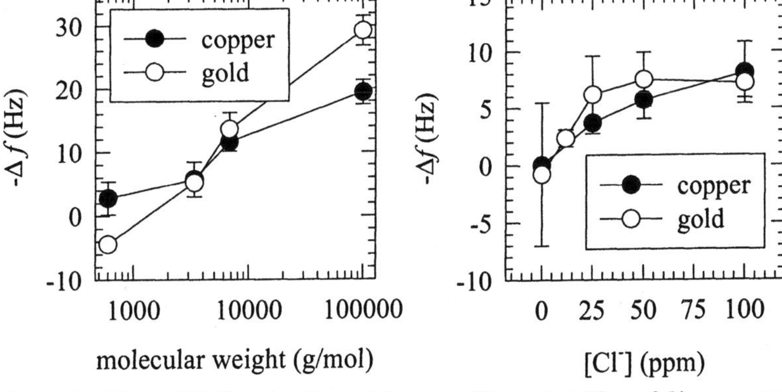

Figure 4. Effect of PEG molecular weight on frequency shifts on a QCM at potentials of -0.450 V and -0.600 V for gold and copper surfaces, respectively. All points are an average of at least 5 measurements with 300 ppm PEG and 50 ppm Cl^-; error bars represent the standard deviation of these measurements.

Figure 5. Effect of Cl^- concentration on frequency shifts on a QCM. All points are an average of at least 5 measurements for 300 ppm 3350 PEG and various Cl^- concentrations. The error bars represent the standard deviation of these measurements.

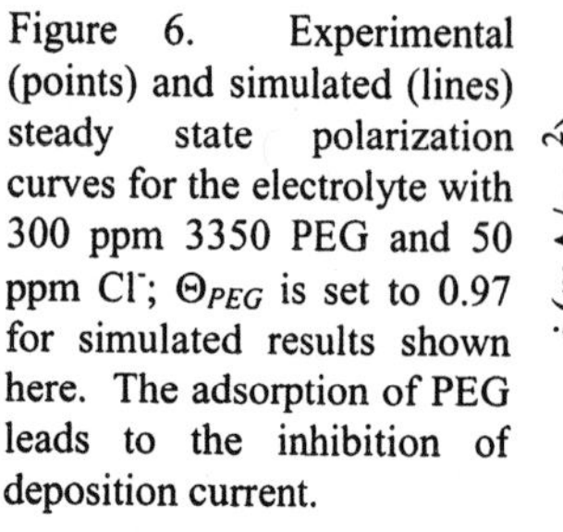

Figure 6. Experimental (points) and simulated (lines) steady state polarization curves for the electrolyte with 300 ppm 3350 PEG and 50 ppm Cl^-; Θ_{PEG} is set to 0.97 for simulated results shown here. The adsorption of PEG leads to the inhibition of deposition current.

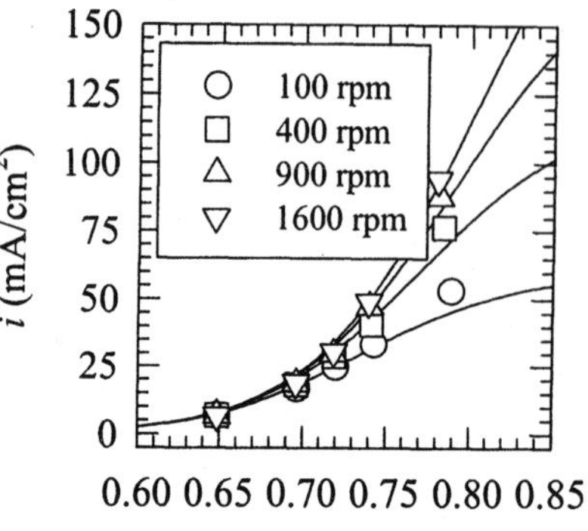

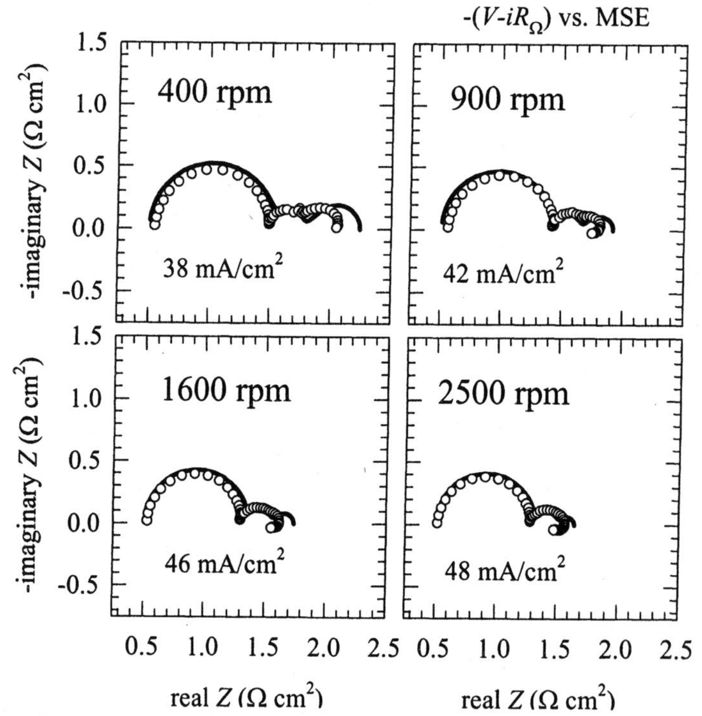

Figure 7. Comparison of experimental (points) and simulated (lines) impedance spectra for the electrolyte with 300 ppm 3350 PEG and 50 ppm Cl^-. $\Theta_{PEG} = 0.97$, as in Figure 6. The emergence of a third loop is predicted by the model for low rotation speeds.

ELECTROLYTE COMPOSITION MONITORING FOR COPPER INTERCONNECT APPLICATIONS

Thomas Taylor, Thomas Ritzdorf, Fred Lindberg, Brad Carpenter
Semitool, Inc., 655 W. Reserve Dr., Kalispell, MT 59901
Mark LeFebvre
Shipley Co., 455 Forest Street, Marlborough, MA 01752

ABSTRACT

Electroplating, widely used for deposition of metal films in electronics applications, is one viable method for effecting copper metallization of semiconductor devices. The ultrafine features of advanced integrated circuits, and the film property tolerances which must be maintained to ensure consistent and reliable device performance, demand that critical parameters of the electroplating process be controlled as tightly as possible. Electrolyte composition plays a significant role in deposit film properties. Evaluations of CVS and CPVS analyses were conducted for two acid copper electrolyte types, and the results were used to develop methods to replenish plating bath additives depleted while depositing copper on semiconductor substrates under conditions simulating high-volume wafer fabrication. HPLC analysis was used to establish bath replenishment methods for one of the electrolyte types. Advantages and drawbacks of the analytical methods were noted, and a topology for a closed-loop electrolyte monitoring and control system is proposed.

INTRODUCTION

The phenomenal growth exhibited by the semiconductor industry over the past three decades has been due in large part to the ability of manufacturers to provide a 25-30% per year cost reduction per function throughout this period. Interconnect architecture improvements now rank among the most intense areas of semiconductor process and integration development, and are anticipated to remain so for the foreseeable future.[1]

One of the enhancements to interconnect structures anticipated to see rapid adoption is the replacement of aluminum and tungsten signal transmission lines by copper. Copper is not expected to be a direct replacement for the conventional materials however; its introduction also drives associated technological developments such as the implementation of effective diffusion barrier materials, and establishment of a patterning sequence which reduces the complexity and cost of multilayer interconnect fabrication. The *de facto* standard for copper interconnect is referred to as Damascene, which involves the deposition of the metal into high aspect-ratio barrier-lined features etched into an underlying dielectric film. Copper interconnect will be introduced using electrochemical deposition (ECD) processes, which feasibility studies have shown to be well matched to the demands of Damascene patterning.[2,3]

Of particular importance is measurement and control of proprietary organic compounds which serve to modify the deposit properties through adsorption onto and desorption from the cathode surface during plating, affecting the diffusion rate of copper cations to nucleation and growth sites. These compounds are typically delivered as multicomponent packages from plating chemistry vendors. One of the most important functions of the additive packages is to influence the ability to fill submicron vias and trenches. Organic additives have also been shown to have dramatic effects on mechanical film properties.[4,5,6,7,8,9] Detection and quantification of these important bath constituents is complicated by the fact that they are effective at very low concentrations in the electrolyte, at several ppm or less.

A number of analytical techniques have been employed with reasonable success in routine assessment of plating bath composition in PWB applications. Of particular importance are those based on electroanalysis and liquid chromatography.

Electroanalytical Methods

Faradaic electroanalysis is attractive as an investigative analytical method principally because what is studied is the electrochemical activity of the bath sample under applied electrical stimulus; the measured responses are related in a fundamental way to the metal deposition process itself. Electroanalysis further offers the opportunity to study the mechanisms and kinetics of the plating process, and the influences the various bath components exert on plating rate suppression and acceleration.

The most widely applied electroanalytical techniques for plating bath analysis are *cyclic voltammetric stripping* (CVS) and *cyclic pulsed voltammetric stripping* (CPVS). Both techniques depend on controlling the voltage between working and auxiliary electrodes with a potentiostat such that the working electrode is cycled between cathodic and anodic potentials while in contact with the electrolyte. A metal film is alternately reduced on the working electrode surface and stripped by anodic dissolution. The potentiostat cycle is defined so that the current can be integrated over time during the stripping period, allowing quantification of the electric charge transferred during the time required for complete film dissolution. The charge is directly related to the molar quantity of metal stripped (and therefore to the amount initially deposited) by Faraday's law. The stripping charge is monitored rather than the charge transferred during film deposition because it is less sensitive to changing electrode surface state and it proceeds to a well-defined endpoint.

The current/voltage/time relationship during analysis is extremely sensitive to variations in electrolyte composition and measurement conditions such as temperature. If sufficient care is taken in methods development and measurement technique then stripping voltammetry can be employed to generate calibration curves with which subsequent analyses can be compared to yield reasonably accurate quantification of analyte composition.

Although the voltammetric stripping techniques are of potential utility in quantitating a number of plating bath components, in practice they are most often employed for evaluating levels of organic additives such as suppressing and brightening agents. Brighteners in particular are among the compounds most prone to breakdown under heavy electrolyte use encountered in production plating conditions, and for which variations in concentrations can have dramatic effects in deposit film properties. A number of calibration methods have been put into practice with good results.[10,11]

The volumes of analyte required for voltammetric stripping analysis are rather small (less than 50 ml) and in the usual titration methods no reagents that are foreign to the plating electrolyte itself are required. This is one of the outstanding benefits of the voltammetric techniques, as the waste streams that are generated can be discharged to the same facility that is used to collect spent or recyclable plating baths. With certain precautions, it is possible that the contents of the analysis cell could be routed directly to the active plating reservoir following measurement, reducing waste to virtually nil.

Potentially the most problematic shortcomings of voltammetric analyses are their sensitivities to so-called 'matrix effects'. Many plating bath components and their breakdown products can display convoluted electrochemical interactions, hence the stripping charge responses can be ambiguous if several constituents have undergone significant concentration change simultaneously. Conversely, compounds and ions that are not electrochemically active over the ranges of potentials explored, and which therefore do not directly affect the deposition or stripping rates under measurement conditions, cannot be detected. For instance, bath impurities that may be occluded in the deposit, or components which affect film properties without appreciably altering plating rate, are likely to escape notice.

These attributes make the voltammetric techniques less than ideal if the user's goal is certain speciation of all bath components. Nonetheless, under conditions where the principal requirement is to detect and correct changes in one or two electrochemically active components which have a rapid rate of consumption, the voltammetric methods are promising candidates for incorporation in closed-loop control systems.

Chromatographic Methods

Chromatographic separation has been a standard technique for quantitative analysis for years, and has seen application in many chemical analysis and control disciplines. As practiced in electroplating, liquid chromatography has utility for determination of virtually all important electrolyte species, including inorganic ions, transition metals and heavy metals, chelating agents and complexes, organic additives and their breakdown products, and others.[12]

The chromatographic methods of greatest interest to electroplaters are *high-performance liquid chromatography* (HPLC) and *ion chromatography* (IC). Features in common between these two methods include the general architecture of the apparatus and the stages of analyte separation and detection. The chromatograph consists of an eluent delivery module including provisions for sample introduction, a separation module, and a detection module. (Fig. 1) The eluent is a carrier solvent which transports small sample volumes through the separation and detection modules. The eluent stream carries the sample through a separation column packed with resin beads which, dependent on the intended analysis, may be coated with insoluble anionic or cationic films or left uncoated. The constituent species of the sample display differing affinities for adsorption onto the bead materials and are hence separated into discrete bands before entering the detector module.

Post-column detection of the time-separated bands can be accomplished in a number of ways, depending on the chemical, electronic, and optical properties of the materials of interest. In ion chromatography, the common detection method is based on the changing conductivity of the separated eluent/sample stream. Other detection methods include UV/VIS absorption and a form of amperometric detection. The former is similar to other spectrophotometric methods, and the later akin to the electroanalytical techniques.

In all embodiments of chromatography, the output response of the detector is plotted against elapsed time, with the resulting chromatogram displaying one or more response peaks separated in time. When compared to chromatograms generated during calibration with samples of known composition, the identity and concentration of the species in an unknown sample can be established: identity by elution time (position on the chromatogram), and concentration by peak area or height.

Shortcomings of the liquid chromatography methods include waste generation, particularly the significant volumes of eluent necessary to perform frequent and successive analyses (approximately 100 ml per measurement in continuous operation). As opposed to the voltammetric methods described earlier, the waste streams differ in makeup from the plating electrolyte, and cannot simply be routed to a common electrolyte collector or discharged into the active plating reservoir. Secondly, the time required to perform the analyses can be long, particularly if dilute eluents are employed with the aim of improving resolution of sample species of similar mobility through the separation phase. The repeatability of the sample injection apparatus directly impacts the precision of the analyte concentration measurements. For certain analyses (e.g., measurement of highly concentrated species, such as the major metal cation in the electroplating bath), the sample may need to be diluted by factors of 100:1 or 1000:1 before injection in the eluent stream to avoid saturating the detector. The precision of the measurement is again related to the precision of the pre-analysis dilution step.

In the other extreme, organic additives in the electrolyte, which are frequently of greatest interest are usually present in extremely low concentrations. Quantitating these materials with sub-ppm accuracy can be extremely challenging with UV absorption. Since these materials often induce a large electrochemical response because of their tendency to concentrate on electrode surfaces, a promising technique to improve signal strength involves the employment of an electrochemical cell as the detector. As noted earlier, amperometric detectors are available for this purpose from instrument suppliers, and methods development for high-resolution separation and measurement have been reported.[13]

EXPERIMENTAL

Commercially supplied acid copper sulfate plating electrolytes were examined, representing two types of commonly available additive packages. From Enthone-OMI, the CUBATH M® bath chemistry reportedly uses a brightening agent based on polyether sulfide (PES). Shipley Company supplied samples of their ELECTROPOSIT 1100® acid copper bath, which from examination of voltammetric behavior appears to use an additive package based on sulfonium alkane sulfonates (SAS). In their work on CVS methods development,[4,14] Haak, Ogden, and Tench deduced significant differences in the deposition kinetics between electrolytes with these general additive types.

Electrochemical deposition (ECD) of copper on 200 mm silicon wafers was performed using Semitool's LT-210® ECD systems, or in some cases a manually operated version of this ECD reactor of the same configuration. Both CVS and CPVS were used to develop analytical methods for the PES- and SAS-type acid copper electrolytes. CVS analyses were performed on an EG&G Potentiostat/Galvanostat Model 263A®, or on a Qualiplate 4000® bath analyzer from ECI. Typical operating parameters were 100 mV/s sweep rate and 2500 rpm on the working electrode. The CPVS analyses were performed using Shipley Company's ELECTROPOSIT Bath Analyzer®.

An HPLC method was developed for the PES-based electrolyte using chromatographic separation and UV/VIS detection. All ion chromatography and HPLC analyses were performed on a Dionex DX 500® system equipped with an LC20 enclosure, AD20 absorbance detector and IP20 isocratic pump or an ED40 electrochemical detector with a conductivity cell, anion self-regenerating suppressor (ASRS-1), and a GP40 gradient pump. The capabilities of the analytical techniques are compared to assess their utility for incorporation into fully automated, closed-loop control of ECD processes for copper interconnect metallization.

Cyclic Voltammetric Stripping

The factors that are available to optimize a CVS technique are the potentiostat parameters (sweep rate, cathodic voltage extreme and anodic voltage extreme), as well as the portion of the voltammetric response used to determine the integrated area of the stripping peak. In addition, the quantitative measurements must be correlated to results obtained from samples of known concentrations in order to determine a calibration curve. A typical CVS technique employed to calculate the concentration of organic additives in acid copper plating baths makes use of a titration method to measure changes in the electrochemical activity of the sample with cumulative volume of successive additive-containing aliquots.

Figure 2 shows information from a series of voltammograms taken from an acid copper plating solution with the PES-type additive system. The curves in this figure represent the electrolyte solution without any additives, the electrolyte with the carrier additive component (M D®), the electrolyte with the brightener/leveler component (M LO 70/30 Special®), and the electrolyte containing both components in recommended use concentrations. Suppression and enhancement of the plating rate can be seen in the cathodic portion of the curves, and is also represented by the area of the stripping peak for each of the voltammograms. Similar data is represented in Figure 3 for the Shipley ELECTROPOSIT 1100 additive system.

For convenient measurement of multicomponent mixtures, one would prefer to define a measurement method such that one component exerts a strong influence on stripping peak area, while others have little or no effect over the concentration range of interest. In such a situation, it would be possible to directly analyze one of the additive components, with no matrix effect from the other component. The normalized areas of the stripping peaks from cyclic voltammograms with various concentrations of the two additive types used in Enthone-OMI's CUBATH M exhibit strong interactions between the organic additives over most of the region of interest. As there is no independent effect of one additive in the concentration range that is used in the plating bath, an alternative technique must be employed. Noting that there is a strong effect of the M LO additive in the dilute additive concentration range, a titration analysis method may be useful in determining the additive concentration. Following the method of ECI[15] a method of titrating small amounts of the bath being analyzed into an electrolyte solution containing no additives has been used for analysis.

A representative plot of normalized peak area versus volume of plating bath (containing additive) added to an electrolyte sample is shown in Figure 4. The repeatability of the measurement is approximately ±5% of the normalized peak area. Figure 5 includes curves displaying the ability of the technique to differentiate varying additive concentrations. Volume of bath sample (or additive) necessary to suppress stripping peak area by a preselected amount (70%) can be correlated to additive concentration, to produce a calibration curve. Figure 5 shows the matrix (or interaction) effect between the two additive components.

The above technique was used to make multiple measurements of plating bath solutions with various additive concentrations to quantify the effects of additive concentration on the reported *BRIGHTENER* and *SUPPRESSOR* values reported by ECI's Qualiplate® analysis system. The *BRIGHTENER* analysis correlates directly to the concentration of Enthone-OMI's M D additive (Fig. 6), while the *SUPPRESSOR* value correlates somewhat to the volume of M LO additive (Fig. 7). The concentration of the M LO component can be correlated to a linear regression of both the *BRIGHTENER* and *SUPPRESSOR* values, as reported by the ECI analyzer. As the CVS techniques were used to control bath concentration over an extended period of time, we observed that the reported *SUPPRESSOR* value would continue to climb, although no additions of the M LO additive were made to the bath. As seen in Figure 8, this suggests that the *SUPPRESSOR* analysis is affected by the MD additive.

Cyclic Pulsed Voltammetric Stripping

The CPVS technique has also been used to measure organic additives in acid copper sulfate plating baths. The technique investigated has been optimized by Shipley Company for use with their ELECTROPOSIT series copper baths and the brightening agents contained in them, and used to determine the brightener levels in these baths for PWB production purposes.[12] In Shipley's implementation, the stripping charge is correlated to brightener concentration through calibration, and is thereafter reported as *Total Brightener Analysis* (TBA) units in evaluation of subsequent samples.

The effects of some of the controlling variables on the values obtained using this technique are seen in Figures 9 and 10 from Fisher and Pellegrino[16] (used with permission). The Shipley CPVS method is said to be consistent within ±5% of the electrolyte component concentrations, although it is very sensitive to temperature. As long as the temperature is tightly controlled during the measurement, however, it is a simple matter to measure the additive concentration in a bath sample without altering the sample in any way. An example analysis shows the dependence of the TBA value on the brightener concentration in the electrolyte for three different concentrations of the carrier additive in Figure 11. It is apparent from these data that the carrier concentration does not significantly interfere with the TBA value, allowing measurements that are representative of the brightener concentration. There is some indication, however, that over very extended periods, bath changes occur that do affect the CPVS measurement. (See Figure 12, again from Fisher and Pellegrino)

CPVS data taken from a prolonged run of an ELECTROPOSIT 1100 plating bath are seen in Figure 13. These data represent a run in which the variation of the brightener component due to run-time parameters was determined through calibration undertaken at the beginning of the run. After modeling additive depletion, the bath was replenished to maintain a consistent brightener concentration range.

The CPVS technique was also applied to the PES-type additive, without modification or optimization of the measurement parameters, for comparative purposes. As seen in Figure 14, the technique is relatively sensitive to one of the two additive types used in this bath. As long as the carrier concentration is relatively close to its recommended level, it is a simple matter to estimate the level of the brightener/leveler additive mixture.

HPLC (High Performance Liquid Chromatography)

The parameters that are normally considered in methods development for HPLC are the eluent composition and concentrations, eluent flow rate, and detector parameters such as wavelength for a UV/VIS spectrophotometric detector. Various detectors exist that may be used in order to maximize the signal for the species of interest while separating it from other components with similar elution times.

Eluent concentration affects the retention time of the various solution components in the column(s). The effect of changing the eluent concentration on the peak separation for a sample of Enthone-OMI MD additive is seen in Figure 15. It can be seen from this figure that as the eluent's acetonitrile concentration (in sulfuric acid) is decreased from 8% to 2%, the lower concentration resolves separate peaks where only one was observable at higher concentration. As the eluent concentration is decreased, the retention time is increased, which translates to longer measurement times and increased eluent consumption. Measurement time can be reduced somewhat by increasing the flow rate of eluent to decrease the elution time, but only at the expense of measurement sensitivity and further eluent waste. This effect is seen in Figure 16.

The ability of HPLC equipment to measure the concentration of a species in an electroplating bath is dependent on its ability to separate the peak of interest from other peaks, and to accurately determine the peak area. The peak area (or height, or other feature of interest) may be maximized by tuning the detector to optimize the response for the particular species being analyzed. Figure 17 shows an example for which the measurement signal can be maximized by selecting a wavelength of approximately 254 nm for the UV/VIS detector's monochrometer.

Examples of HPLC spectra for fresh CUBATH M electrolyte and a bath that has been used for a period of time are shown in Figures 18 and 19, respectively. These data were accumulated after optimizing the eluent concentration and detector settings in order to resolve separate peaks and maximize signal-to-noise with UV/VIS detection.

These HPLC measurement parameters were used to monitor CUBATH M additive concentration during a marathon run of an LT-210 ECD system. These data, plotted in Figure 20, indicate that as the peak labeled "MD" is held constant through bath replenishment techniques, another unidentified peak increases markedly with time, and a third peak (which appears only after running the bath) maintains an equilibrium value. We hypothesize that these unidentified peaks correspond to two or more organic byproducts whose concentrations change with time as the bath ages. This tends to support the observation made during CVS analyses of the CUBATH M electrolyte under extended-run conditions in which increasing rate suppression was presumed to result from the appearance and build-up of breakdown products from the brightening agents in the PES-based additive package.

CONCLUSIONS

Several analytical methods exist which may be appropriate for inclusion in a closed-loop, feedback-controlled replenishment system to stabilize the compositions of acid copper electrolytes employed for semiconductor interconnect metallization.

Electroanalytical techniques such as CVS and CPVS have the potential to provide rapid measurements with low hardware investment, minimum reagent cost and little added waste disposal burden. However, since organic additive packages may include components which interact with each other (or the breakdown products which accumulate over time), it may be difficult to precisely correlate observed changes of electrochemical activity to variations in individual bath components over extended periods of bath use. Liquid chromatography offers the ability to speciate a wide variety of electrolyte components with less ambiguity, though with drawbacks including greater overhead in methods development, measurement time, waste generation, and system/operating cost.

With continued refinement of the hardware and further efforts expended in methods development, one or more of the techniques examined may mature to the point where the demands of closed-loop bath monitoring and control for semiconductor applications are fully met. At this point in time, however, it appears that the most workable approach is one in which both voltammetric and chromatographic methods are integrated into a hierarchical analytical approach which exploits their relative strengths while minimizing their shortcomings. Such a multi-level approach might include CVS or CPVS systems integrated into individual ECD systems themselves, dedicated to performing frequent bath analyses tracking changes in concentrations of organic additives. The results of these voltammetric measurements are processed by an on-board CPU, which performs trend analysis and makes replenishment calls to a replenishment unit (which may be dedicated to the tool, or 'multiplexed' to a number of ECD systems). Frequent analysis is possible because of the relatively short measurement times associated with the voltammetric methods, the minimal waste involved (and the ease of waste handling, if any), and the fact that the hardware cost is low enough to allow each plating system to be served by a dedicated analysis unit.

Since the onset of bath changes which are unresolvable by voltammetry may be anticipated even with frequent, dedicated analysis, a chromatographic system would be configured to accept sample streams from a number of plating systems. HPLC or IC analysis would take place on a less frequent basis, gated either by a periodic sampling schedule or episodically as dictated by excursions in individual plating reservoirs' composition trends. The chromatography results may be employed to modify the transfer functions which link the systems' on-board voltammetry-based analyses to organic additive replenishment events, initiate less-frequent dosing of inorganic species, or to indicate the need for electrolyte treatment or replacement.

Establishment of such a sophisticated closed-loop compositional control system demands the coordinated efforts of ECD system designers, electrolyte manufacturers, analysis system vendors and, not least, the semiconductor manufacturers who as the end users are most directly affected by the reliability and performance of electroplating processes and the methods used to monitor and control them.

ACKNOWLEDGEMENTS

The authors wish to thank the engineers and technicians of Semitool's Advanced Technology Group laboratory, the chemists and technologists at Shipley Company, and the members of SEMATECH's advanced interconnect team focused on copper metallization and integration. Special thanks are due to Traci Deglow and Jennie Hollingsworth of Semitool for manuscript production. This paper would not have been possible without their unstinting contributions of time and effort. Further, we would like to acknowledge the employees of Dionex Corporation, ECI, EG&G, and Enthone-OMI for their material and intellectual contributions to advancing the state-of-the-art in copper electroplating for microelectronics applications.

REFERENCES

[1] K. David and M. Bohr, "Interconnect Scaling - Future Trends and Requirements," *SEMI Symposium on Chip Interconnection Digest of Technical Papers*, pp. 1-17 (1996).

[2] D. Edelstein et al., "Full Copper Wiring in a Sub-0.25 μm CMOS ULSI Technology," *Proc. IEEE IEDM*, pp. 773-776 (1997).

[3] E.M. Zielinski et al., "Damascene Integration of Copper and Ultra-Low-k Xerogel for High Performance Interconnects," *Proc. IEEE IEDM*, pp. 936-938 (1997).

[4] R. Haak, C. Ogden and D. Tench, "Cyclic Voltammetric Stripping Analysis of Acid Copper Sulfate Plating Baths, Part 1 : Polyether-Sulfide-Based Additives," *Plat. & Surf. Fin.*, April 1981.

[5] E.K. Yung, L.T. Romankiw and R.C. Alkire, "Plating of Copper into Through-Holes and Vias," *J. Electrochem. Soc.*, pp. 206-215 **136**, 1 (1989).

[6] T. Pearson and J.K. Dennis, "Effect of Pulsed reverse Current on the Structure and Hardness of Copper Deposits Obtained from Acidic Electrolytes Containing Organic Additives,"*Surface and Coatings Tech.*, **42**, pp. 69-79 (1990).

[7] V.A. Lamb and D.R. Valentine, "Physical and Mechanical Properties of Electrodeposited Copper: I. Literature Survey," *Plating*, pp. 1289-1311, Dec. 1965.

[8] H.J. Wiesner and W.P. Frey, "Some Mechanical Properties of Copper Electrodeposited from Pyrophosphate and Sulfate Solutions," *Plat. & Surf. Fin.*, pp.51-56, Feb. 1979.

[9] D. Anderson, R. Haak, C. Ogden, D. Tench and J. White, "Tensile Properties of Acid Copper Electrodeposits," *J. Appl. Electrochem.*, pp. 631-637, **13** (1985).

[10] W. Freitag, C. Ogden, D. Tench and J. White, Determination of the Individual Additive Components in Acid Copper Plating Baths," *Plat. & Surf. Fin.*, v70, (10), 55, (1983).

[11] P. Bratin, "New Developments in Use of CVS for Analysis of Plating Solutions," *Proceedings of AES Analytical Methods Symposium*, Chicago, IL (1985).

[12] K. Haak, "Ion Chromatography in the Electroplating Industry," *Plat. & Surf. Fin.*, September 1983.

[13] E.A.M.F. Dahmen, *Electroanalysis: Theory and Applications in Aqueous and Nonaqueous Media and in Automated Chemical Control*, p. 346, Elsevier Science Publishers B. V., Amsterdam (1986).

[14] R. Haak, C. Ogden and D. Tench, "Cyclic Voltammetric Stripping Analysis of Acid Copper Sulfate Baths, Part 2 : Sulfoniumalkanesulfonate-Based Additives," *Plat. & Surf. Fin.*, March 1982.

[15] Qualiplate QP-4000 User's Manual, V. 1.4, January 26, 1993.

[16] G.L. Fisher and P.J. Pellegrino, "The Use of Cyclic Pulse Voltammetric Stripping for Acid Copper Plating Bath Analysis," *Plat. & Surf. Fin.*, June 1988.

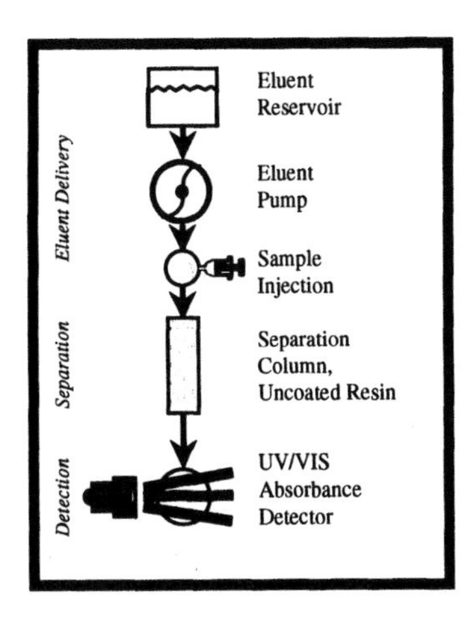

Figure 1. HPLC Schematic

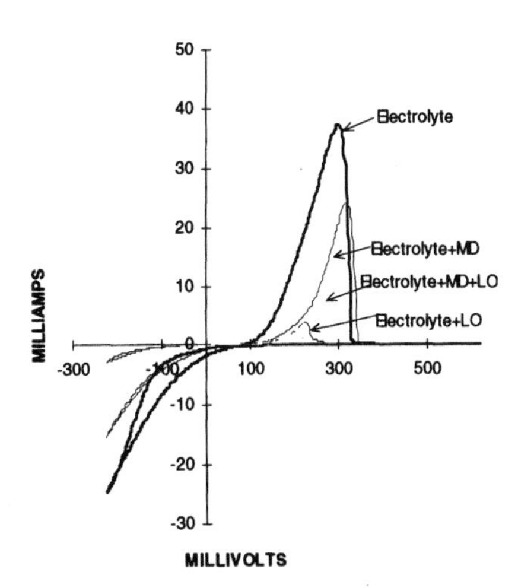

Figure 2. CVS Voltammograms of Enthone Cubath M®

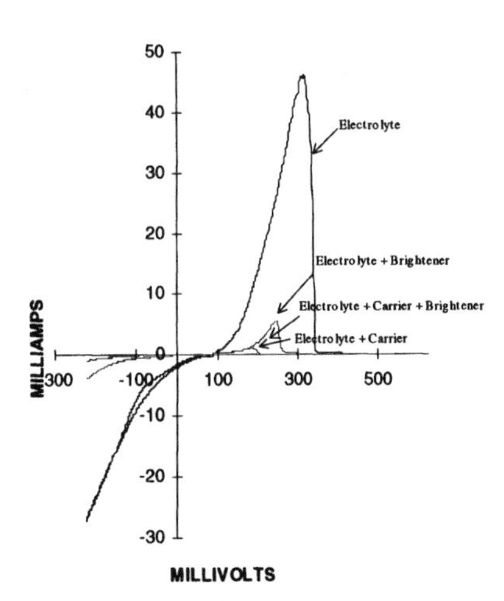

Figure 3. CVS Voltammograms of Shipley Electroposit 1100®

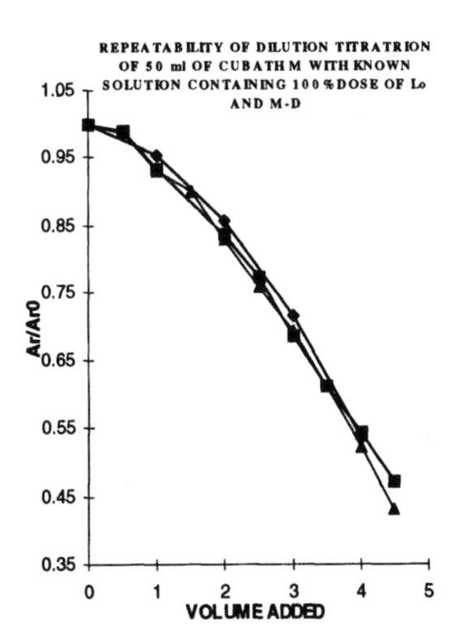

Figure 4. CVS Repeatability

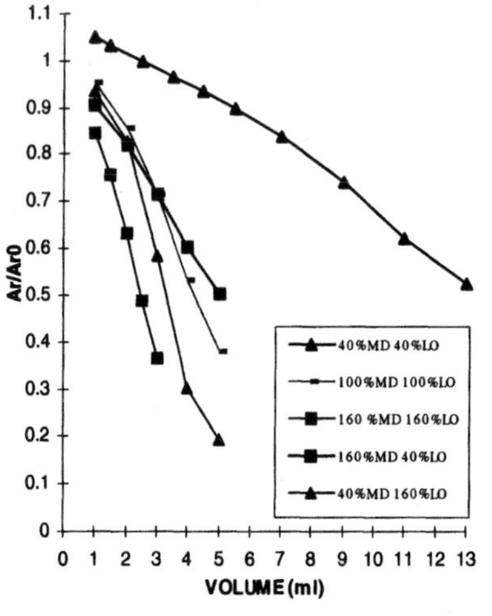

Figure 5. Normalized Stripping Peak Area vs. Titration Volume

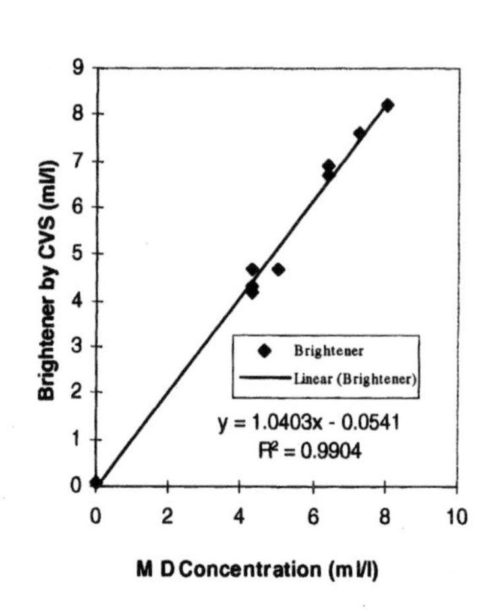

Figure 6. Brightener Analysis vs. M D Concentration

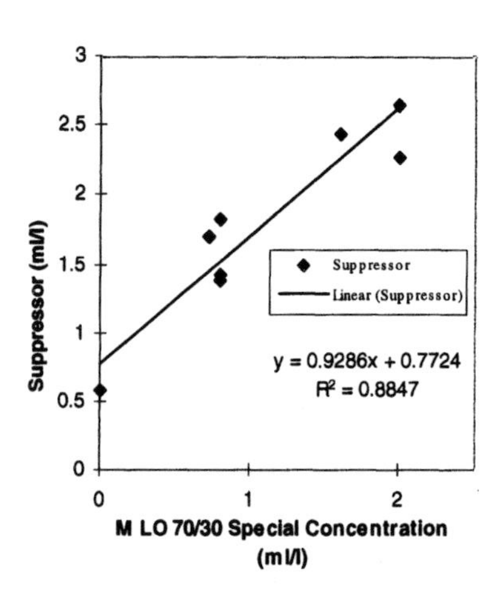

Figure 7. Suppressor vs. M LO Concentration

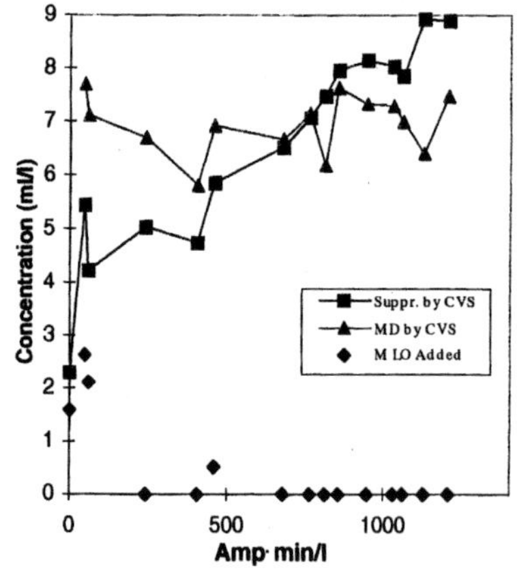

Figure 8. Organic Analysis vs. Bath Life

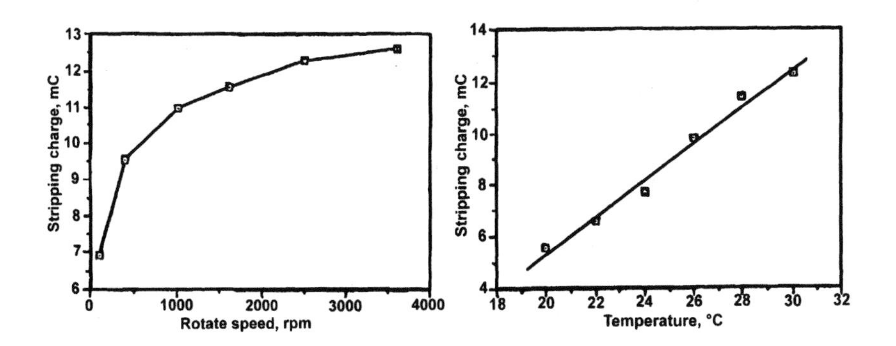

Figure 9. CPVS RDE Speed Dependence

Figure 10. CPVS Temperature Dependence

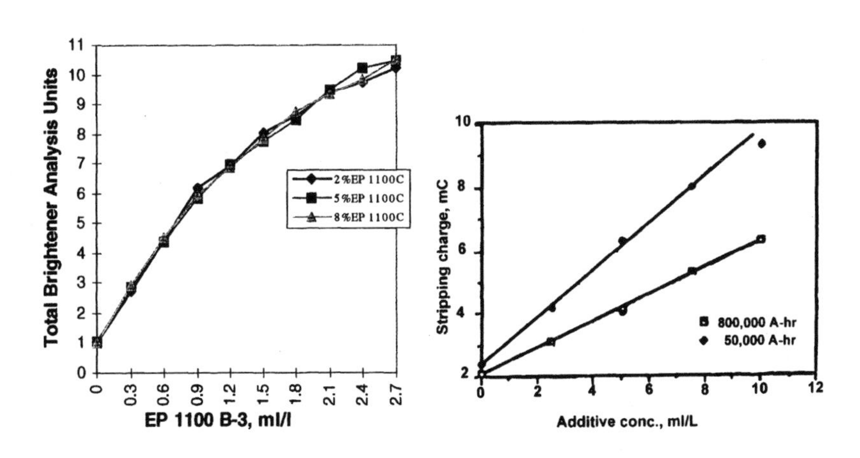

Figure 11. TBA Value as a Function of Organic Additives

Figure 12. Shipley Bath Aging Effects

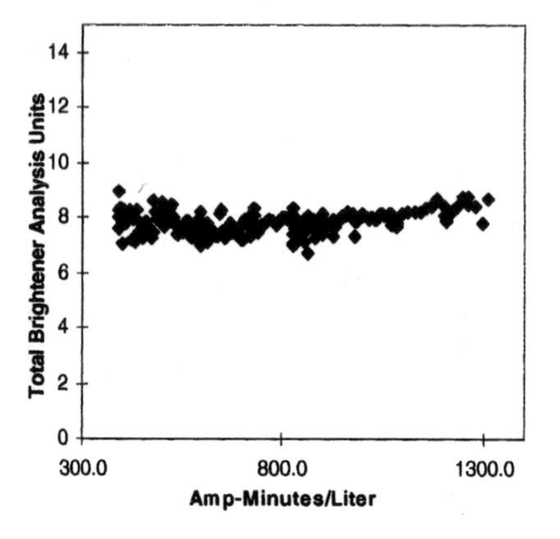

Figure 13. Brightener Analysis (TBA) vs. Bath Life

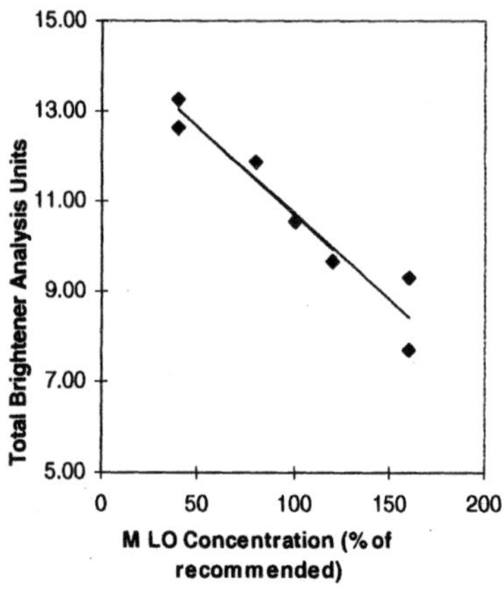

Figure 14. CPVS Analysis of CUBATH M

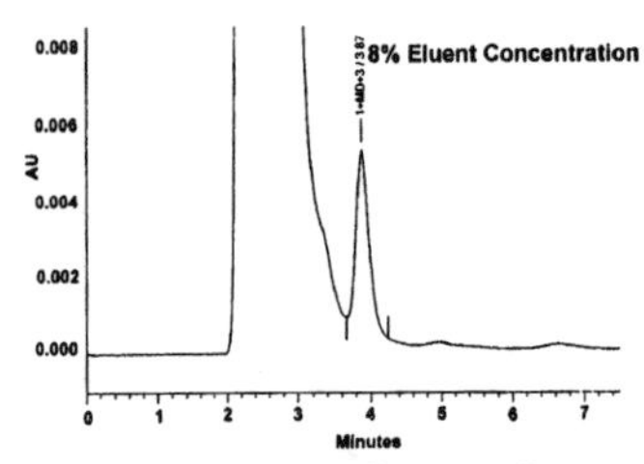

Figure 15a. Peak Separation vs. Eluent Concentration

2% Eluent Concentration

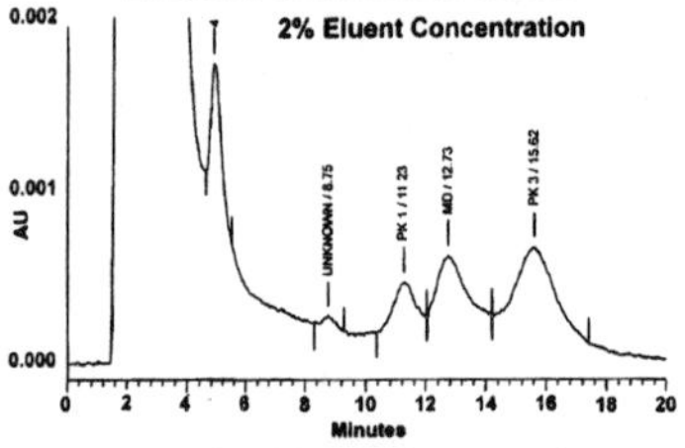

Figure 15b. Peak Separation vs. Eluent Concentration

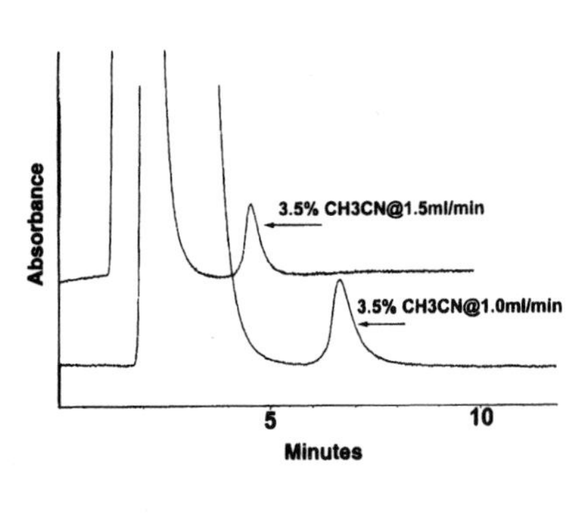

Figure 16. HPLC vs. Flow Rate

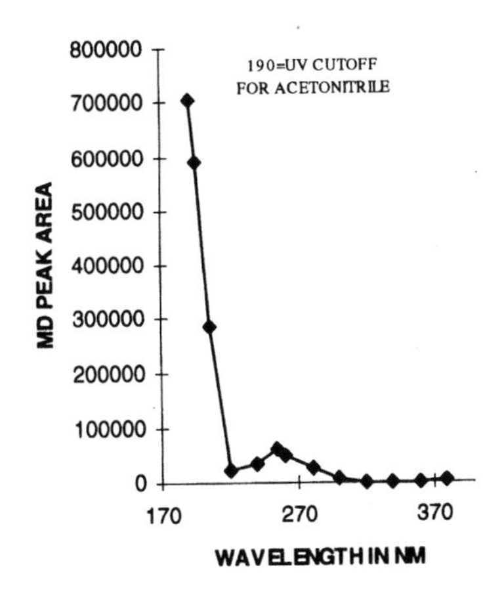

Figure 17. MD at Various UV Wavelengths

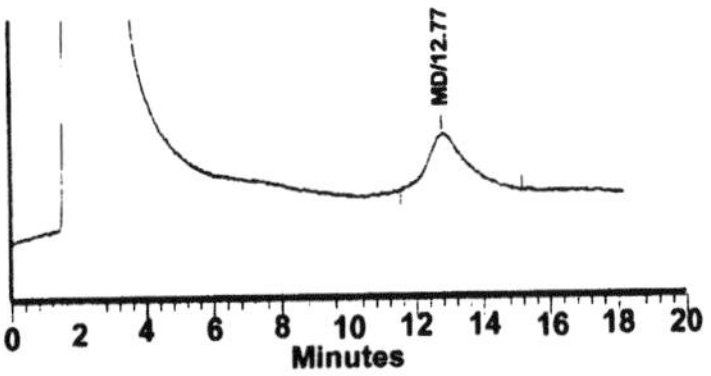

Figure 18. HPLC Chromatogram of Fresh Acid Copper Plating Bath with Organic Additives

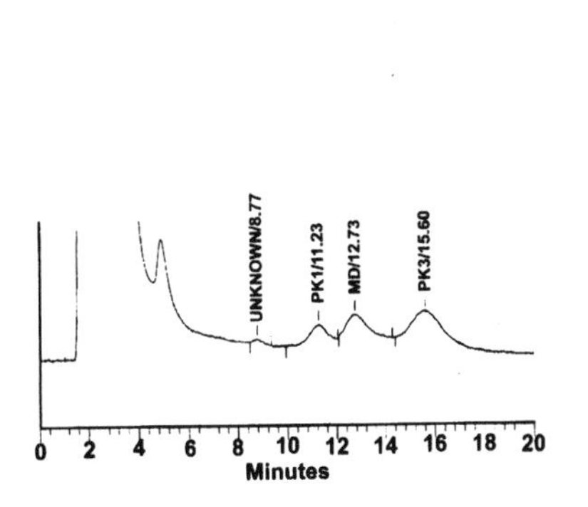

Figure 19. HPLC Chromatogram of Acid Copper Plating Bath Showing Organic Reaction Byproducts

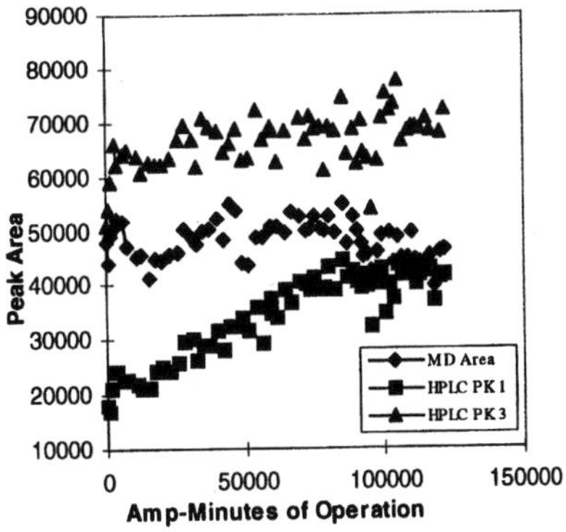

Figure 20. HPLC Analysis of Marathon Run

DAMASCENE COPPER ELECTROPLATING

FOR CHIP INTERCONNECTIONS

P.C. Andricacos, C. Uzoh, J.O. Dukovic, J. Horkans and H. Deligianni

IBM T.J. Watson Research Center
Yorktown Heights, N.Y. 10598

ABSTRACT

Damascene copper electroplating for on-chip interconnections, a process that we conceived and developed in the early 1990s, makes it possible to fill submicron trenches and vias with copper without creating a void or a seam and has thus proven superior to other technologies of copper deposition. We discuss here the relationship of additives in the plating bath to superfilling, the phenomenon that results in superconformal coverage, and we present a numerical model which accounts for the experimentally observed profile evolution of the plated metal.

INTRODUCTION

On-chip interconnections comprise a multilevel structure of fine wiring located on top of the transistor circuitry of logic or memory chips whose role is to connect circuits together. To avoid significant degradation of circuit speed, on-chip interconnections should permit rapid signal transmission among the various parts of the circuitry. Ever since the development of the integrated circuit about 40 years ago, the most pervasively used materials for the fabrication of the wiring structure have included aluminum as the conductor (or more recently an aluminum-copper alloy for better reliability [1]) and silicon dioxide as the insulator. The transition to copper as the conductor and a better insulator began with IBM's announcement in September 1997 [2], and signals one of the most important changes in materials that the semiconductor industry has experienced since its creation. Copper metallization was implemented first since significant gains can be obtained by copper alone.

It has been stated that the three major challenges in the implementation of copper have been i) the method of depositing and ii) patterning the metal and iii) finding a suitable barrier material that prevents the copper metal from diffusing into the insulator material. Although these generalizations neglect the enormous integration and manufacturing challenges, they serve to highlight the fact that a combination equivalent to blanket sputter deposition and Reactive Ion Etching technology used to deposit and pattern aluminum was not available for copper. It is precisely this technological challenge that electrodeposition met in combination with an integration approach called Damascene [3].

In IBM, various form of Physical Vapor Deposition (including sputtering, dep-etch, Electron Cyclotron Resonance, reflow), Chemical Vapor Deposition (CVD), and electroless plating were all being examined initially as potential methods of depositing copper. Electrolytic copper was simply not on the horizon. It only existed as a possibility in the minds of a very small group of researchers who had conceptualized the phenomenon of superfilling (to be discussed later in the article), a phenomenon that makes it possible to fill submicron Damascene structures without defects [4]. It turns out that superfilling is an attribute unique to electrodeposition and thus played the key role in the success of the technology. The first Damascene - copper wafers were electroplated towards the end of 1989 and the beginning of 1990. It only took a few months for the copper integration team to realize that electroplated copper was the way to go since it delivered the best quality copper of all methods that were being evaluated at the time and offered potential for throughput improvement and cost reduction. The summer of 1991 was a period of intense plating effort; this led to the formal adoption of plating later that year as the fabrication method of a bipolar device that was part of IBM's semiconductor product plans at the time.

The eclipse of bipolar technology in favor of CMOS technology lead to a delay in the implementation of the copper interconnect program, which was revived in 1994 [5] and was implemented in manufacturing by the end of 1997. During this period efforts in electroplating development focused on i) the infrastructure for plating tools capable of handling a high throughput of 200 mm wafers and ii) the detailed integration of electroplating technology together with all other fabrication technologies to produce high reliability interconnect structures. Fabrication of copper / silicon dioxide wiring structures and performance of resulting devices were described in detail in 1997 [6]. Copper on-chip interconnection technology entered high - volume manufacturing in 1998 at IBM's plant at Burlington, Vermont.

ADVANTAGES OF COPPER INTERCONNECTIONS

These have been described in detail by Edelstein [8 - 10]. He examines the factors that contribute to *logic gate delay* and points out the need for "*hierarchical wiring*" structures where the *lower* wiring levels are at a minimum possible pitch and thickness to minimize capacitance and maximize wiring density, while the *higher* wiring levels are scaled horizontally and vertically to maintain a constant capacitance while reducing resistance. In addition to the low - capacitance (C) and low - resistance (R) wires, the hierarchy allows for low - RC wires of intermediate dimensions. Typical width for the low - lying wires is about 0.25 microns, while it can increase by an order of magnitude for high - lying wires.

A reduction in the wiring resistance by as much as 45 % that can be accomplished with copper as the core interconnection material, thus has a significant impact not only on the resistance of the "fat" wiring but also (indirectly) on the capacitance of the "thin" wiring whose dimensions can be scaled downward. This allows an additional benefit in reduction of crosstalk, which is difficult to quantify in general, but has already led to significant benefits in noise-related chip design and performance results.

It is important to note that wire resistance must take into account not only the core material of the interconnect (Al(Cu) or Cu) but also any other (usually higher resistivity) materials used to surround the core material. Edelstein makes the comparison between Al(Cu) wires made by RIE that have bottom and top Ti layers (which, during sintering, produce $TiAl_3$

with a resistivity of about 35 μΩ–cm and 3.5 times the original Ti thickness), and Cu wires made by a Damascene process (described below) that leaves cladding on the bottoms and the sides of the lines. The cladding also has high resistivity but does not react with Cu and serves as a diffusion barrier and an adhesion layer.

In addition to resistance reduction, Edelstein cites two other primary advantages of copper interconnection technology: high electromigration resistance and amenability to dual Damascene processing. Electromigration is the diffusive transport of conductor material as a result of the momentum transfer by passage of extremely high electron current densities. Atoms are regarded as being driven in the direction of the "*electron wind*" causing the cathode end of the wire to become depleted and ultimately resulting in the formation of voids. Scaling down of low-lying wires to reduce capacitance and to increase wiring density as mentioned above as well as the inexorable trend to higher performance and transistor current-drives leads to increased current densities. For this reason, an extendible conductor metallurgy must survive stringent electromigration tests. In addition to electromigration, stress voiding, mass transport resulting from thermally - induced stress gradients, is another reliability concern. Electromigration tests of 0.3 μm lines conducted at an elevated temperature (295 °C) and a stress current of 2.5 million amps per square centimeter showed a T_{50} lifetime (the time interval needed for 50% of the devices tested to fail) that was more than 100 times longer for Cu than Ti/Al(Cu)/Ti lines. Excellent results were obtained for stress migration testing [6]. Both sets of results argue strongly in favor of the extendibility of copper interconnections to higher usable current densities and better overall performance.

The last primary advantage of copper interconnections in Edelstein's analysis, namely amenability to dual Damascene processing (and its concomitant process cost reduction and other advantages), relates to the manner that electroplating is integrated together with other deposition, removal, patterning, and lithographic processes used in the fabrication of the device. Integration places tremendous requirements on the way electrodeposition is carried out.

INTEGRATION OF ELECTROPLATING PROCESSES

Because current must be conducted to the entire wafer surface for electroplating to occur at all points, a seed layer is usually deposited in a blanket fashion prior to the electroplating step by a process such as sputtering or equivalent (Figure 1). In Damascene processing, the seed layer is deposited on top of the patterned substrate and therefore electroplating occurs everywhere, inside and outside features, from the bottom upwards and from the sides inwards. There are good reasons for adopting a Damascene process of integration. A barrier is usually required to prevent interaction between the conductor and the insulator and to provide good adhesion between the two. A via level and a trench level can be metallized and planarized simultaneously (dual Damascene process) thus saving in fabrication costs. The requirement of a barrier layer and amenability to dual-level fabrication are some of the advantages of a Damascene process of integration.

Damascene electroplating as applied to the process of making copper on-chip interconnections, represents a deviation from another popular method of integrating electroplating called "through-mask plating" (Figure 1). Here, the seed layer is deposited over the wafer surface *prior* to patterning. Electroplating occurs only in those lithographic areas that are not covered by the masking material. Through-mask plating has been used extensively in the fabrication of inductive recording heads and in other applications [10 - 12].

Both through - mask and Damascene electroplating require post - electrodeposition steps for the metallization process to be completed. In through - mask plating, the seed layer remains in place after completion of the plating step and must be removed by a dry or wet etching process. In a Damascene process, excess material is deposited on top of the useful wiring structure and must be removed by a planarization process called Chemical Mechanical Planarization (CMP). CMP is a process with electrochemical character whose significance in the fabrication of copper wiring cannot be overemphasized. However, a CMP discussion is well beyond the scope of the present article.

The foremost challenge in Damascene plating is to fill via holes, trenches, and their combinations completely, without voids or seams. Figure 2 shows possible ways for the profile of plated copper to evolve in time. In conformal plating, deposition at a uniform rate at all points of a feature leads to the creation of a seam, or, if the shape of the feature is reentrant, a void. Subconformal plating leads to the formation of a void even in straight-walled features. Subconformal plating results when substantial depletion of the cupric ion in the plating solution inside the feature leads to significant concentration overpotentials which, in turn, cause the current to flow preferentially to more accessible locations outside the feature. Also, if the feature depth is large (say in excess of 50 μm), the ohmic drop in the electrolyte may cause nonuniformity in the distribution of the current in favor of external feature locations. For defect-free filling, a deposition rate that increases with depth along the sides and is highest at the bottom of the feature is desired.

As early as 1990 in IBM we discovered that plating from certain plating solutions that contain additives leads to this superconformal deposition behavior that eventually produces void-free and seamless structures (Figure 3). We call this behavior "superfilling" [4].

SUPERFILLING

Superfilling can be understood qualitatively by comparing deposition rates at different points along the feature profile, as shown in Figure 2 (bottom left). The most noticeable difference may occur between points A and E. However, we consider the difference between points B and C, i.e., two points at different elevations on the side wall, to be a more fundamental determinant of superfilling, especially in high-aspect-ratio cavities. Since any two points are electrically shorted by the seed layer in the solid phase, the difference in total overpotential, $\Delta\eta = \eta^B - \eta^C$, must be equal to 0. The difference in total overpotential can be written as the sum of the difference in ohmic drop in the electrolyte, $\Delta\phi_\Omega$, plus the differences in concentration and surface overpotentials, $\Delta\eta_C$ and $\Delta\eta_S$, respectively. Thus, $\Delta\phi_\Omega + \Delta\eta_c + \Delta\eta_s = 0$. Because the feature is small (typically submicron), the difference in ohmic drop in the electrolyte $\Delta\phi_\Omega$ is negligible (on the order of one microvolt). The situation here is very different from the one that prevails in through-hole plating, where the ohmic drop in the electrolyte is substantial because the holes are typically tens of microns in depth. Assuming further that the difference in the cupric ion concentration is negligible (an assumption whose validity obviously diminishes with increasing aspect ratio and decreasing feature size), the difference in concentration overpotential becomes negligible, $\Delta\eta_c \sim 0$. One finally obtains that $\Delta\eta_s \sim 0$. Writing the surface overpotential in terms of a Tafel expression, it follows that $\frac{RT}{aF}\ln\frac{i^B}{i_0^B} = \frac{RT}{aF}\ln\frac{i^C}{i_0^C}$, where $\imath_0$is the exchange current density, a is the cathodic transfer coefficient, and $\imath$ is the current density (proportional to the rate of copper deposition by Faraday's law). *Assuming that the exchange current density is lower the higher the flux of additives,* it follows that $\imath_0^B < \imath_0^C$simply because point B is more accessible to additive diffusion than point C. It follows that $\imath^B < \imath^C$, i.e. the rate of copper deposition is higher at point C.

With the aim of improving our understanding of shape-change behavior in Damascene plating through a quantitative framework, we undertook a numerical modeling effort in 1991. We adapted a numerical model that had been applied to leveling in conventional electroplating (13) and to shape evolution in through-mask plating (14).

Physically, the essential characteristics of the model are as follows (Figure 4). The local rate of copper deposition is proportional to the local current density i by Faraday's law. The current distributes itself so as to take the path of least resistance as it approaches the trenched electrode surface. Transport of the metal ion M (in this case, the cupric ion, Cu^{2+}) and of an inhibiting additive A is dominated by convection except within a concentration boundary layer that extends several tens of microns from the electrode surface. We treat this zone as stagnant, with each species moving only by diffusion. At the outer edge of the boundary layer, we assume that the cupric ion and the inhibitor are at their well-mixed bulk concentrations. Since the feature dimensions are much smaller than the boundary-layer thickness, we take i to be uniform at the boundary-layer edge. The current encounters a voltage barrier or overpotential at the electrode surface. Since the barrier becomes higher as current density increases (according the Tafel kinetic expression (16)) there is no reason for point A to receive a higher current density than point B unless one of the the following cases applies: **1.** The ohmic pathway to point A is significantly more favorable than to point B; **2.** The metal ion has been depleted to a significantly lower concentration at point B than at point A (difference in concentration overpotential); or **3.** The rate constant for electrodeposition, i_0, is higher at point A than at point B as the result of differential inhibition or catalysis. We can rule out case **1**, since ohmic drop in the plating solution is negligible at the length scale of 1 micron. Case **2** only applies as the current density approaches the transport-limited current density i_L, which is nearly always avoided. (It is noteworthy that neither effect **1** nor **2** could cause superconformal plating; rather, each would result in subconformal coverage.) We are left with case **3**. It is well known that i_0 can be strongly influenced by adsorbed inhibitors. The surface concentration of adsorbate would have no reason to vary along the profile unless it were influenced by the diffusive transport of the inhibitor A. It must be recognized that diffusion cannot have a sustained effect unless the adsorbate is *consumed* (either by reaction or by incorporation into the deposit). The simplest and strongest case of diffusion influence is diffusion *control*; hence we assume, for simplicity, that the inhibitor concentration $c_A$drops to zero adjacent to the electrode surface. Under this assumption, the flux of the inhibitor, N_A, is easy to compute from boundary-value problem corresponding to Fick's second law of diffusion. From the nature of the Laplace equation, we know that strong field effects driven by the profile geometry can arise, causing strong variations in N_A along the profile. Taking the view that the surface concentration of the adsorbates responsible for retarding electrocrystallization is determined by a dynamic balance between the arrival of fresh additive and its consumption by reaction or by incorporation in the deposit, we relate the degree of kinetic inhibition directly to the flux of inhibitor, N_A. We do this simply by multiplying the rate constant for electrodepositon by an inhibition factor ψ, which ranges between 1 and 0, decreasing monotonically with the dimensionless inhibitor flux N_A^*. The form of the expression $\psi(N_A^*)$ is discussed below.

A simple area-blockage treatment of inhibition (15) has been employed in a shape-change simulation to model classical leveling with some success (16). An equivalent description of inhibited kinetics was used in Reference 13, where the inhibition factor had the following form: $\psi = \frac{1}{1+K_{LEV}\frac{N_A^*}{N_M^*}}$. However, we found that such a treatment was not adequate to describe the shape-change behavior that we refer to as superfilling. In particular, the area-blockage model can describe differences in local kinetics necessary to cause slower plating outside the cavity than inside; but it cannot generate the magnitude of rate differentiation *within* the cavity that permits the rounding of the internal corners and the prevention of seam formation.

We found it necessary to use an inhibition expression $\psi(N_A^*)$ for which ψ varies gradually over a very wide range of N_A, *i.e.*, several orders of magnitude. Some experimental support for this finding is furnished by the observation that, in the plating bath with all components at standard concentration except for one inhibiting additive, the plating potential jumps significantly when the inhibitor concentration is raised from 2% to 4% of its nominal value, and this sensitivity extends over roughly two orders of magnitude in concentration. The expression we adopted was $\psi = \frac{1}{1+b\,N_A^{*\,p}}$. The fractional exponent p was introduced, somewhat empirically, to widen the dynamic range of fluxes over which differential inhibition can occur. Values of p = ¼ and b = 10 were chosen mainly to capture the corner rounding and general shape-change behavior observed experimentally.

The mathematical system is summarized in dimensionless form in Figure 4. All equations and nomenclature correspond directly to Reference 14, with three exceptions. First and most important, the present model uses a different expression for the inhibition factor ψ, as noted above. A second difference, of minor consequence, is that in the present model, we neglect the anodic or reverse-reaction term of the Butler-Volmer kinetic expression, Reference 14, Eqn. 19, leaving the simpler Tafel expression. A third difference is that the mean current density $\bar{i}$ in the present model (which enters the dimensionless groups Wa_T and Sh) is based on the superficial area rather than the topographic area of the trenched electrode.

An account of the problem statement of Figure 4 follows. Within a laterally symmetric section of the concentration boundary layer, there are three field variables, which all obey the Laplace equation: the dimensionless potential ϕ^*, the dimensionless metal-ion concentration, c_M^*, and the dimensionless additive concentration, c_A^*. The surface-normal derivatives $\nabla^*\phi^* \cdot n^*$, $\nabla^* c_A^* \cdot n^*$, and $\nabla^* c_M^* \cdot n^*$ (abbreviated in the Figure as $\phi^{*\prime}$, $c_A^{*\prime}$, and $c_M^{*\prime}$) are constrained to zero at the symmetry boundaries. (i.e., there are no fluxes across symmetry lines.) At the top of the boundary layer the potential gradient is taken to be uniform, $\phi^{*\prime} = 1$, and the metal ion and inhibitor are at their bulk concentrations, $c_M^* = 1$ and $c_A^* = 1$. It is only at the electrode surface that the three field variables, ϕ^*, c_M^*, and c_A^*, are coupled. Here, we impose c_A^* =0, in accordance with the assumption that the inhibitor is consumed under mass-transfer control. The resulting flux profile, $N_A^* = \nabla c_A^* \cdot \boldsymbol{n}$, enters the expression for ψ in the kinetic expression $\phi^{*\prime} = k\,\psi\, c_M^{*(\gamma+\frac{\alpha_c}{n})}\, e^{\frac{\phi^*}{Wa_T}}$, which relates the field variables ϕ^* and c_M^* (where k is a dimensionless rate constant, $k = i_0^{\infty, c_{A=0}} / \bar{i}$). The potential and the metal-ion concentration are also related by a flux-matching condition, namely $c_M^{*\prime} = Sh\ \phi^{*\prime}$.

The solution depends on Wa_T, Sh, $\gamma + \frac{a_c}{n}$, b, and p. (The rate constant k does not affect the current distribution under Tafel kinetics.) The parameters Wa_T, Sh, and $\gamma + \frac{a_c}{n}$ are not freely adjustable, but are determined from handbook constants and process conditions.

The numerical method (quadratic boundary element method) was the same as that of Ref 14, and the scheme for repositioning the nodes to represent profile evolution is essentially that used in Reference 13, with some improvements.

Figure 5 compares a cross-sectional SEM of a partially plated trench (top) with the model simulation (bottom). The trench width is 1.0 micron and the pitch spacing is 2.25 micron. The corresponding dimensionless parameter values are $Wa_T = 13,000$; $Sh = 0.008$; and $\gamma + \frac{a_c}{n} = 0.85$. Values for p and b in the expression for the inhibition factor were ¼ and 10, respectively. The match between experiment and simulation, though not perfect, is fairly good and indicates that the model, based on differential inhibition caused by diffusion-controlled additives, can describe superfilling behavior.

Extendibility of electroplating has been demonstrated down to 0.1 micron trenches [17]. It appears that superfilling is operational over a wide range of dimensions.[1]

SUMMARY

Copper on-chip interconnections represent not only a change in materials but also a new way of integration and a new way of depositing the conductor metal. Copper interconnections are superior to Al(Cu) interconnections because of the decreased resistance, improved reliability, and reduced process complexity. Electrodeposition has played a key role in making implementation of the technology possible since it can deposit copper in Damascene structures without defects such as seems or voids. This unique property of electrodeposition is due to a phenomenon we call "superfilling", in which the rate of the copper deposition reaction increases as one goes down into a feature as a result of the differential inhibition of the reaction kinetics by the additives present in the plating solution.

REFERENCES

1. I. Ames, F.M. d'Heurle, and R.E. Horstmann, "Reduction of Electromigration in Aluminum Films by Copper Doping," *IBM J. Res. Develop.*, **14**, 461(1970); C.-K. Hu, K.P. Rodbell, T.D. Sullivan, K.Y. Lee, and D.P. Boulder, "Electromigration and Stress - Induced Voiding in Fine Al and Al - Alloy Thin - Film Lines," *IBM J. Res. Develop.*, **39**, 465(1995).

2. L. Zuckerman, "IBM to Make Smaller and Faster Chips - Second Breakthrough in A Week Has Wide Uses," *The New York Times*, page D1, Monday, September 22 ,1997.

3. M.M. Chow, J.E. Cronin, W.L. Guthrie, W. Kaanta, B. Luther, W.J. Patrick, K.A. Perry, and C.L. Standley, "Method for Producing Coplanar Milti-level Metal/Insulator Films on a Substrate and for

[1] Superfilling is distinctly different from leveling. Leveling reduces the roughness of a surface and smooths defects such as scratches; superfilling produces void-free and seamless deposits inside lithographically defined cavities with vertical walls and high aspect ratios.

Forming Patterned Conductive Lines Simultaneously with Stud Vias," United States Patent 4,789,648 (Dec. 6, 1988).

4. P.C. Andricacos, C. Uzoh, J.O. Dukovic, J. Horkans, and H. Deligianni, "Damascene Copper Electroplating for Chip Interconnections," *IBM J. Res. Develop.*, **42**, 567 (1998).

5. B. Luther, J.F. White, C. Uzoh, T. Cacouris, J. Hummel, W. Guthrie, N. Lustig, S. Greco, N. Greco, S. Zuhoski, P. Agnello, E. Colgan, S. Mathad, L. Saraf, E.J. Weitzman, C.K. Hu, F. Kaufman, M. Jaso, L.P. Buchwalter, S. Reynolds, C. Smart, D. Edelstein, E. Baran, S. Cohen, C.M. Knoedler, J. Malinowski, J. Horkans, H. Deligianni, J. Harper, P.C. Andricacos, J. Paraszczak, D.J. Pearson, and M. Small, "Planar Copper - Polyimide Back End of the Line Interconnections for ULSI Devices," *Proceedings of the 10th International IEEE VLSI Multilevel Interconnection Conference*, 1993, p. 15.

6. D. Edelstein, J. Heidenreich, R. Goldblatt, W. Cote, C. Uzoh, N. Lustig, P. Roper, T. McDevitt, W. Motsiff, A. Simon, J. Dukovic, R. Wachnik, H. Rathore, R. Schulz, L. Su, S. Luce, and J. Slattery, "Full Copper Wiring in a Sub-0.25 μm CMOS ULSI Technology," *Tech. Digest IEEE Int. Electron Devices Mtg.*, 1997, p. 773.

7. D.C. Edelstein, G.A. Sai-Halasz, and Y.-J. Mii, "VLSI On-Chip Interconnection Performance Simulations and Measurements," *IBM J. Res. Develop.*, **39**, 383 (July 1995);

8. D.C. Edelstein, "Advantages of Copper Interconnects", *Proceedings of the 12th International IEEE VLSI Multilevel Interconnection Conference*, 1995, p.301.

9. D.C. Edelstein, J. Heidenreich, R. Goldblatt, W. Cote, C. Uzoh, N. Lustig, P. Roper, T. McDevitt, A. Stamper, W. Motsiff, A. Simon, J. Dukovic, R. Wachnik, H. Rathore, P. McLaughlin, T. Katsetos, R. Schulz, L. Su, N. Rohrer, and S. Luce, "Sub-0.25 μm CMOS ULSI Technology with Multilevel Copper Interconnections," Realize, Inc., Tokyo, 1998, p. 207.

10. P.C. Andricacos and L.T. Romankiw, "Magnetically Soft Materials: Their Properties and Electrochemistry," in *Advances in Electrochemical Science and Engineering*, H. Gerischer and C.W. Tobias, Eds., volume 3, pp. 227-321, VCH, New York, 1993.

11. M. Datta, R.V. Shenoy, C. Jahnes, P.C. Andricacos, J. Horkans, J.O. Dukovic, L.T. Romankiw, J. Roeder, H. Deligianni, H. Nye, B. Agarwala, H.M. Tong, and P. Totta, "Electrochemical Fabrication of Mechanically Robust PbSn C4 Interconnections," *J. Electrochem. Soc.*, **142**, 3779 (1995).

12. P.C. Andricacos, J.H. Comfort, A. Grill, D.E. Kotecki, V.V. Patel, K.L. Saenger, A.G. Schrott, "Plating of Noble Metal Electrodes for DRAM and FRAM," United States Patent 5,789,320 (Aug. 4, 1998).

13. J.O. Dukovic and C.W. Tobias, "Simulation of Leveling in Electrodeposition," *J. Electrochem. Soc.* **137**, 3748(1990).

14. J.O. Dukovic, "Feature-scale Simulation of Resist-patterned Electrodeposition," *IBM J. Res. Develop.*, **37**, 125(1993).

15. S.I. Krichmar, "Theory of the Leveling Effect in the Electrochemical Behavior of Metals," *Sov. Electrochem.*, 1 (7), 763 (1965) (translated from *Elektrokimiya*, 1 (7), 858 (1965)).

16. K.G. Jordan and C.W. Tobias, "The Effect of Inhibitor Transport on Leveling in Electrodeposition," *J. Electrochem. Soc.*, **138**, 1251 (1991).

17. C.-K. Hu, K.Y. Lee, L. Gignac, S.M. Rossnagel, C. Uzoh, K. Chan, P. Roper, and J.M.E. Harper, "Extendibility of Cu Damascene to 0.1 μm Wide Interconnections," in *Advanced Interconnects and Contact Materials and Processes for Future Integrated Circuits,* S.P. Murarka, M. Eisenberg, D.B. Fraser, R. Madar, and R. Tung, Eds.,Materials Research Society Symposium Proceedings Volume **514**, 1998, p. 287.

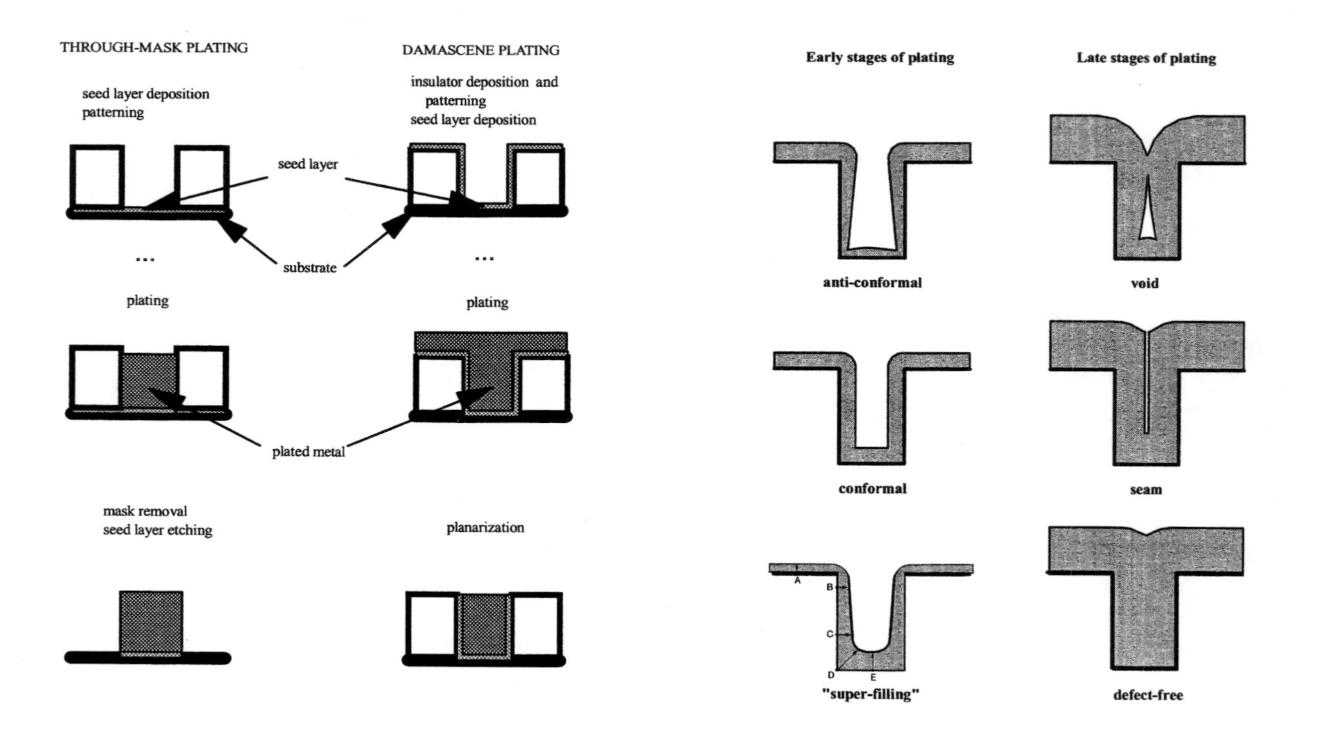

Figure 1. Integration approaches of electrodeposition; through-mask (left), Damascene (right)

Figure 2. Types of profile evolution in Damascene plating

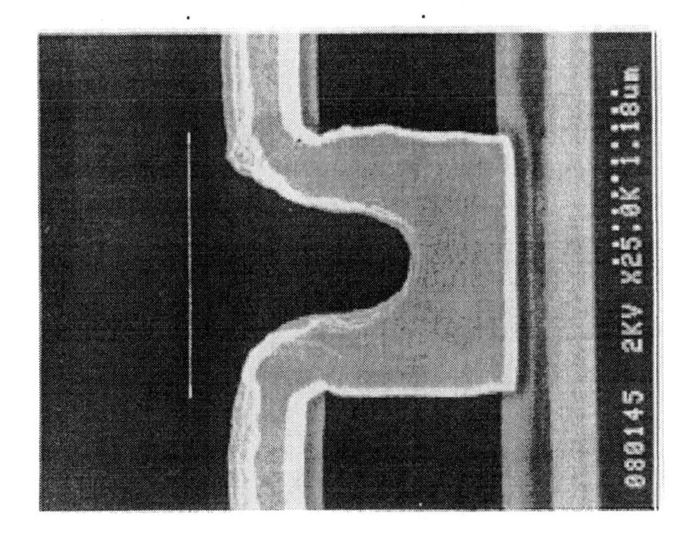

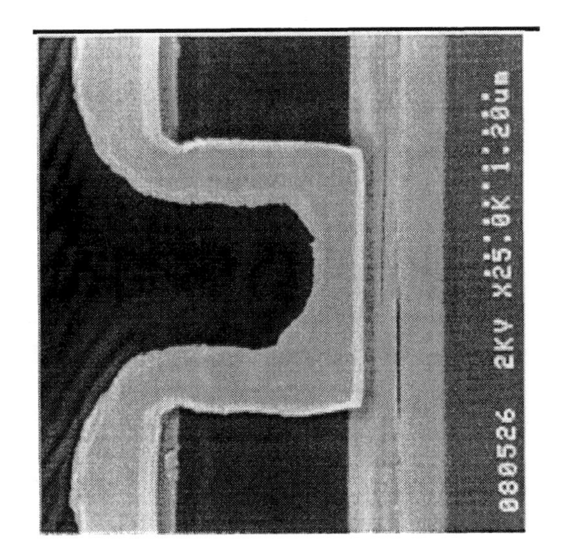

Figure 3. Cross section of plartially filled lines showing: superfilling from a complete set of additives (left); conformal plating with edge roundning from incomplete set of additives (right)

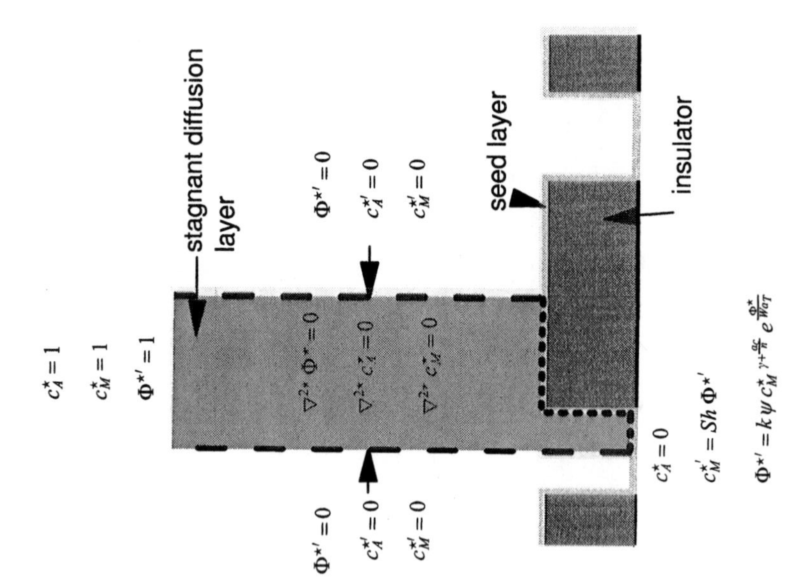

Figure 4. Model of superfilling

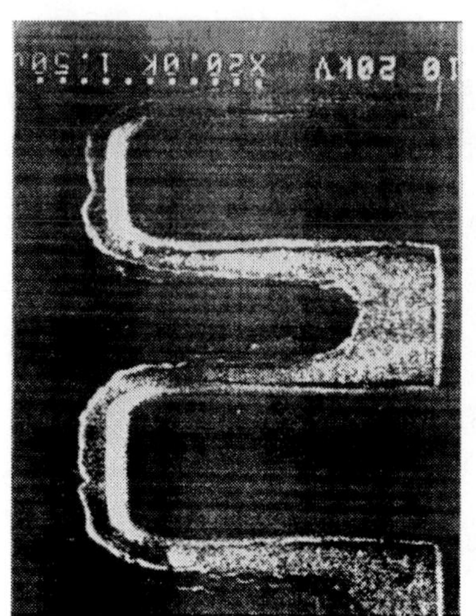

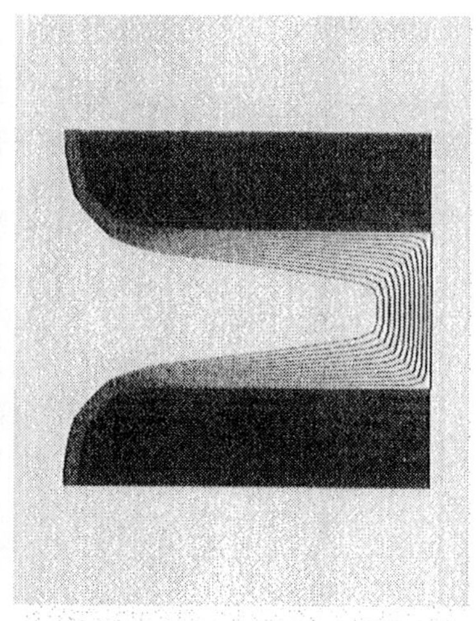

Figure 5. Comparison of suerfilling as observed experimentally (top) with model prediction (bottom)

CONSIDERATIONS FOR INTEGRATION OF ELECTROPLATED COPPER ONTO SEMICONDUCTOR SUBSTRATES

Cindy R. Simpson, David M. Pena, J. Vernon Cole

Advanced Products Research and Development Laboratory, Motorola, Inc.
3501 Ed Bluestein Blvd., Austin, TX 78721

ABSTRACT

The use of copper as a back-end metallization scheme has been receiving a fair amount of attention lately, fueled strongly by announcements by Motorola and IBM. Several methods currently exist to deposit and fill the vias and trenches, however, electroplating has been shown to be one of the more viable of these choices. A thin seed layer is first deposited, followed by the electroplated copper to fill, which is then removed in the field by CMP. Since the fill and CMP of copper are relatively new to the industry, the integration aspects need to be considered in order for a viable product, with good yields, to be produced.

This paper will discuss the factors which influence the uniformity of the electrodeposited copper films on semiconductor substrates. The seed thickness, anode size, diffuser, flow rates and rotation rates have all been varied. A general factorial DOE was used and the analysis done with JMP. Non-uniformity values at 1 sigma, as well as 49 point contour plots have been used to determine the important variables in the control of the uniformity for copper plating applications.

INTRODUCTION

As microelectronics begin to scale, the need for a lower resistivity material, combined with a more reliable material, will become an increasingly important consideration. Copper has a bulk resistivity of 1.68 μohm-cm as compared to an aluminum resistivity of 2.65 μohm-cm. In addition, the electromigration lifetimes have been reported to be improved when moving to a copper interconnect (1-3).

However, copper has a number of obstacles to overcome in order for it to be used as a production material. First, the process scheme requires an inlaid approach, whereby the dielectric material is patterned, followed by the copper fill and finally a chemical mechanical polish (CMP) step to remove the residual copper from the field. The deposition of the copper, therefore, must follow the needs of the CMP process. If the CMP tool polishes edge fast, then the plating must deposit edge heavy and vice versa for the case of edge light. It is, therefore, necessary to know the parameters which govern the uniformity of the plating , or fill step.

Since copper electroplating has been in use in the printed wiring board industry for many decades, an enormous amount of work has already been done to determine the parameters which have the most affect on the uniformity of the deposited films. One of the key parameters has always been the distance of the anode to the cathode, or plating surface. However, in a production tool, this distance is usually set by the chamber dimensions. The anode size is something that is easily changed as is the seed thickness, rotation rate of the wafer, the flow rate of the solutions, and the diffuser panel. The relative effect that each of these has on the uniformity has been investigated and will be reported.

EXPERIMENTAL

Depositions of electroplated copper were carried out on a Semitool Equinox EQH444PR system, which is a commercially available electroplating tool consisting of dedicated plating chambers and dedicated spin/rinse/dry chambers. Copper sulfate based plating bath was used with the components held at a constant value throughout the experiment. The solution in this system flows upward from a large reservoir, around the outside of the anode and then through the diffuser. The solution then touches the wafer and is drawn to the outside via capillary action. The solution then flows over an overflow and below tot he reservoir. The chamber is shown schematically in Figure 1.

The wafers were blanket seed having 500Å, 2000Å, or 5000Å of a copper seed layer. The underlying barrier was not varied during the DOE so as to not have this have an effect on the uniformity.

The plating rate was held constant throughout the experiment so that it also did not enter in as a variable. The final film thickness was always 1μm +/- 200Å. The plating times were adjusted for each seed thickness so as to have the constant 1μm film.

All rotation rates and flow rates were adjusted via the software on the tool. Rotation rates studied were 0, 20, and 50 rpm. The maximum was chosen so as to avoid too much turbulence at the wafer. The flow rates studied were 1, 3, and 6.5 gpm. The maximum was chosen due to the limitations of the tool. These two conditions were not used in the DOE but rather, separate, using a standard set of conditions. For variations in the rotation rate, a flow rate of 6.5 gpm was used. Conversely, for the variable flow rate study, a constant 20rpm of the wafer was used. No diffuser was used in this section of the study so as to remove any effects that could be caused by either the solution flow through a diffuser, or solution turbulence between a diffuser and the wafer. Two barrier/seed combinations were tested using the variable rotation rate. The thickness of the copper seed layer was held constant.

Copper anodes were milled to radii of 8", 6", 4", and 2". Each anode contained the copper plating solution vendors' recommended levels of phosphorous (≤0.1%). The anodes were backed with a Teflon panel so as to minimize the flow effects on the anode film. In addition, the anodes were seasoned similarly to provide a consistent anode film.

There were three diffusers, in addition to no diffuser, used for this study. The first was the standard diffuser recommended by the tool vendor. It consisted of a random array of holes about an approximately 8" diameter. Another diffuser was doughnut shaped with a 0.75" ring of polyethylene about the outside edge. There were no holes in the polyethylene area of this diffuser. The final diffuser was another doughnut shape with a 1.25" ring of polyethylene about the outside edge.

The DOE was a general factorial design with all combinations of levels for the three factors. A list of 48 experiments were determined. The factors were the anode size, the diffuser design, and the seed thickness.

Data were obtained on a KLA/Tencor NC-110 non-contact film thickness tool. The resistivity of the film was assumed to be 1.9 μohm-cm for all analyses so that thickness values were the output. A 49 point contour map was obtained for all conditions. Two wafers were run for each condition. An average value for the non-uniformity is reported.

RESULTS AND DISCUSSION

A) Flow Rate Effects

Variations in the flow rate were found to be quite dramatic with the increased flow yielding an increase in the non-uniformity of 40% as is seen in Figure 2. It was expected that the potential drop across the wafer would be the dominant source of non-uniformity and the sensitivity to parameters affecting mass transfer would be weak. However, for this study, a very low current waveform was employed which is expected to increase the importance of these secondary effects.

Modeling work done at Motorola, using a simulation of the flow in the chamber geometry, provided quantitative insight into the cause of the observed trends. Without the diffuser plate, the solution flows up near the chamber wall and then out past the wafer edge into an overflow area. As the flow rate increases, more of the solution follows a direct path from the inlet to the overflow exit. Therefore, there is a good supply of replenished solution at the wafer edge but diffusion radially through the chamber is the only mechanism for transporting the copper to the majority of the wafer surface. These mass transfer effects are believed to lead to the observed edge heavy plating and increased non-uniformity with increased flow rate.

B) Rotation Rate Effects

The variations in the uniformity as brought about by the rotation rate of the wafers were tested on two different underlying barriers. The uniformity of the barrier/seed layer before plating was measured and determined not to have significant differences, i.e. less than 2%, 1 sigma. The contour maps of the two samples also showed similar edge and center thickness values.

The results indicate that the trend is similar between the two different barrier/seed combinations. However, a difference is seen in the absolute uniformity values as seen in Figure 3. By viewing the contour plots as shown in Figure 4a it is seen that at a rotation rate of 0 gpm, a very edge heavy plate is seen with little contour variations seen in the center of the wafer. The edge is approximately 13 kÅ and the center is 9 kÅ. With the 20 gpm rotation rate sample, Figure 4b, once again the edge heavy plating is seen but now a slight mound appears in the center. The edge is 12 kÅ and the center is 9.5 kÅ. This trend, caused by the rotation driven flow effects pulling solution to the center of the wafer and pushing the inlet stream away from the edge, reduces the non-uniformity value as is seen in Figure 3. Finally for the rotation rate of 50 gpm essentially no plating is observed at the center of the wafer. The contour plot, seen in Figure 4c, is consistent with a vortex being formed under the wafer center while the rotation driven flow pushes fresh solution farther from the edge. The minimum thickness is seen to be approximately 2 kÅ whereas the edge is approximately 11kÅ.

The sensitivity to variations in the rotation rate can be minimized by placement of a diffuser between the anode and wafers, however, only gross variations were sought for this study.

C) Designed Experiment

Seed Layer Thickness Effects. It is widely accepted that the seed layer thickness should improve the non-uniformity of the electroplated copper. This was tested using the 500Å, 2000Å, and 5000Å seed layers. This was tested for all anode sizes and all diffusers. Figure 5 is a plot of the wafer uniformity vs. the thickness of the copper seed layer using no diffuser. For the case of the 2", 6", and 8" anodes, the trend is clear with the thicker seed layer giving a more uniform deposit. However, for the 4" anode, there does not seem to be any apparent gain in uniformity with increased thickness.

For all types of diffusers, this random uniformity is seen, i.e. there is no apparent improvement in uniformity with the increased seed thickness, although most of the data trends in that direction. It is thought that this is due to the combination effects of the diffusers, anodes and the seed thickness.

Anode Size. The size of the anode was investigated with a 2", 4", 6", and 8" soluble copper anode. Figure 6 is a plot of the uniformity vs. the anode size using no diffuser. It is clear from this plot that as the anode size increases, that the non-uniformity also increases.

A reason that this is seen is shown in Figure 7 where the contour plots for the four anode sizes are shown. It is seen that when the anode is small, the 2" and 4" case, the edge plate is very light but a very heavy plate is seen in the center. This is to be expected from the current distribution lines that can be drawn from such a scenario. Conversely, for the 6" and 8" cases, the edge now plates heavy and the center becomes less pronounced. Once again, the current distribution dictates that the edge now have very high distribution and therefore the edge heavy plating.

Diffuser Design. There were four shapes for the diffuser in this study, including the no diffuser factor. The standard diffuser was a set of random holes in an 8" polyethylene disk. This diffuser was recommended by the plating tool manufacturer as one which provides the best uniformity under standard plating conditions, i.e. seed thickness and waveforms. In addition, there were two ring shaped diffusers. The first had a ring approximately 0.75" in width extending from the outside portion of the cup. The second was a ring with an approximate width of 1.25". This also is from the outside of the cup.

As can be seen in Figure 8, in a plot of the uniformity variations with diffuser type for a 2000Å seed layer, the standard diffuser shows the best uniformity value using an 8" anode. The ring which measure 0.75" was also very good with the overall uniformity values at both the 8" and the 6" anode sizes. The case of no diffuser behaved as discussed above, while the 1.25" ring had the best uniformity at 6", followed closely by the 8" anode uniformity value.

It should be noted that in cases where the current is shielded between the anode and the wafer surface, the use of a diffuser yields data that trends in an expected manner, i.e., the smallest anode size gives the worst uniformity values while the larger anode gives the better performance.

CONCLUSIONS

The variation in anode size, diffuser shape, and seed thickness have been evaluated for the effects that each have on the uniformity of the electroplated copper film. In addition a separate study on the effects that the flow rate and rotation rate of the wafer have on uniformity has also been conducted. It has been shown that the flow rate has a dramatic effect as does the rotation rate. The flow was shown to be turbulent at the edges leading to the edge heavy plate seen. The rotation rate was shown to have created a vortex which did not allow for efficient plate in the center of the wafer and therefore a high non-uniformity.

The designed experiment was run to look at the effects of the three factors mentioned above. It was shown that there are rather dramatic effects which must be considered when designing a plating chamber. Much of the design will be dependent on the subsequent processing steps and the requirements therein.

It is duly noted that changes in the conclusions may occur when secondary interactions in the DOE have been analyzed. However, several experiments need to be performed in order to do this comaprison.This will be subject of a follow on paper.

ACKNOWLEDGMENTS

The authors wish to thank the folks in the Advanced Materials bay for their support in obtaining these data. We also acknowledge useful discussions with Prof. Jacob Jorne from the University of Rochester and Tom Ritzdorf and his team at Semitool.

REFERENCES

1. P. Singer, Semiconductor International, 17,**52** (1994).
1. S. Venkatesan, et al., IEDM, December 1997.
3. C.H. Ting, et al., Extended Abstract 544, Electrochemical Society, Volume 97-2, 1997.

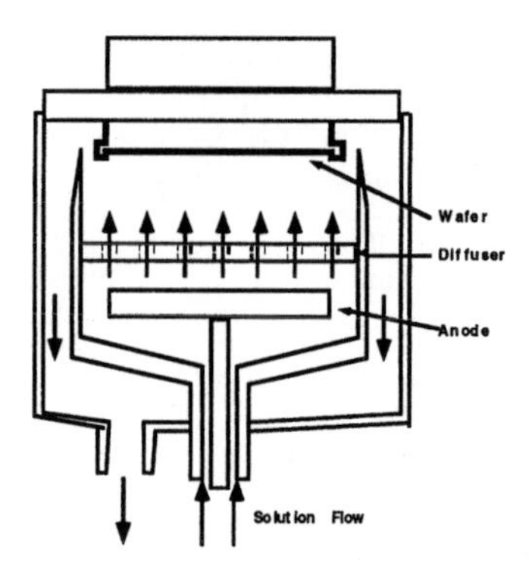

Figure 1: Schematic of plating chamber.

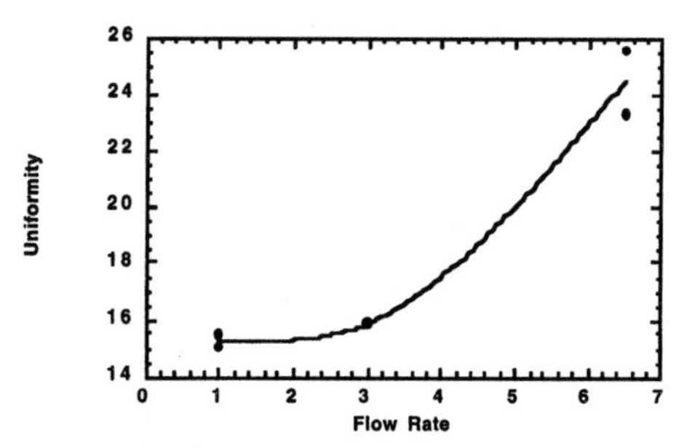

Figure 2. Plot of Uniformity vs. Flow Rate showing dramatic differences as the flow rate is increased.

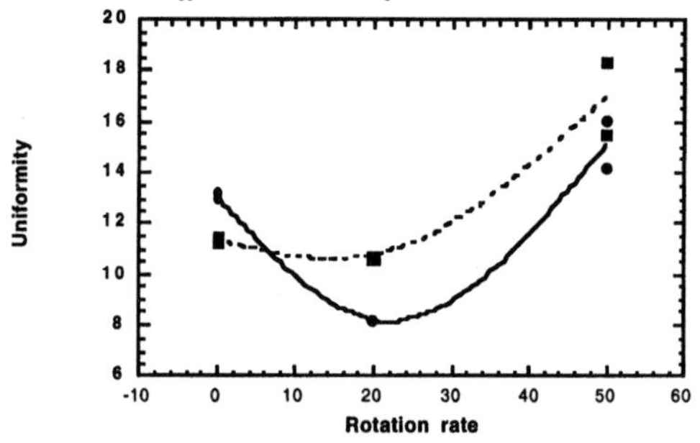

Figure 3: Change in uniformity with variable rotation rate for two different barrier/seed combinations. The seed thickness remained constant.

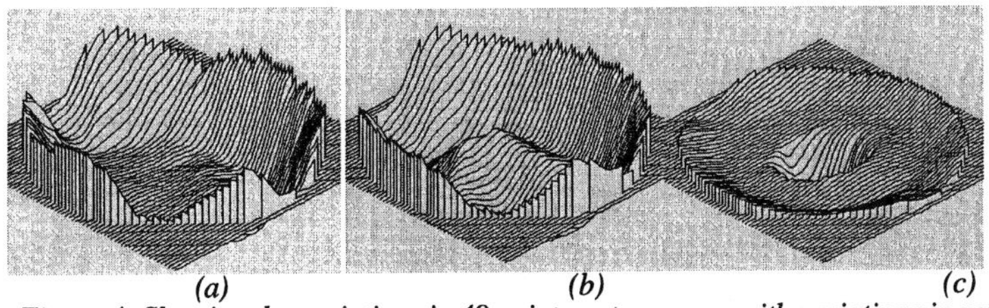

Figure 4: Showing the variations in 49 point contour maps with variations in rotation rate. Figure 4a is for no rotation of the wafer, 4b is shown for 20 rpm, and 4c is for 50 rpm.

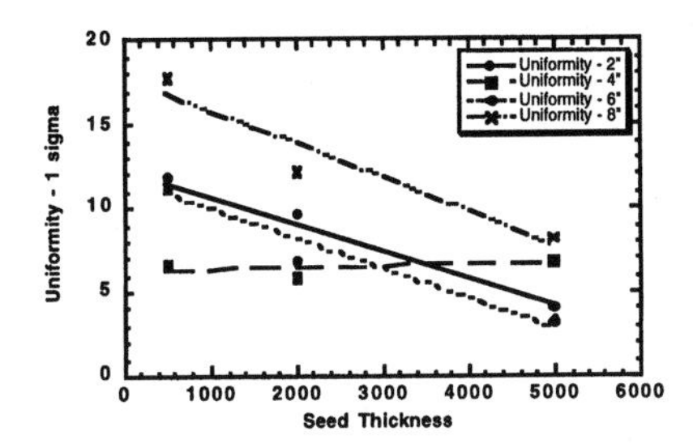

Figure 5: Plot of Uniformity vs. Seed layer thickness for all anode sizes tested.

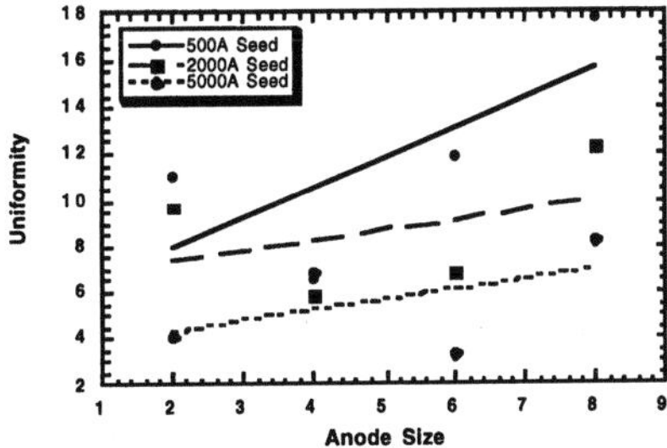

Figure 6: Plot of the uniformity vs. the anode size for the no diffuser case. A trend towards a more non-uniform plate is seen with increased anode size.

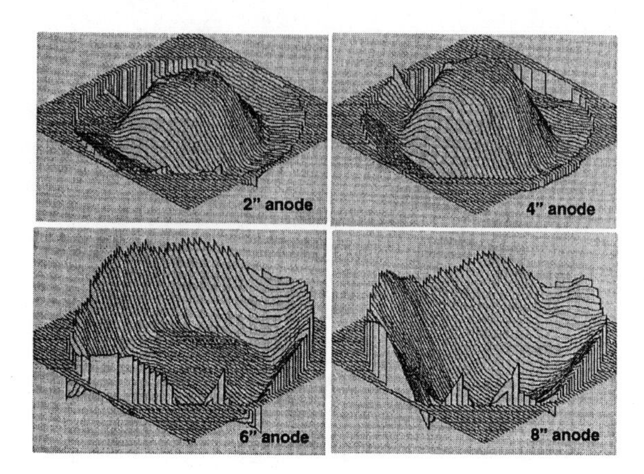

Figure 7: Contour plots for variable sized anodes.

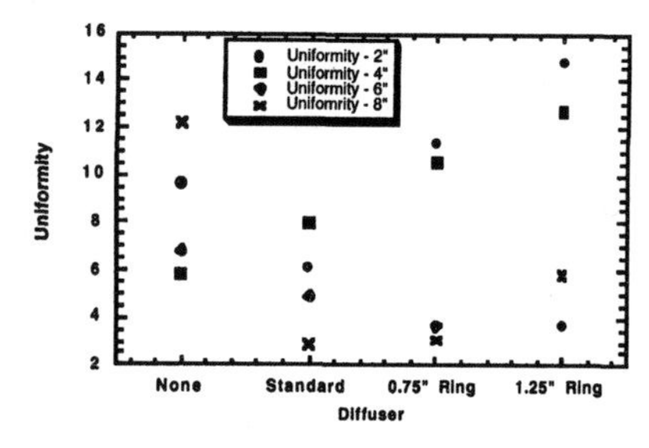

Figure 8: Plot of Uniformity vs. Diffuser shape for various anode sizes. In all cases the seed thickness was 2000Å.

Electroless Plating on Semiconductor Wafers

R. Aschenbrenner, A. Ostmann* and H. Reichl

Fraunhofer-Institut für Zuverlässigkeit und Mikrointegration
Gustav-Meyer-Allee 25
13355 Berlin, Germany
Tel. (+49) 30 46403 160 Fax (+49) 30 46403 161

* Technische Universität Berlin
Forschungsschwerpunkt Technologien der Mikroperipherik
TIB 4/2-1, Gustav-Meyer-Allee 25
13355 Berlin, Germany
Tel. (49) 30 314 72873 Fax (49) 30 314 72835

1. Abstract

In this work, we will describe the concepts and results of electroless plating techniques on silicon wafers for two different applications:

- Electroless bumping
- Electroless Plating for VLSI circuits

Flip chip technology requires the formation of bumps on semiconductor devices. Traditional bumping methods need expensive equipment for sputtering, photolithography and electroplating or evaporating. In contrast to the common techniques the cost for a maskless wet-chemical bumping process is significantly lower.

A chemical bumping technology developed and implemented at TUB/IZM will be presented. The maskless process is based on electroless nickel deposition. Batches of 25 wafers of 150 mm diameter can be processed in a 30 l tank. The process time is determined by the nickel plating which has a rate of 20 µm/hour. The uniformity of bumps is better than 1 µm for 20 µm bumps on 100 mm wafers.

For plating of very fine metal patterns such as contact holes and interconnections, we have developed a process based on electroless copper.

The mechanical properties of copper deposits and the kinetics of electroless copper plating were analyzed for various types of baths.

Selective copper plating introduces some new problems in general, like the compatibility with integrated circuits materials. There are also some particular problems that are associated with the technique, which are described here.

2. Electroless Bumping

2.1 Electroless Nickel Bumping

Flip chip technology arrived an increased level of acceptance for many different applications. The first driving force for the introduction of this technology was the need to achieve increased speed and performance along with higher I/O count. A breakthrough, however, will be the use of flip chip due to cost reduction. An important impact for this will be the implementation of lower cost bumping processes since the established methods need expensive equipment for metal sputtering and photolithography. Chemical bumping processes based on electroless nickel plating have been presented from several authors /1-10/ as a low-cost alternative. The advantages of chemical bumping are:

- very low cost
- not dependent on wafer size
- capacity for high volume production
- applicable for flip chip soldering or adhesive joining

In this paper the chemical bumping process of the TUB/IZM will be presented. Regarding first investigations with a recently implemented equipment some process and cost issues will be discussed. Different types of test wafers from 100 mm to 150 mm diameter were used for parameter optimization. 13 types of functional wafers from 7 manufacturers, mostly with AlCu metallizations of different compositions, have been successfully bumped. Figure 1 shows chemically plated nickel bumps on a test wafer with an area pad layout.

Figure 1: Ni/Au Bumps on a CMOS wafer

The chemical bumping process is fully wet-chemical and maskless, therefore no expensive facilities for sputtering or masking are necessary. The recently at TUB/IZM installed equipment is similar to that for wet-chemical processing. For the different chemical bathes heatable tanks are used. The main part of the bumping line is a tank for the electroless Ni bath. It has a volume of 30 l and can take 25 wafers of 150 mm diameter in standard wafer carriers. Moreover single wafers up to 200 mm can be plated. The bath is constantly pumped through a 1 µm filter. The operating temperature is stabilized ± 1 °C. Analytical control and replenishment of the bath is done manually. In the near future an automatic control system will be installed permitting continuous bath operation. The rinsing after each process step is performed with de-ionized water in resistance-controlled rinsers. The equipment is installed in a cleanroom lab. Quality of processed wafers is regularly controlled by optic microscopy, profilometer measurements and shear tests. For additional analysis SEM, EDX and infrared microscopy are used.

Process Steps

Only commercially available bathes and chemicals were used in the process, making it easy transferable. The process steps for the chemical bumping are:

- Al cleaning
- Al activation
- electroless Ni deposition
- immersion Au coating

Each process step is followed by a careful rinsing in de-ionized water.

The investigations on the process steps were performed using test wafers with a 1 µm AlSi1% metallization and 1 µm oxide passivation. The size of bondpads was 100x100 µm². Bumps were characterized by their mechanical adhesion (shear tests), height (profilometer measurements), resistance (four-point measurements). The shear strength measurement gives an important information about the quality of the interface Al/Ni. Low adhesion of bumps can lead to delamination and consequently to contact failures of assembled flip chip devices. Additionally weak adhesion is correlated with high contact resistance /4/. Sufficient shear strength of Ni bumps (100x100 µm² size) for flip chip interconnection should be 100 cN or more, corresponding to a contact resistance below 1 mW. The most important influence on adhesion and resistance has the combination of Al cleaning and activation. Optimal cleaning time has to be selected depending on parameters like Al alloy, oxide thickness and Al grain size.

Al Cleaning

In a slightly alkaline Al cleaner thick oxide layers are removed while the Al surface is roughened. The roughness enables the formation of a uniform, fine-grained zinc layer. The cleaning time has a strong influence on bump adhesion. A longer cleaning enhances adhesion but also leads to stronger depletion of Al

Al Activation

Different treatments to activate Al for electroless Ni bumping have been discussed. Activation by Pd solutions generally has the problem of selectivity /2,7/. A careful selection of chemicals and process parameters is

necessary to obtain a high yield. Most commonly zincate treatment has been applied to activate Al /1,3,4,6,8,10/. The advantage of zincating is the very good selectivity. Zinc is plated only on Al surfaces resulting in a stable and reliable process.
A commercially available, alkaline zincate solution is used for Al activation. Suitable times are between 20 to 40 s enabling a safe handling even of large wafers .

Electroless Ni Plating
For the electroless Ni plating a commercial bath based on sodium hypophosphite is used. The rate of Ni deposition is 20 µm/h at a temperature of 90 °C. The plated Ni contains 10 % P and has a resistivity of 70 µWcm.

A careful selection of the bath is important in order to get a safe and uncomplicated processing. Unsuited bathes or improper process parameters can easily cause a variety of bump defects. Insufficient stability of the bath leads to random growth of Ni particles which can stick to the wafer surface. Also distorted bumps can be observed under some conditions. This effect, which has been reported by Van der Putten et al. /11/, is caused by a poisoning of the catalytic Ni surface. For this reason several commercial bathes can only plate thin Ni layers on the bondpads.

The internal connections of semiconductor devices can lead to different electrical potentials of the Al bondpads in chemical solutions. The effect was first reported by Sard et al. /12/ for the electroless plating of Au beam leads. On bondpads which are connected to the Si bulk no proper zinc layer and no Ni bumps will be deposited due to the formation of a galvanic couple Al/Si. Therefore a protection of the wafer backside by a foil or a resist is necessary. Another source of electrical bias is the photovoltaic effect across illuminated p-n junctions. It causes a shift of the electrical potential on specific bondpads. Ni deposition rate can be reduced or increased according to the polarity of the photovoltage.

Immersion Au Coating
A final Au coating on the Ni is necessary to prevent oxidation and enables long-time solderability of bumps. Since the Au plating stops if the Ni is completely covered a maximum thickness of 0.25 µm can be achieved. The standard plating time is 10 min. corresponding to 0.15 µm Au.

Wafer Design Rules
Al bondpads thickness should be 1 µm or more in order to have sufficient Al after cleaning and activation. AlSi, AlSiCu and AlCu metallizations in different compositions were investigated. All types have been processed with good results. Deep probe marks in the Al can lead to complete dissolution at some locations on the pad.

The achievable bump thickness is limited since the Ni growth to the sides can cause short circuit between adjacent bondpads. The maximum height is half the distance between two bondpads. For example on a die with 200 µm pitch and 100 µm pad diameter Ni bumps up to 40 µm height can be obtained.
The passivation must be free of defects. Cracks cause a growth of Ni which can produce short circuit. This effect will also occur on parts of a wafer surface which were scratched by improper handling.

Ni also grows on Si which is not covered by an oxide or the passivation. Unprotected Si in the wafer scribe line will cause plating of a Ni layer with very low adhesion. Therefore the scribe line should be almost insulating except for defined test structures.
The wafer backside has to be protected during the process. For this purpose laminated adhesive foil or spin-on resist coating has been used.
A summary of the wafer design rules are shown in table 1.

metallization	AlSi, AlCu, AlSiCu
Al thickness	³ 1 µm
bump height	< 1/2 pad distance
passivation	defect-free nitride, oxide, polyimide
scribe line	insulating
wafer backside	protected (foil, resist)

Table 1: Summary of wafer design rules for chemical bumping.

2.2 Electroless Copper Bumping

We have developed the possibility of a new bumping method based on electroless nickel and electroless copper plating. As an alternative to the electroless Ni bumps, the use of Ni/Cu bumps was investigated [13]. Selective copper plating introduces some new problems in general, like the compatibility with integrated circuits materials.

A direct copper deposition on the aluminum bondpads is by experience of the authors not possible, because of the high alkaline content of the solution. Therefore, before copper plating is carried out, a sufficiently thick nickel layer is deposited on the pads to protect them (Fig. 2). Approximately 5 - 7 µm of nickel are necessary to seal and protect the aluminum bondpad from the high alkaline copper bath. Nickel and copper plating is isotropic, hence bumps grow laterally and overlap the passivation. When the nickel layer is too thin, OH^- ions can move between the passivation and the nickel, reaching the aluminum and damaging it.

The shear strength of nickel copper bumps deposited on 100 * 100 μm^2 aluminum pads was measured with a shear tester. The average value of the shear force was measured at 180 cN. The measurements of the contact resistances of bumps (size: 100 * 100 μm^2) were better than 2 mΩ.

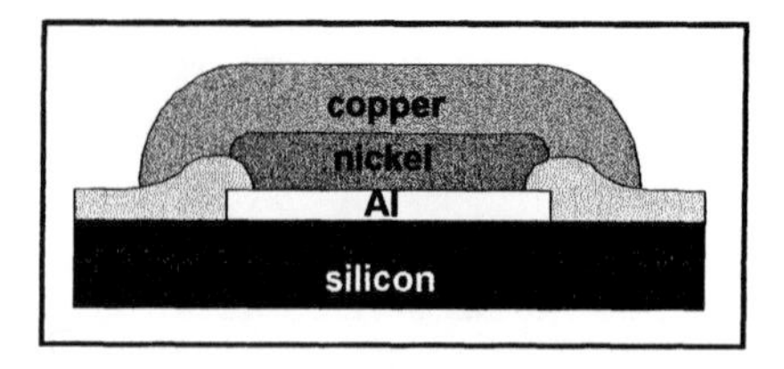

Fig 2: Electroless nickel/copper bump

A strong non-selectivity is clearly recognized in conjunction with high plating rates. In extreme cases this leads to a short circuit and to irregular bump surfaces. Lower plating rates have a slower start and long exposure periods. Electroless copper solutions with tartrate and quadrol as the complexing agent produced the best results. Ideal growth conditions for this type of bath are in the range of 1.5 - 3.5 µm/h. In this range a copper layer of 5 µm can be plated in 1.5 hours without external deposition and with no degradation of the integrated circuit.

3. Electroless Plating for VLSI circuits

The reason for all the interest in copper for VLSI is simply that it has a lower resistivity than aluminium. Cu has a very low resistivity (1.7 µΩ·cm) and expected good reliability, therefore, it is an obvious choice for long and narrow interconnections /14,15/.

The electroless process requires a conductive surface for deposition. The best results for the first metall layer were achived with TiW as seed layer for Pd activation. The fabrication process for the VLSI interconnection structures is shown in Table 2. The sample with TiW is first cleaned and prepared in different aqueous solutions for the Pd activation. This step is done in order to clean and to increase the roughness of the TiW, creating a suited surface for Pd activation and Cu deposition. The copper was deposited from aqueous solution in a bath with room temperature. The last step is the top coating which is performed by depositing a thin nickel film on top of the copper layer. Ni coating reduces the oxidation for the films (Fig. 3).

The excellent selictivity achieved by this process is illustrated in Fig. 4. In this picture we present 1.2 µm Cu lines with a space of 1.0 µm. The SEM picture shows the good coverage of the TiW and the smooth surface. The resolution is limited by the lateral growth of the Cu film.
The specific resistivity of this Cu film was measured by a four-point probe and calculated from the thickness to 2.1 µΩ·cm.

Process Flow
Preparation of the TiW surface Activating the TiW surface with Pd Electroless Cu patterning Electroless Ni-coating

Tab. 2: Typical process flow for electroless Cu deposition on TiW

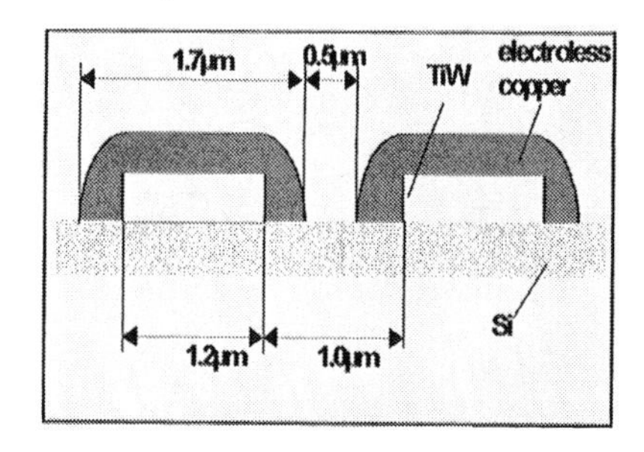

Fig 2 Schematic representation of electroless copper deposition on TiW lines

Fig. 3: Example of Cu-film on TiW (1.2µm line and 1.0µm space)

4. References

/1/ K. Wong, K. Chi, and A. Rangappan, „Application of Electroless Ni Plating in the Semiconductor Microcircuit Industry“, Plat. and Surf. Finishing, July 1988, pp. 70-76.

/2/ K. Yamakawa, M. Inaba, and N. Iwase, „Maskless Bumping by Electroless Plating for Thin and Low Cost Microcircuits“, Proc. ISHM, Baltimore 1989, pp. 620-626.

/3/ J. Simon, E. Zakel, and H. Reichl, „Electroless Deposition of Bumps for TAB Technology“, Metal Finishing, October 1990, pp. 23-26.

/4/ A. Ostmann, J. Simon, and H. Reichl, „The Pretreatment of Al Bondpads for Electroless Nickel Bumping“, Proc. IEEE MCM Conf., Santa Cruz 1993, pp. 74-78.

/5/ M. Uchida, K. Nozawa, and Y. Karasawa, „Interconnective technology with metallization for LCD“, Proc. IEMT, Japan 1991, pp. 97.

/6/ A. Aintila, A. Björklöf, E. Järvinen, and S. Lalu, „Electroless Ni/Au Bumps for Flipchip-on-Flex and TAB Applications“, Proc. IEEE Int. Electronic Manufacturing Techn. Symp. 1994, pp. 160-163.

/7/ C. Lin, I. Yee and B. Nelson, „Bumping Process on Integrated Circuits by Electroless Plating“, Proc. American Electropl. and Surf. Fin. Soc Atlanta 1992, pp. 531-540.

/8/ J. Liu, „Development of a Cost -effective and Flexible Bumping Method for Flip-Chip Interconnections“, Hybrid Circuits No. 29, September 1992, pp. 25-31.

/9/ C. NiDheasuna, A Mathewson, J. Barrett, G. Bruton, G. O'Riordan, D. L. Burke, and T. Spalding, „Electroless Plating Techniques for Microelectronic Packaging“, DVS 158 pp. 247-250.

/10/ J. Audet, L. Belanger, G. Brouillette, D. Danovitch, and V. Oberson, „Low Cost Bumping Process for Flip Chip“, Proc. ITAB Symp., San Jose 1995, pp. 16-21.

/11/ A. M. T van der Putten and J. W. G. de Bakker, „Geometrical Effects in the Electroless Metallization of Fine etal Pattwerns“, J. Electrochem. Soc.,Vol. 140 No. 8, pp. 2221-2228, (1993).

/12/ R. Sard, Y. Okinaka, and H. A. Waggener, „Electroless Beam Lead Plating“, J. Electrochem. Soc., Vol. 1 No. 1, pp. 62-66 (1974).

/13/ R. Aschenbrenner, A. Ostmann, U. Beutler, J. Simon and H. Reichl, „Electroless Nickel/Copper Plating as a New Bump Metallization“, IEEE Transaction CPMT -Part B, Vol. 18, NO. 2, 1995, pp. 334 - 338

/14/ Pei-Lin Pai and Chiu H. Ting, „Selective Electroless Copper for VLSI Interconnection“, IEEE Electron Device Letters, Vol. 10, NO. 9, 1989, pp. 423 – 425

/15/ Roger Palmans and Karen Maex, „Feasibility study of electroless copper deposition for VLSI“, Applied Surface Science 53, 1991, pp. 345 - 352

INTERACTION BETWEEN THE ELECTROLESS COPPER DEPOSITION SOLUTION AND THE LOW-K FLUORINATED DIELECTRICS

D. T. Hsu, H. Y. Tong and F. G. Shi
Department of Chemical & Biochemical Engineering and Materials Science
University of California, Irvine, CA 92697-2575

S. Lopatin and Y. Shacham-Diamand
School of Electrical Engineering and the Cornell Nanofabrication Facility
Cornell University, Ithacha, NY 14853-5401

Bin Zhao and M. Brongo
Rockwell Semiconductor Systems
4311 Jamboree Road, Newport Beach, CA 92660

P. K. Vasudev
SEMATECH
2706 Montopolis Drive, Austin, TX 78741

The compatibility of fluorinated poly(arylethers) (FLARE™1.0 & FLARE™1.51) and fluorinated polyimides (FPI-45M & FPI-136M) with the electroless Cu deposition solution was investigated. Fourier transform infrared (FTIR) and ellipsometry were employed to investigate possible chemical and physical property changes in two classes of low-k materials before and after their electroless Cu deposition solution treatments for various solution temperatures and treatment times. Our FTIR results demonstrate that FLARE™1.51 can react with the electroless Cu deposition solution, resulting in a disappearance of the 1726 cm^{-1} absorption band associated with one of the C=C stretching vibration modes of either the benzene ring or the aromatic R group. For fluorinated polyimides, the chemical reaction induced by the electroless Cu deposition solution was detected for FPI-45M based on the FTIR spectra. It was found that two new absorption bands were formed at 2852 cm^{-1} and 2891 cm^{-1} under certain electroless solution treatment conditions.

INTRODUCTION

The semiconductor industry has recently placed a great emphasis on the interlayer dielectric (ILD) polymers with low dielectric constant (low-k) as well as low moisture absorption and high thermal stability. Many new polymers (1-4) have been introduced with smaller dielectric constant than that of the conventional SiO_2, a conventional material currently used by the industry. The low-k polymers offer many advantages for circuit performance in reducing RC delays, inter-line capacitance and cross talk noise (5-7). In order to use these low-k polymers in IC manufacturing, however, their compatibility with existing and future multilevel interconnect processes (including the copper scheme) should be investigated. It is generally agreed that copper is the most promising candidate that can replace the aluminum based metallization as the interconnection metal because of its low resistivity and good electromigration resistance. Among various approaches including plasma vapor deposition (PVD), chemical vapor deposition (CVD), and electrochemical deposition (ECD) processes, electroless Cu deposition has been demonstrated to be promising for the sub-quarter-micron interconnect ULSI applications (8,9) Since the interconnect plays a vital role in terms of the performance considerations, the interconnection process node which uses Cu-based metallization in conjunction with the low-k materials is considered as a promising combination for integrated circuit fabrication. Thus, there is a critical need for investigation of the compatibility between the low-k materials and the electroless Cu deposition process. Among the several approaches for Cu metallization, the electroless copper metallization is promising because of it's low processing temperature and low cost. Consequently, a study of the compatibility of the low-k materials with the electroless copper deposition process used for copper metallization is urgently needed.

The objective of the present work is focused on the chemical compatibility between the low-k material and the electroless Cu deposition solution developed for the sub-quarter-micron interconnect applications (8,9). The low-k materials, fluorinated poly(arylethers) and fluorinated polyimides, have dielectric constants lower than that of the conventional dielectric material SiO_2. Different formulations for fluorinated poly(arylethers) (FLARE™1.0 and FLARE™1.51) (10,11) and for fluorinated polyimides (FPI-45M and FPI-136M) (12) have been investigated to examine their compatibility with the electroless Cu deposition solution. Fourier transform infrared (FTIR) spectroscopy and ellipsometer were employed to investigate possible chemical and physical property changes of the low-k materials before and after electroless Cu solution treatments at different solution temperatures and treatment times.

EXPERIMENTAL PROCEDURE

The electroless Cu deposition solution contained cupric sulfate, ethylenediaminetetraacetic acid (EDTA) and formaldehyde. The pH value of the solution was in the range of 12.3-12.7, adjusted by tetramethylammonium (TMAH). Stabilizer and surfactant were also added to the solution to increase the solution stability and to decrease the surface tension of the solution (9). The detailed composition of the electroless Cu deposition solution is shown in Table I. The low-k materials, fluorinated poly(arylethers)

(FLARE™1.0 and FLARE™1.51) and fluorinated polyimides (FPI-45M and FPI-136M) were used as samples for the investigation of their compatibility with the electroless Cu deposition solution. The dielectric samples were immersed in the electroless Cu solution at two different temperatures, 57°C and 77°C, for 5, 10, 15, minutes. During the immersion in the electroless Cu deposition under the real deposition conditions, the deposition of Cu did not occur because there was no seed layer initially deposited to the sample films. However, the backside of the samples were not protected by any other film and Cu deposition was observed, indicating that the treatment condition is suitable for electroless Cu deposition. After being treated by the solution, samples were rinsed in de-ionized water for 10 min, then were dried in N_2 flow for 1 min.

Table I Composition of Electroless Cu Deposition Solution

Chemicals	Concentration (g/l)
$CuSO_4$	8
EDTA	14
HCHO	5
TMAH	23
2,2'-dipyridyl	0-0.2
Triton®	0-0.2
RE-610	0.02

Infrared spectra were collected using a MIDAC (PRS-102) FTIR spectrometer capable of scanning from 400 to 4000 cm^{-1}. The FTIR spectra were collected at 4 cm^{-1} resolution with 32 scans co-added, and were represented by plots of transmission versus wavenumber. All figures of FTIR are spectra of untreated samples and samples treated by electroless Cu solution. Some spectra contain the absorption bands at around 2300 cm^{-1}, due to CO_2 from the background spectrum of the air. A Rudoplh Research ellipsometer (Auto EL IV) was used to investigate the change of thickness and refractive index before and after the solution treatments. In ellipsometry, monochromatic light with a wavelength λ=632.8 nm and known polarization is incident under a fixed angle of incidence ϕ=70°C.

RESULTS AND DISCUSSION

Characterization of chemical change for fluorinated poly(arylethers)

Fluorinated poly(arylethers) based on perfluorobiphenyl (FLARE™) were developed by AlliedSignal (10,11). The generalized formula for the fluorinated poly(arylether) (13) is presented in Figure 1. As shown in Figure 1, the generalized FLARE™ polymer is filled with fluorine and the difference between FLARE™1.0 and

FLARE™1.51 lies in the different nature of aromatic R groups and the fluorine content. Figures 2 and 3 present FTIR spectra of thin FLARE™1.0 films treated by the electroless Cu solution for different solution temperatures and for different treatment times. Although the overall intensities of absorption bands increase with the deposition time at a solution temperature of 57°C, this change is due to the copper deposition on the back of silicon substrate. The relative band intensities did not change with increasing deposition time, indicating that no interaction between FLARE™1.0 and the electroless Cu solution. In addition, there is no peak occurrence or disappearance for treated FLARE™1.0 in comparison with its untreated sample. This again indicates that there is no reaction occurring between FLARE™1.0 and electroless Cu deposition solution under the present experimental conditions.

As for thin FLARE™1.51 films, the overall intensity change was also a result of the copper deposition on the silicon substrate. Since there is no appearance of any new band, it can be concluded that there is no formation of new functional groups. However, in comparison with the untreated sample, as shown in Figures 4 and 5, a disappearance of the band at 1726 cm^{-1} was observed in the spectra after the FLARE™1.51 films were treated by the electroless Cu solution for different solution temperatures and treatment times.

The disappearance of the 1726 cm^{-1} band shown in Figures 4 and 5 indicates a significant structural change in FLARE™1.51 after the electroless Cu deposition solution treatments. Since the FLARE™ polymer contains no carbonyl group, this 1726 cm^{-1} band can be assigned either to the stretching vibration mode of C=C double bond for the benzene ring or the C=C stretching for the aromatic R group. The different nature of aromatic R groups is one of the factors determining the various properties of FLARE™ polymers. Thus, there is no doubt that the aromatic R group plays an important role for the observation outlined above.

The disappearance of the 1726 cm^{-1} absorption band can be attributed to two possible reaction mechanisms since the band at 1726 cm^{-1} can be associated with one of the C=C stretching vibration modes of either the benzene ring or the aromatic R group. The first possible mechanism can be directly linked to the aromatic R group in FLARE™1.51. This aromatic R group may have reacted with the electroless Cu deposition solution, which is then reflected as the disappearance of the 1726 cm^{-1} absorption band. Without breaking C-O bonding, the C=C double bond of the aromatic R groups may be reduced to the C-C single bond due to the interaction with the electroless Cu deposition solution, resulting in the disappearance of absorption band at 1726 cm^{-1}. The formation of C-C, however, could not be virtually observed in the spectra. The other possible reaction mechanism we should consider is that some fluorine atoms bound to the benzene ring may interact with the electroless Cu deposition solution, leading to the structural change of the benzene ring. This structural change is reflected by the disappearance of 1726 cm^{-1} absorption band. However, the intensity of C-F absorption band seems to remain the same before and after the electroless Cu deposition solution treatments, which means the total concentration of fluorine does not change with the electroless Cu solution. Therefore it is possible that the disappearance of absorption band at 1726 cm^{-1} is mainly due to the possible reaction between the aromatic R group and the electroless Cu deposition solution. Whereas the exact interaction mechanism still remains to be elucidated.

Characterization of chemical change for fluorinated polyimides

Fluorinated polyimides have been increasingly used for microelectronic applications. For instance, fluorinated polyimides can be utilized as interlayer dielectric materials in integrated circuits due to their superior properties, i.e., low dielectric constant, high glass transition temperature, low moisture absorption and high thermal stability. In the prsent work, two fluorinated polyimide , namely FPI-45M and FPI-136M, developed by DuPont (12) were utilized for investigation. The generalized structures for FPI-45M and FPI-136M are schematically shown in Figure 6 (12). Figures 7-10 illustrate the FTIR spectra for FPI-45M and FPI-136M films before and after the treatment by the electroless Cu solution for different solution temperatures and for different treatment times. As shown in Figures 7 and 8, the overall intensities of absorption bands for FPI-136M change due to the copper deposition on the back of silicon substrate. The relative band intensities, however, do not change with increasing deposition time, which indicates that no interaction between FPI-136M and the electroless Cu solution under treatment conditions. Additionally, no occurrence or disappearance of absorption bands was observed, indicating that there is no chemical reaction occurring between the FPI-136M and the electroless Cu deposition solution.

In the case of FPI-45M, there is no chemical reaction observed from the FTIR spectra, after the electroless Cu solution treatment at 57°C, as illustrated in Figure 9. Whereas, Figure 10 shows an observation of new absorption bands for the FPI-45M films treated by the electroless Cu solution at solution temperature of 77°C in comparison with the untreated sample. For deposition time of 5 min, there is a reduction of overall intensity, but the relative peak intensities do not vary with the increasing deposition time. As illustrated in Figure 10, two small bands appear at 2852cm^{-1} and 2891cm^{-1} in the spectrum for 10 min at 77°C, and they become more evident in the spectrum for 15 min. These new absorption bands may be associated with the C-H stretching of C-C-H groups or aldehydes. The formation of new peaks results from the chemical reaction between the FPI-45M and the electroless Cu deposition solution. Thus, the chemical reaction between FPI-45M and the electroless Cu deposition solution can occur at relative high solution temperatures and the relatively long duration of solution treatments.

Glass transition temperature and Coefficient of thermal expansion (CTE)

The samples treated in the solution at 57°C for 20 min and their corresponding untreated counterparts were further investigated to determine if the aforementioned chemical change induced by electroless solution treatments are reversible or irreversible after thermal cycle annealing. An ellipsometer equipped with a hot stage was employed for the measurements. The sample was first heated up to 400°C and then cooled down to the room temperature. The glass transition temperature and coefficient of thermal expansion were thus measured as a function of annealing temperature, respectively. As shown in Table II, the change in the glass transition temperature for FLARE™1.51 is observed after the electroless Cu solution treatment. The variation of the glass transition temperature of FLARE™1.51 is expected in view of the fact that its chemical structure has been altered by the electroless solution. It is also expected that FLARE™1.51 and FPI-45M exhibit a significant change in their CTEs (see Table II) due to the solution-induced change.

Table II Changes in the T_g and CTE

	Untreated		Treated at 57°C 20 min		Change	
	T_g (°C)	CTE ppm/°C	T_g (°C)	CTE ppm/°C	T_g %	CTE %
FLARE™1.0	247	41	250	n/a	1	n/a
FLARE™1.51	280	90	234	102	-16	13
FPI-45M	325	8	n/a	4	n/a	-50
FPI-136M	315	7	n/a	n/a	n/a	n/a

CONCLUSIONS

This work presents preliminary results on the compatibility between the electroless Cu deposition solution and low-k fluorinated poly(arylethers) (FLARE™1.0 and FLARE™1.51) as well as fluorinated polyimides (FPI-45M and FPI-136M). It has been found that there are no changes observed in IR spectra for FLARE™1.0 and FPI-136M after the electroless Cu deposition solution treatments. However, significant structural changes in FLARE™1.51 and FPI-45M have occurred after their solution treatments. In the case of FLARE™1.51, our results demonstrate that FLARE™1.51 can react with the electroless Cu deposition solution, resulting in a disappearance of the 1726 cm^{-1} absorption band associated with one of the C=C stretching vibration modes of either the benzene ring or the aromatic R group. Based on the FTIR spectra of FPI-45M, two new absorption bands at 2852 cm^{-1} and 2891 cm^{-1} have been observed due to the chemical reaction induced by the electroless Cu deposition solution under a certain condition of the electroless solution treatment (77°C, 15 min). It has been found that these new absorption bands may correspond to the C-H stretching of the C-C-H group or aldehyde. To determine if the aforementioned chemical change induced by electroless solution treatments are reversible or irreversible, thermal cycle annealing experiments have been performed. As expected, the glass transition temperature of FLARE™1.51 is changed, resulting from the chemical structural change induced by the electroless Cu deposition solution treatments. CTEs of both FLARE™1.51 and FPI-45M also exhibit solution-induced changes. However, the detailed reaction mechanisms between these low-k materials and the electroless Cu deposition solution still remain to be elucidated further.

ACKNOWLEDGMENTS

This work at UCI was supported by SEMATECH, MICRO program and Rockwell. Semiconductor Systems.

REFERENCES

1. N. H. Hendricks, Mat. Res. Soc. Symp. Proc., 443, 3 (1997).
2. T. Ramos, K. Roderick, A. Maskara and D. M. Smith, Mat. Res. Soc. Symp. Proc., 443, 91 (1997).
3. C. Jin, S. List, S. Yamanaka, W. W. Lee, K. Taylor, W. -Y. Hsu, L. Olsen, J. D. Luttmer, R. Havemann, D. Smith, T. Ramos and A. Maskara, Mat. Res. Soc. Symp. Proc., 443, 99 (1997).
4. A. Grill, V. Patel, K. L. Saenger, G. Jahnes, S. A. Cohen, A. G. Schrott, D. C. Edelsten and J. R. Paraszczak, Mat. Res. Soc. Symp. Proc., 443, 155 (1997).
5. S. P. Murarka, Solid State Technology, 83 (March, 1996).
6. J. Wary, B. Olson and W. Beach, Semiconductor Int'l., 211 (June, 1996).
7. B. Zhao, S. -Q. Wang, S. Anderson, R. Lam, M. Fiebig, P.K. Vasudev and T.E. Seidel, Mat. Res. Soc. Symp. Proc., 427, 415 (1996).
8. Y. Shacham-Diamand, V. M. Dubin, M. Angyal, Thin Solid Films, 262, 93 (1995).
9. V. M Dubin, Y. Shacham-Diamand, B. Zhao, P. K. Vasudev and C. H. Ting, J. Electrochem. Soc., 144, N3 898 (1997).
10. Trademark of AlliedSignal.
11. F. W. Mercer and R. Sovish, US Patent No. 5,114,780 and 5,115,082 (assigned to AlliedSignal).
12. B.C. Auman, Mat. Res. Soc. Symp. Proc., 381, 19 (1995)
13. N. H. Hendricks, K. S. Y. Lau, A. R. Smith and W. B. Wan, Mat. Res. Soc. Symp. Proc., 381, 59 (1995).

FIGURES

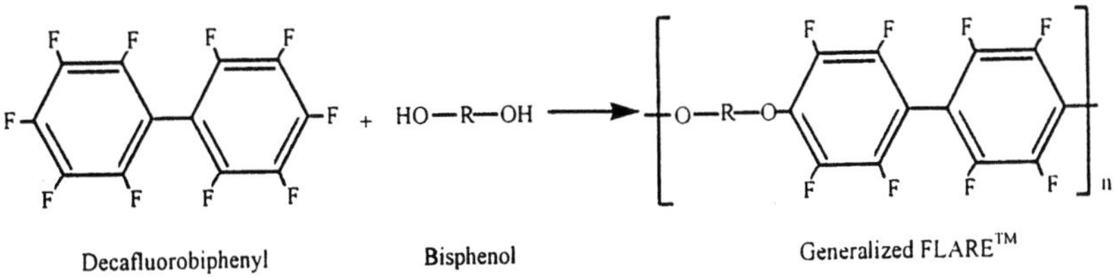

HO—R—OH

Decafluorobiphenyl Bisphenol Generalized FLARE™

Figure 1. The generalized synthesis of FLARE™.

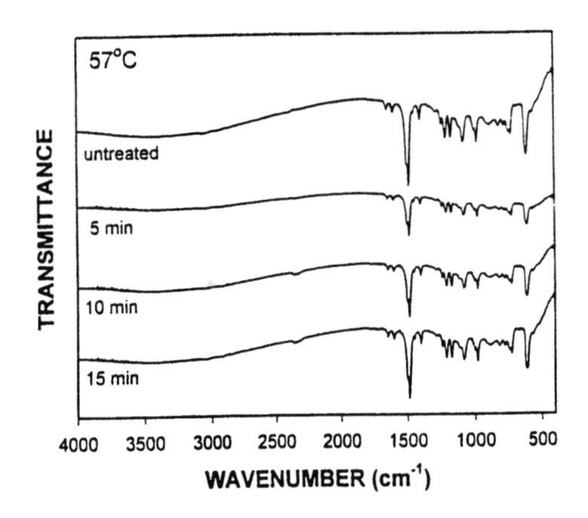

Figure 2. FTIR spectra of FLARE™1.0 treated by the electroless Cu deposition solution at 57°C for 5, 10, 15 min.

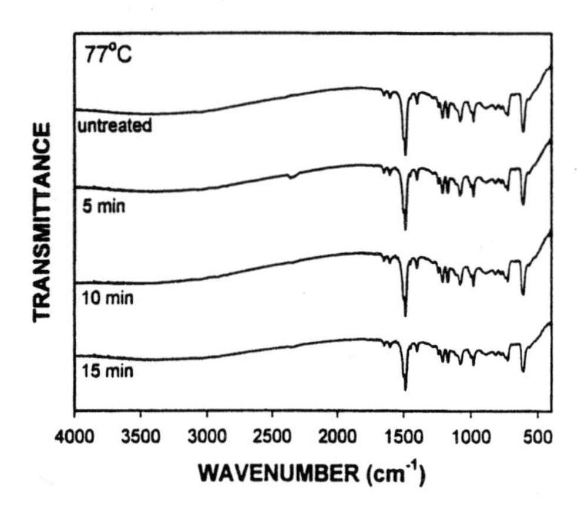

Figure 3. FTIR spectra of FLARE™1.0 treated by the electroless Cu deposition solution at 77°C for 5, 10, 15 min.

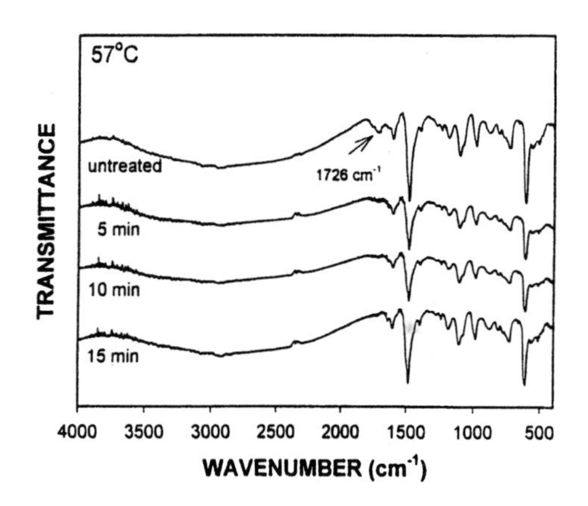

Figure 4. FTIR spectra of FLARE™1.51 treated by the electroless Cu deposition solution at 57°C for 5, 10, 15 min.

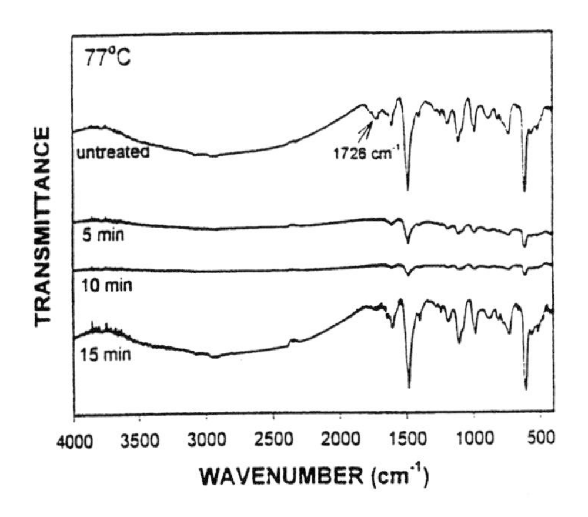

Figure 5. FTIR spectra of FLARE™1.51 treated by the electroless Cu deposition solution at 77°C for 5, 10, 15 min.

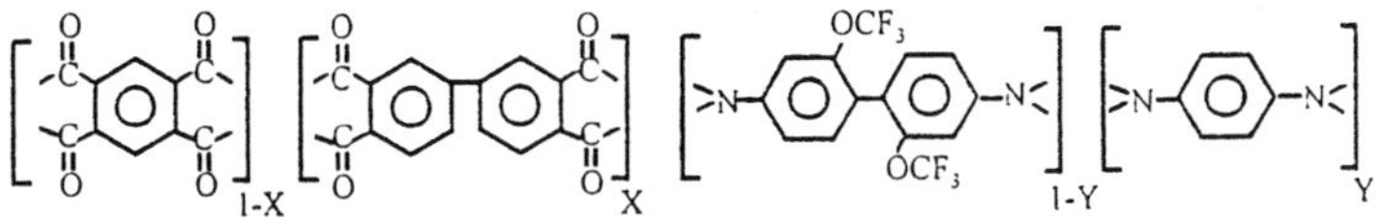

(a)

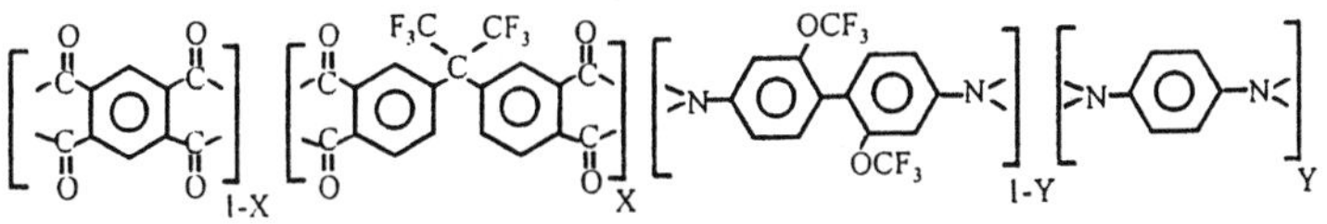

(b)

Figure 6. The general structures for (a) FPI-45M and (b) FPI-136M.

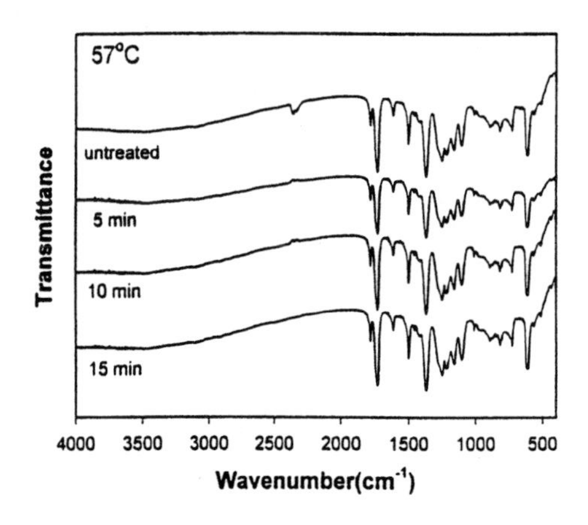

Figure 7. FTIR spectra of FPI-136M treated by the electroless Cu deposition solution at 57°C for 5, 10, 15 min.

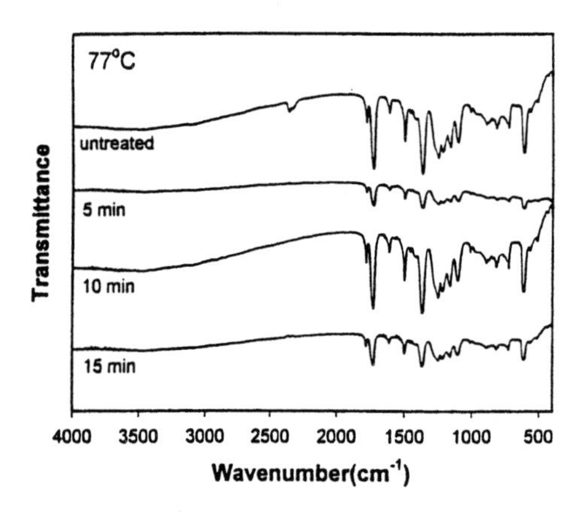

Figure 8. FTIR spectra of FPI-136M treated by the electroless Cu deposition solution at 77°C for 5, 10, 15 min.

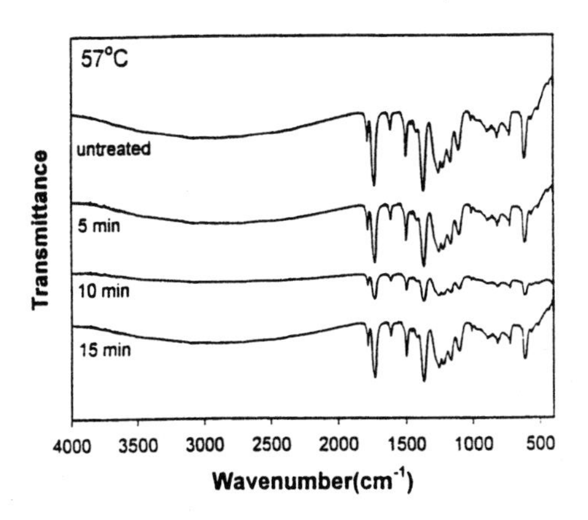

Figure 9. FTIR spectra of FPI-45M treated by the electroless Cu deposition solution at 57°C for 5, 10, 15 min.

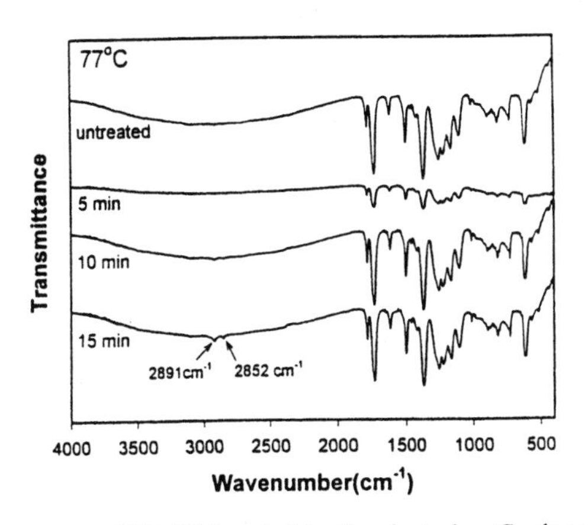

Figure 10. FTIR spectra of FPI-45M treated by the electroless Cu deposition solution at 77°C for 5, 10, 15 min.

ELECTROCHEMICAL DEPOSITION OF COPPER ON n-Si/TiN

Gerko Oskam, Philippe M. Vereecken, John G. Long, Brendan J. Moran, and Peter C. Searson

Department of Materials Science and Engineering
The Johns Hopkins University
Baltimore, MD 21218

In this paper we report on the electrochemical deposition of copper onto n-type silicon with a TiN barrier film. We show that the nucleation and growth kinetics follow instantaneous nucleation followed by three dimensional, diffusion limited growth. The nucleus density was dependent on the applied potential and increased from 9×10^7 cm^{-2} at -0.25 V to 5×10^8 cm^{-2} at -0.7 V.

INTRODUCTION

In integrated circuit manufacturing, the semiconductor industry is changing from aluminum to copper interconnect technology since copper has a lower resistivity and a higher maximum current density for electromigration [1,2]. These properties are sufficient to provide the necessary increase in performance and reliability for the next generation of ULSI devices. In addition, with dual-damascene processing [2], vias and interconnect lines can be fabricated at the same time thereby decreasing the number of deposition steps.

Since copper is soluble in silicon it is necessary to deposit a thin diffusion barrier onto the silicon before deposition of the copper metallization. Candidate materials for diffusion barriers include transition metals, transition metal alloys, transition metal silicides, polycrystalline metal nitride, and ternary amorphous alloys [3].

Copper can be deposited by a number of techniques, including chemical vapor deposition, physical vapor deposition, electroless deposition, and electrochemical deposition. Of these techniques, electrochemical deposition appears to be the leading candidate due to its inherent advantages in filling high aspect ratio structures with complex geometries. Due to the difficulties in depositing directly onto most diffusion barriers, a copper seed layer is vapor deposited prior to electrochemical deposition

In this paper, we report on the nucleation and growth kinetics of copper onto unpatterned n-type silicon wafers with a TiN barrier layer. The goal of this work is to explore the possibility of electrochemical deposition of high quality copper films without the need for a seed layer. We show that under appropriate experimental conditions, a high density of copper nuclei can be obtained at the surface. Furthermore, continuous copper films can be obtained by subsequent kinetically limited growth of the clusters.

EXPERIMENTAL

All experiments reported were performed on n-Si (100), $N_D = 1 \times 10^{15}$ cm^{-3}, with a 30 nm TiN barrier layer. In order to avoid problems associated with the sheet resistance of the TiN film, an ohmic contact was made to the back side of the silicon wafer using In/Ga eutectic. The aqueous 50 mM Cu^{2+} solution was prepared from 25 mM $CuCO_3 \cdot Cu(OH)_2$ with 0.32 M H_3BO_3 and 0.18 M HBF_4. The pH of the solution was about 1.4.

The experiments were performed using a conventional three electrode cell with Ag/AgCl (3 M NaCl) as reference and platinum gauze as counter electrode. The reference electrode was placed close to the Si/TiN working electrode with the aid of a Luggin capillary. All potentials are given with respect to the Ag/AgCl (3 M NaCl) reference (0.22 V versus NHE). The experiments were performed in air and at room temperature.

RESULTS AND DISCUSSION

Figure 1 shows a current - potential curve for n-Si/TiN in 50 mM Cu^{2+} solution at a scan rate of 10 mV s^{-1}, starting from the open circuit potential, OCP = 0.20 V. On the first cycle, the onset of Cu^{2+} reduction occurs at about 0 V with a characteristic diffusion limited growth peak at -0.15 V. After the deposition peak, the current again increases at a potential of about -0.75 V due to the reduction of protons at the n-Si/TiN surface, which is partially covered with copper.

The reverse scan in Figure 1 shows a diffusion limited deposition current of about 5 mA cm^{-2}. At potentials positive to 0.02 V, a stripping peak is observed. Note that if copper is electrodeposited onto a silicon surface, stripping of copper cannot occur due to the 0.63 eV barrier height of the Si/Cu Schottky junction [4]. The n-Si/TiN contact, however, is ohmic so that stripping can take place on the reverse scan.

On subsequent cycles, the deposition peak is shifted to about 0.00 V since the copper is not completely removed during the stripping wave. This shift of the deposition peak indicates that a nucleation overpotential is required for the deposition of copper onto n-Si/TiN. The second and third sweeps are essentially the same, and the shape of the voltammograms suggests that deposition and dissolution of copper on n-Si/TiN/Cu is a reversible process.

Cyclic voltammograms were recorded at scan rates from 25 mV s^{-1} to 2025 mV s^{-1}. Prior to each measurement, the electrodeposited copper from the previous scan was stripped at 0.2 V. Figure 2 shows the maximum current density at the peak, i_{peak}, of the first scan versus the square root of the scan rate, ν, and Figure 3 shows the potential at the current maximum, U_{peak}, versus the logarithm of the scan rate. From Figures 2 and 3 it follows that i_{peak} is proportional to $\nu^{1/2}$, and that U_{peak} is proportional to $\log(\nu)$ characteristic of simple diffusion limited charge transfer reactions [5]. Note that at a scan rate of 2 V s^{-1}, U_{peak} has shifted considerably to about -0.4 V. The observation that the peak potential shifts to more negative potentials upon increasing the scan rate indicates that the kinetics of charge transfer are not sufficiently fast to maintain steady-state conditions upon changing the potential.

Figure 4 shows a series of current transients for copper deposition on n-Si/TiN in the 50 mM Cu^{2+} solution for deposition potentials in the range from -0.25 V to -0.70 V. All the transients exhibit an initial current spike due to charging of the double layer. The nucleation and growth process is characterized by a current peak; the deposition current first increases due to nucleation of copper clusters and three dimensional diffusion limited growth, followed by a decrease as the diffusion zones overlap resulting in one dimensional diffusion limited growth to a planar surface. At long times, the transients at potentials between -0.4 V and -0.7 V converge on a curve where the current is linear with $1/\sqrt{t}$, in accordance with the Cottrell equation [5,6]. For the transients recorded at potentials more positive than -0.4 V, the current densities at 0.3 s are smaller than the Cottrell value, suggesting a partial kinetic control of the growth process. Furthermore, the current at -0.25 V is smaller than at -0.35 V in agreement with this explanation.

Experimentally, the mechanism of nucleation and growth can be obtained from analysis of the transient current response. The time dependent deposition current density (normalized to the geometric surface area), i(t), for instantaneous nucleation followed by three dimensional diffusion limited growth is [7,8]:

$$i(t) = \frac{z F D^{1/2} c}{\pi^{1/2} t^{1/2}} \left[1 - \exp\left(- N_\infty \pi D t \left(\frac{8 \pi c M}{\rho} \right)^{1/2} \right) \right] \qquad \{1\}$$

where D is the diffusion coefficient, c is the bulk metal ion concentration, M is the molar weight of the deposit, ρ is the density of the film, and N_∞ is the final nucleus density. For progressive nucleation, the time dependent deposition current density is given by [7,8]:

$$i(t) = \frac{z F D^{1/2} c}{\pi^{1/2} t^{1/2}} \left[1 - \exp\left(- \frac{2}{3} A N_\infty \pi D t^2 \left(\frac{8 \pi c M}{\rho} \right)^{1/2} \right) \right] \qquad \{2\}$$

The time dependence of the nucleus density, N(t), is given by [7,8]:

$$N(t) = N_\infty (1 - \exp(-At)). \qquad \{3\}$$

where A is the nucleation rate constant. For the case where At << 1, the nucleation is instantaneous and all nuclei are formed at the same time followed by growth; for At >> 1, the nucleation is progressive and new nuclei are formed while existing nuclei are growing. Hence, for instantaneous nucleation, a narrow size distribution of clusters is expected while the distribution can be wide for progressive nucleation.

In order to determine wether nucleation is instantaneous or progressive, the transients are most conveniently analyzed in reduced form in terms of the maximum current, i_{max}, and the time at which the maximum current is observed, t_{max} [7,8]. For instantaneous nucleation:

$$\frac{i^2}{i_{max}^2} = 1.9542\left(\frac{t_{max}}{t}\right)\left[1 - \exp\left(-1.2564\frac{t}{t_{max}}\right)\right]^2 \qquad \{4\}$$

and for progressive nucleation:

$$\frac{i^2}{i_{max}^2} = 1.2254\left(\frac{t_{max}}{t}\right)\left[1 - \exp\left(-2.3367\frac{t^2}{t_{max}^2}\right)\right]^2 \qquad \{5\}$$

Figure 5 shows the deposition transients in Figure 4 replotted in reduced form, along with the theoretical curves for instantaneous and progressive nucleation given by equations {4} and {5}. From this Figure it can be seen that the deposition transients at potentials from -0.4 V to -0.7 V show good agreement with the theoretical curves for instantaneous nucleation. The deviation from the theoretical curve at short times is due to the contribution from the tail of the charging current. In principle, the transients can be corrected for the charging current [9], however, that was not possible here since the sampling interval was not sufficiently fast.

The current transients at -0.35 V and -0.25 V show a deviation from the theoretical curves for diffusion limited growth after the maximum in the dimensionless plots. This may be due to partial kinetic control of the growth process at these small overpotentials, as was concluded from Figure 4. The cyclic voltammogram recorded at 10 mV s^{-1} indicates that the steady-state current is diffusion limited in this potential range. However, Figure 3 shows that the deposition peak shifts to considerably more negative potentials upon increasing the scan rate, indicating that upon fast changes of the applied potential, the reaction rate will be smaller than the steady state (diffusion limited) reaction rate for a certain time period.

For instantaneous nucleation, the density of nuclei can be calculated from t_{max} and i_{max} [7,8]:

$$N_\infty = 0.065\left(\frac{\rho}{8\pi c M}\right)^{1/2}\left(\frac{zFc}{i_{max}\, t_{max}}\right)^2 \qquad \{6\}$$

Figure 6 shows that the nucleus density determined from the current transients increases from about 10^7 cm^{-2} at -0.4 V to about 10^8 cm^{-2} at -0.7 V. Also shown in the Figure are the nucleus densities obtained from scanning electron microscope images for samples where about 10 mC cm^{-2} of charge was passed at a constant deposition potential. In this case the nucleus densities increase from about 9 x 10^7 cm^{-2} at -0.25 V to 5 x 10^8 cm^{-2} at -0.7 V, illustrating reasonable agreement between the values obtained by the two methods.

The results from these experiments were used to determine the conditions for the deposition of high quality copper films on n-Si/TiN surfaces. The approach was to deposit a high density of copper nuclei by applying a potential pulse from the open circuit potential to a potential in the range from -0.7 V to -1 V. The length of the nucleation pulse was optimized in order to prevent coalescence of the nuclei under diffusion limited growth. After the nucleation pulse, the potential was stepped to a potential in the range from -0.1 V to 0.1 V in order to grow the clusters under kinetic control and to obtain continuous copper films. Preliminary experiments showed that this approach results in good quality copper films on n-Si/TiN surfaces.

REFERENCES

1. The New York Times, Sept. 22 (1997).
2. D. Edelstein, J. Heidenreich, R. Goldblatt, W. Cote, C. Uzoh, N. Lustig, P. Roper, T. McDevitt, W. Motsiff, A. Simon, J. Dukovic, R. Wachnik, H. Rathore, R. Schulz, L. Su, S. Luce, and J. Slattery, Proc. IEEE International Electron Devices Meeting (IEDM), **43**, 773 (1997).
3. S.-Q. Wang, MRS Bulletin, **19**, 30 (1994).
4. E. H. Rhoderick and R.H. Williams, Metal-Semiconductor Contacts, Oxford, New York (1978).
5. Southampton Electrochemistry Group, Instrumental Methods in Electrochemistry, Ellis Horwood, New York, (1990).
6. G. Oskam, J.G. Long, M. Nikolova, and P.C. Searson, Materials Research Society, Proc., **451**, 257 (1997)
7. G. Gunawardena, G.J. Hills, I. Montenegro, and B. Scharifker, J. Electroanal. Chem., **138**, 225 (1982).
8. B.R. Scharifker and G.J. Hills, Electrochim. Acta, **28**, 879 (1983).
9. P.M. Vereecken, K. Strubbe, and W.P. Gomes, J. Electroanal. Chem., **433**, 19 (1997)

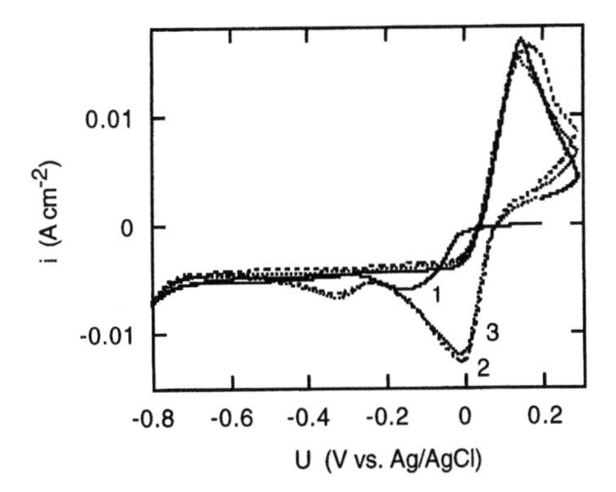

Figure 1. Current - potential curves for n-Si/TiN in 25 mM $CuCO_3 \cdot Cu(OH)_2$ + 0.32 M H_3BO_3 + 0.18 M HBF_4 (pH = 1.4) at a scan rate of 10 mV s^{-1}. The first scan was recorded starting from the open circuit potential at 0.2 V.

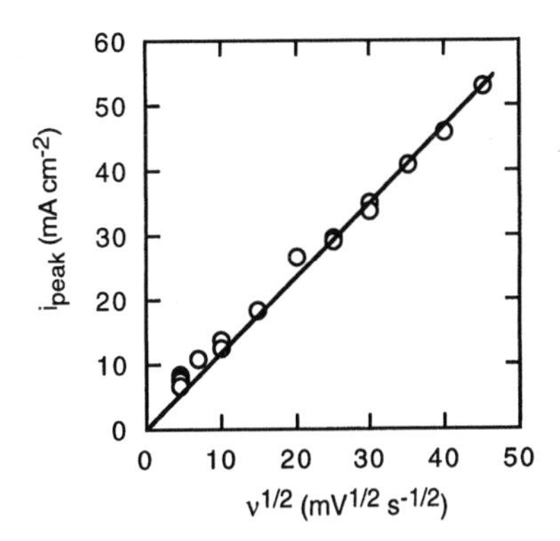

Figure 2. Peak deposition current versus the square root of the scan rate determined from the first scan started from the open circuit potential. Prior to each measurement, the electrodeposited copper from the previous experiment was stripped at 0.20 V.

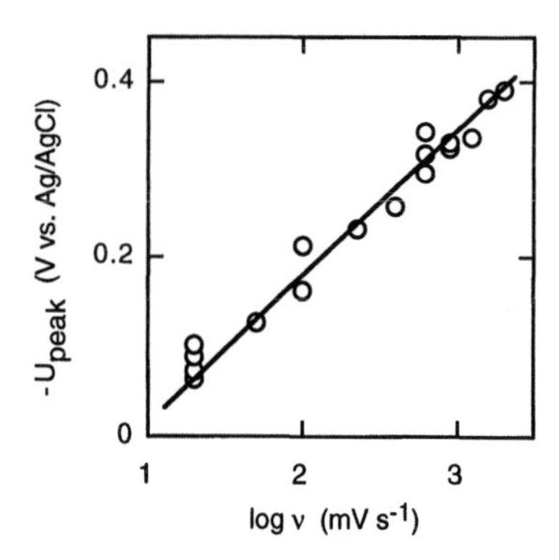

Figure 3. The potential at the current peak maximum versus the logarithm of the scan rate for the same measurements as shown in Figure 2.

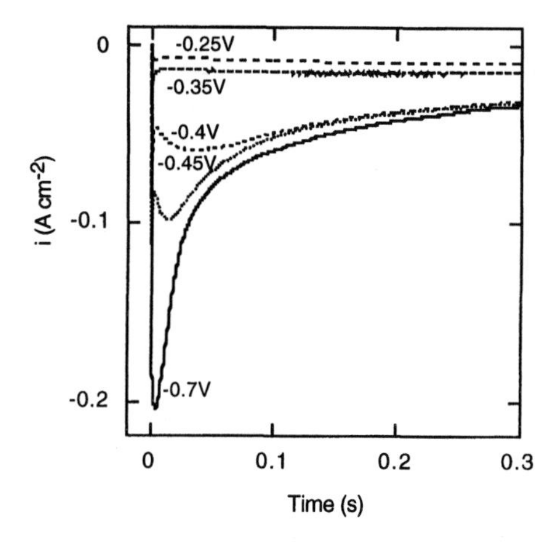

Figure 4. Current transients for the deposition of copper for various deposition potentials. The potential was stepped from the open circuit potential to the potential indicated in the Figure.

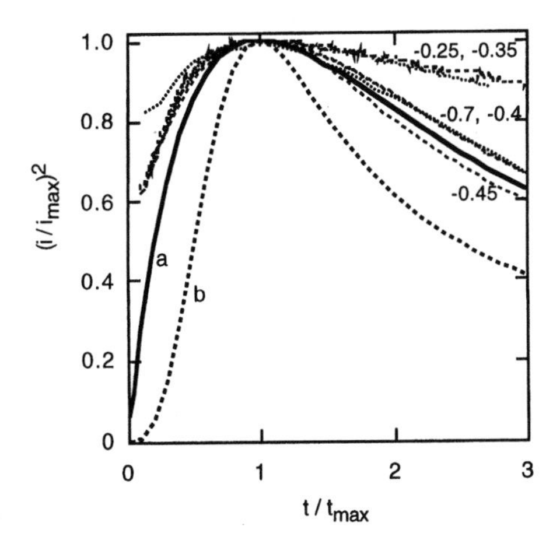

Figure 5. Reduced parameter plots for the current transients for the deposition of copper shown in Figure 4. The deposition potentials are indicated in the Figure. Also shown are the theoretical curves for instantaneous (a) and progressive (b) nucleation.

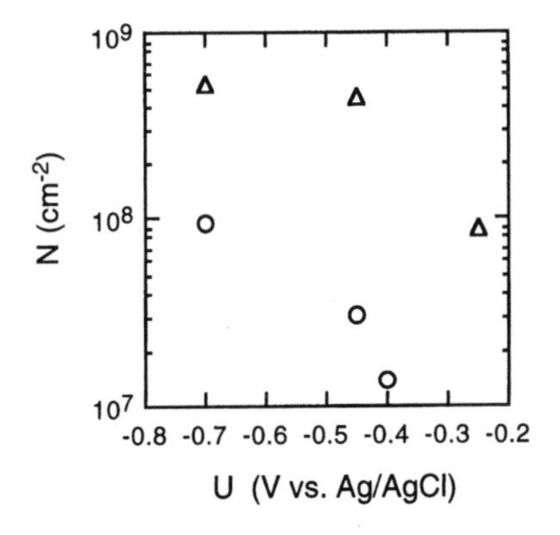

Figure 6. Nucleus density (o) determined from deposition transients in the potential regime where the deposition transients follow the instantaneous nucleation model, and (Δ) nucleus density determined from scanning electron microscope images of samples where about 10 mC cm^{-2} copper was deposited at constant potential.

FABRICATION PROCESS SIMULATION OF A SILICON FIELD ELECTRON EMITTER MOSFET STRUCTURE

Gerald Oleszek and Manuel Rodriquez
University of Colorado
Microelectronics Research Laboratories
Dept. of Electrical and Computer Engineering
Colorado Springs, CO 80933

ABSTRACT

Applications for Silicon Field Electron Emitter Arrays include high density Field Emission Displays and other applications which require the use of highly collimated electron beams. Two-dimensional computer numerical simulation results are presented of the process steps used in the fabrication of a Silicon Field Electron Emitter MOSFET structure. The fabrication processes of silicon and oxide etching, ion implantation and annealing, oxidation, lift off etching and patterning are simulated. The vertical cross-section, shape, and dimensions of the simulated Silicon Field Electron Emitter MOSFET structure are found to be in close agreement with fabricated structures.

INTRODUCTION

Micrometer Field Electron Emitter structures using Silicon micro-fabrication processes have been reported and their electron emission properties have been measured (1-2). Applications for field electron emitter structures include high-density field emission displays, high-resolution electron lithography, and other applications, which require the use of highly collimated electron beams. Field electron emitter structures require that the shape of the emitter be a very fine point (tip). In addition, the emitter tips must be extremely small and spaced in a planar array with micron separations between them. Each emitter shape must be identical to ensure uniform electron emission properties across the planar array. Silicon microchip fabrication technology is thus ideally suited for the fabrication of Field Electron Emitter Arrays.

In this paper, two-dimensional computer simulation results are presented of the process steps used in the fabrication of the Silicon Field Electron Emitter MOSFET structure as reported by Hirano, et al (3). The fabrication processes of silicon and oxide etching, ion implantation and annealing, oxidation, lift off etching and patterning are simulated and the vertical cross-sections associated with sequential processing steps are presented.

SIMULATION MODELS

Two dimensional, physically based computer simulation relies on numerically solving equations at a number of discrete points that represent a cross section of the device being simulated. The collection of points used for this finite element analysis is referred to as the *grid*. Computational nodes are linked to every single point on the grid. Since a computation must be performed at each discrete node, the structure of the grid critically affects the time required to complete a simulation. A coarse grid containing a small number of points is often used to obtain a rough idea of the structure, while a fine grid is used for final results or at critical regions within the structure.

Various numerical models can be utilized in the simulation of the process steps required to complete the device structure. Some of these models trade off computational time for accuracy while others will produce more accurate results depending of the real life processing conditions encountered when fabricating an actual device. The critical processes in simulating the fabrication of the Silicon Electron Field Emitter MOSFET structure are reactive ion etching of the Silicon substrate and thermal oxidation. Other processes used include deposition of evaporated material, diffusion of impurities in Silicon and ion implantation.

Analytical oxidation models in SSUPREM4 are limited to planar, bare Silicon surfaces. To obtain accurate solutions in the simulation of arbitrary structures, including structures with steps and trenches, numerical models must be used. In all models, oxidation equations are solved at the silicon/oxide interface to obtain the growth rate at each node in the grid. As oxide grows, its expansion is determined by the chemical reconfiguration of adjacent materials. The expansion of the material promotes a certain flow, which is simulated by the numerical models in SSUPREM4. The compressible model used in our simulation makes use of the reduced hydrodynamic equation to simulate creeping-flow motion in the material (4). Since a slight compressibility is allowed in the numerical model used, solutions using this method are not completely accurate since the internal stress of the material is not taken into account. In our simulation, this model was utilized to grow the oxide mask and to sharpen the electron field emitter tip.

A Reactive Ion Etching (RIE) model was utilized to etch the gate and drain holes in the simulated structure. This analytical model possesses a varying isotropic etch rate component as well as a varying anisotropic etch rate component. Both the isotropic rate and the anisotropic rate can be independently varied to simulate RIE under various conditions. The exact geometry of this etch is constrained by the number of nodes on the grid that define a line between the specified geometrical coordinates. Etching can only be specified for a particular material; residual etching of mask materials is not readily simulated.

Deposition of an evaporated metal on the surface of the silicon is another process in our simulation. A unidirectional deposition model, simulating the arrival of a vapor stream in one direction to any region not shadowed by overlying material, was used. The growth rate of the deposited film in a shadowed region is equal to zero. The magnitude of the film growth rate is dictated by the cosine distribution law which states that the film thickness grows at a rate proportional to $\cos(\omega)$, where ω is the angle between the vapor stream and the normal surface.

An analytical model is used to simulate the implant profile. The analytical model used in the simulation of implanted profiles relies on the reconstruction of previously measured implant profiles. Spatial-distribution moments are utilized by SSUPREM4 to calculate the profiles of implantation distributions. The method used is based on the theory of "Range Concepts" (5), where an ion-implantation profile is build from a set of previously measured or calculated moments. A Pearson distribution, analytical implant model, is used for the calculation of the implant profile in our simulation.

Diffusion of the implanted profile defines the n^+ emitter/source and drain regions in the simulated structure . Mathematical equations describing diffusion in silicon have beenextensively researched and modeled in the past (6). The concentration and depth of the p^+ diffusion after annealing and/or oxidation steps is determined by analytical calculation of the diffusion equations at each node in the grid.

SIMULATION RESULTS AND DISCUSSION

The most serious problem associated with Field Electron Emitter arrays are with respect to current instability and non-uniformity caused by structural (shape) and work function variations of the emitter tip. It has been reported (7) that current stabilization may be achieved by utilizing a constant current source in series with the emitter tip. In particular, it has been found that emission current could be well controlled and stabilized by the saturated drain current of a MOSFET device, rather than field emission mechanisms (3). A vertical cross-section of this structure is shown below in Fig. 1. In this structure, the gate electrode serves two functions: as an extraction electrode and as a control gate for the drain current supplied by the tip. Additionally, each emitter tip is junction isolated from the substrate.

The Athena process modeling program, available from Silvaco International, was used for the simulations. Simulated vertical cross-sections of the Silicon Field Electron Emitter

MOSFET structure after sequential fabrication process steps are shown in Fig. 2 (a) to (f). A description of the fabrication steps, corresponding to Fig. 2, is provided below.

(a) A p-type (100) wafer with a resistively of 4 ohm-cm is used as the starting wafer. A 300 nm thick oxide layer is thermally grown and then patterned into a 1.5 um length oxide layer on the Silicon surface.
(b) The Silicon is then Reactive Ion etched to form the emitter tip region.
(c) A B+ implantation at 50 KeV and dose of 1E13 cm-2 is performed and a thermal oxidation at 1000 C grows 0.63 um of oxide. Nb is then deposited and patterned.
(d) The oxide is etched which exposes the Silicon emitter tip by a lift off process.
(e) The drain and source regions are formed by a phosphorous implantation and subsequent annealing.
(f) Completed Structure – after Al deposition and patterning for metalization of the gate and source.

Figure 3 shows the two-dimensional grid used for the finite element simulations shown in Fig. 2. A fine grid was used to simulate regions which required very precise numerical computations such as at the surface regions; whereas a course grid, as in regions away from the surface, is used for less precise computations. In this manner computational time is minimized. A comparison of the simulated vertical cross-sectional area and a SEM of a fabricated structure are shown in Fig. 4. It is seen that the vertical cross-section and shape of the simulated structure is in close agreement to the fabricated structure.

The most critical processing steps with respect to emitter tip geometry were found to be the Silicon etch step and the oxidation time after Silicon etch. Figure 5 shows the variation of emitter tip geometry for Silicon etch depths of 0.66, 1.12 and 1.33 um respectively. It is seen that both the emitter tip geometry and the geometric relationship of the emitter tip to the Nb extraction electrode vary considerably with Silicon etch depth. Similarly, Fig. 6 shows the variation of emitter tip geometry and the geometric relationship of the emitter tip to the Nb extraction electrode for oxidation times of 105, 125 and 150 minutes. It is seen in Fig. 6 that as the oxidation time increases the emitter tip recedes below the Nb extraction electrode.

SUMMARY

Two-dimensional computer simulation results were presented of the process steps used in the fabrication of a Silicon Field Emitter Tip MOSFET structure. The fabrication processes of silicon and oxide etching, ion implantation and annealing, oxidation, lift off etching and patterning are simulated. The vertical cross-section, shape, and dimensions of the simulated Silicon Field Electron Emitter MOSFET structure were found to be in close agreement with fabricated structures. The unique features of the fabrication process were the isotropic Silicon etching and the self-aligned B+ ion implantation. Both of these process steps were found to provide precise control of the emitter tip shape and extraction gate electrode.

ACKNOWLEDGEMENTS

The authors would like to acknowledge SILVACO International for the use of Simulation software, Athena and Atlas, and Mr. David Horn, of SILVACO International, for his support and encouragement throughout the course of this work.

REFERENCES

1. T. Asano and J. Yasuda, Jpn. Appl. Phys. Vol. 53 (1996) pp. 6632-6636.
2. T. Asano, IEEE Trans. Electron Devices, 38, 2392, (1991).
3. T. Hirano, S. Kanemaru and J. Itoh, J. Appl. Phys., 35, 6637, (1996).
4. Chin, "Two Dimensional Oxidation, Modeling and Applications". Ph.D Thesis, Department of Electrical Engineering, Standford University, Jun. 1983.
5. J. Lindhard, M. Scharff. and H.E. Schiott, Kgl. Dan. Vid. Selsk. Mat.-fys. Medd., 33 (1963).
6. A.S. Grove, Physics and Chemistry of Semiconductor Devices, Wiley, NY, 1967, pp. 38-39.

7. Y. Kobori and M. Tanaka, U.S. Patent 5 162 704, (1992).

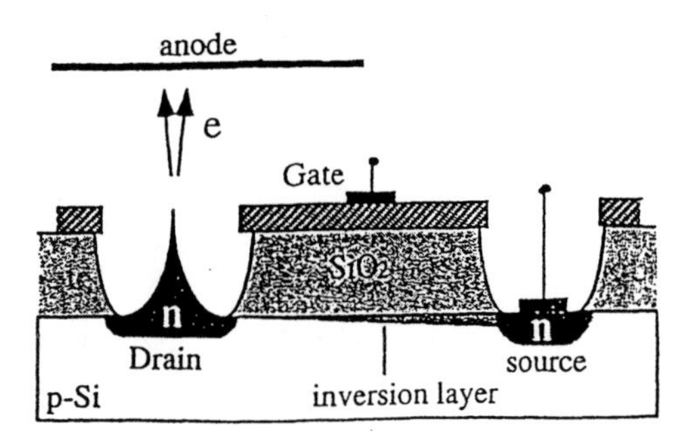

Fig. 1 Vertical cross-sections of a Silicon Field Electron Emitter MOSFET structure.

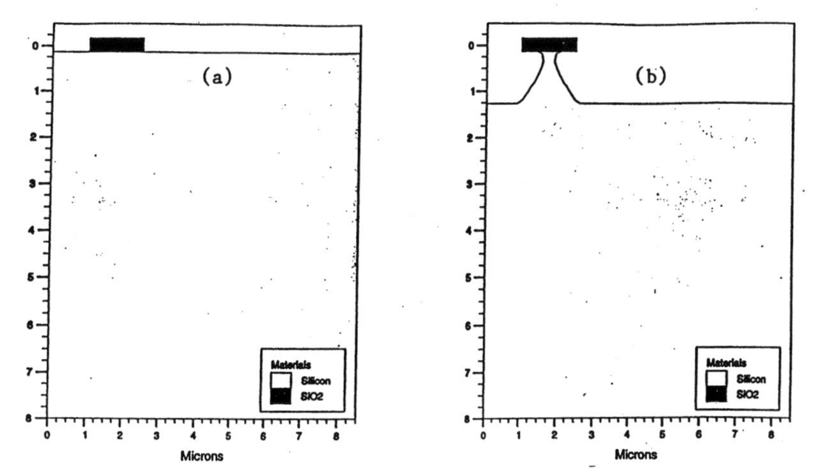

Fig. 2 Vertical cross-sections of a Silicon Field Electron Emitter MOSFET structure
(a) after oxide patterning, (b) after Silicon reactive ion etching.

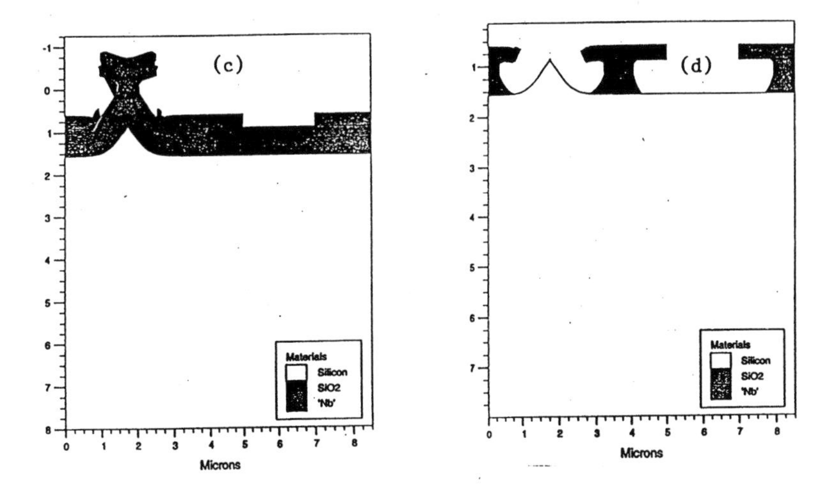

Fig. 2 Vertical cross-sections of a Silicon Field Electron Emitter MOSFET structure
(c) after B+ implant,oxidation and Nb patterning, (d) after oxide etching to form the emitter tip.

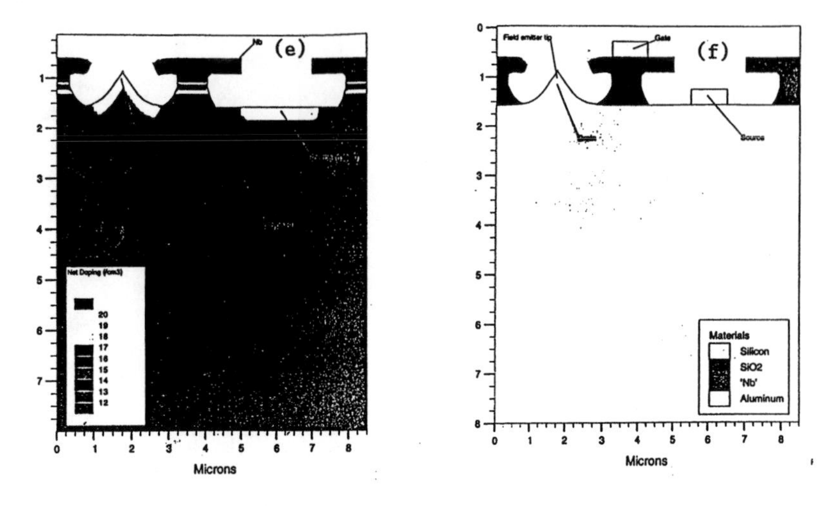

Fig. 2 Vertical cross-sections of a Silicon Field Electron Emitter MOSFET structure (e) after drain-source P implantation and annealing, (f) completed structure.

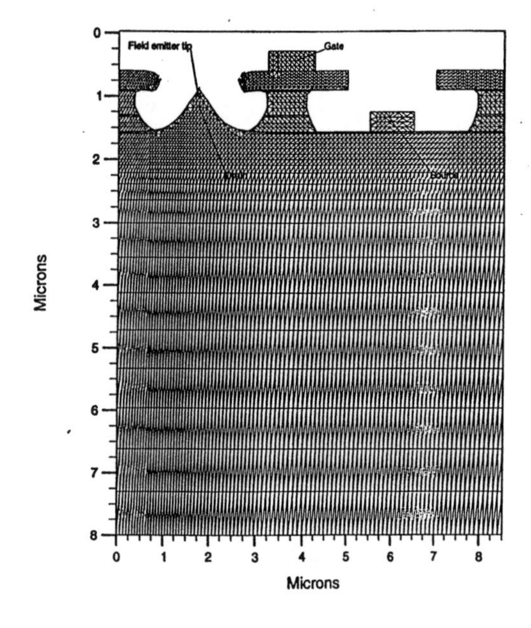

Fig. 3 Grid used for two dimensional finite element simulations.

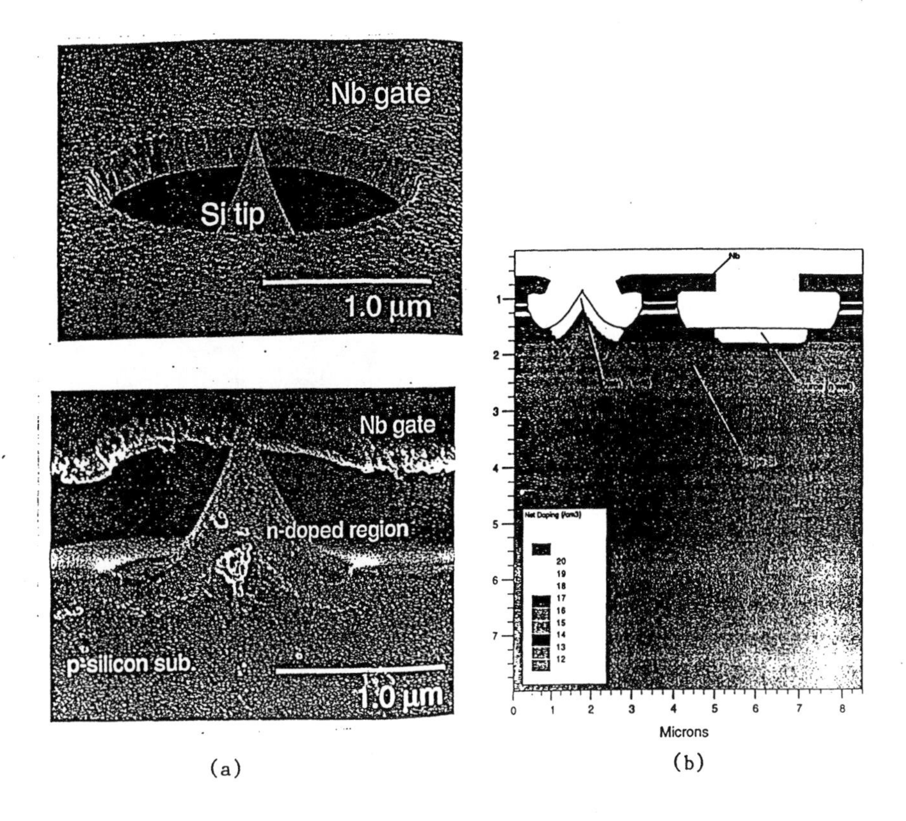

Fig. 4 Comparison of (a) fabricated and (b) simulated Silicon Field Electron Emitter MOSFET structure.

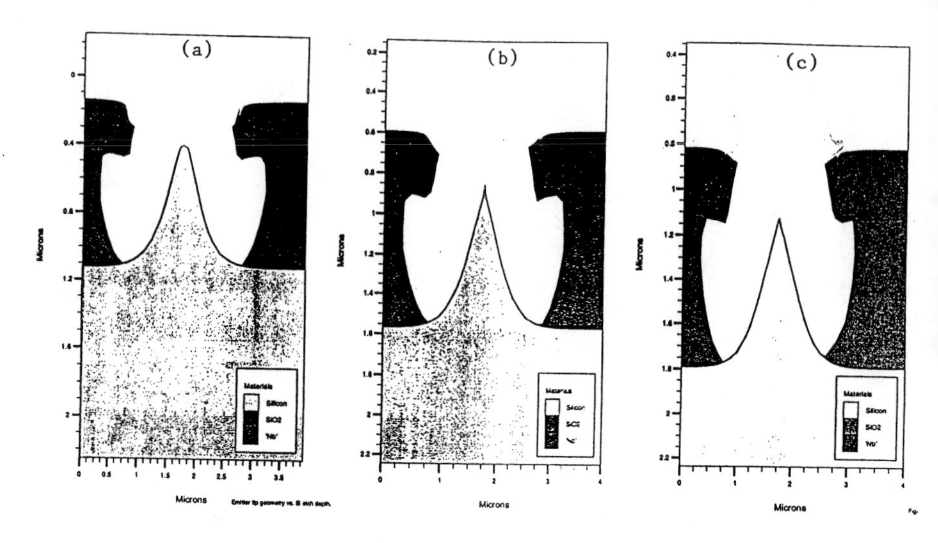

Fig. 5 Variation of Emitter tip geometry with Silicon etch depths of: (a) 0.66 um (b) 1.12 um and (c) 1.33 um.

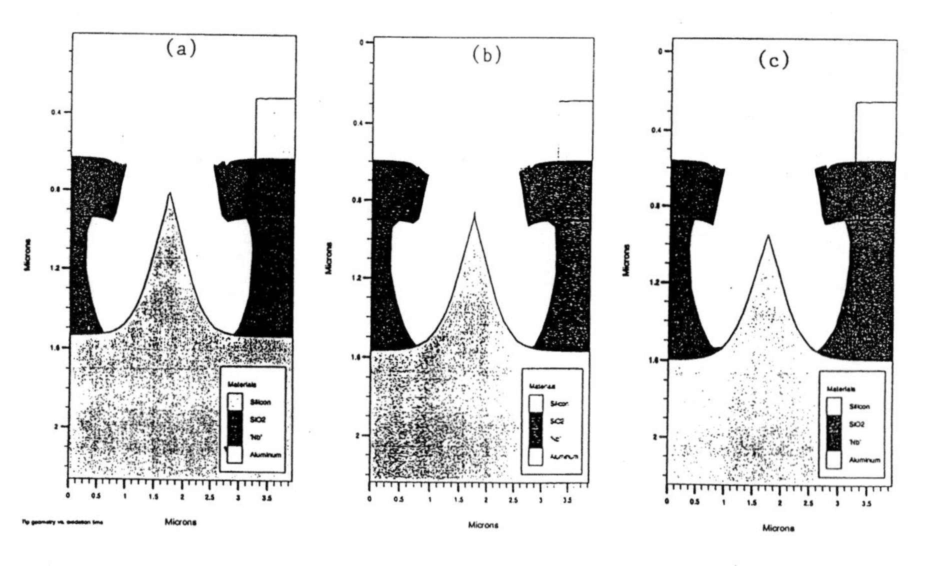

Fig. 6 Variation of Emitter tip geometry with Oxidation time after Silicon etch of: (a) 105 min. (b) 125 min. and (c) 150 min.

SIMULATION OF A SELF ALIGNED LOCAL OXIDATION SILICON FABRICATION PROCESS FOR SUBMICRON FIELD ELECTRON EMITTERS

Gerald Oleszek and Manuel Rodriguez
University of Colorado
Microelectronics Research Laboratories
Dept. Of Electrical and Computer Engineering
Colorado Springs, CO 80933

ABSTRACT

The fabrication of a submicron Field Emitters and its arrays using Silicon micro fabrication processes have been reported in the literature and their emission properties have been investigated. Potential applications of Field Emitter Arrays are for Field Emission Displays, high resolution electron beam lithography as well as other applications which require highly collimated electron beams.

There have been a number of fabrication processes for Field Emitters reported which utilize Silicon micro fabrication processes to produce arrays of emitters. Of importance in each of these processes is the requirement to produce a well defined emitter geometry in the shape of a sharp tip with high packing density. In this paper, computer simulation results are presented for the fabrication of a Field Emitter Array with submicron gate apertures using a self aligned local oxidation Silicon process. It was found that the shape, dimensions and aspect ratio of a Field Emitter with submicron gate apertures obtained by simulation were consistent with cross sections of fabricated Field Emitters reported in the literature.

INTRODUCTION

The microfabrication of field electron emitters is an area that holds much promise for the development of high frequency amplifiers and cold micro-cathodes. Arrays of these emitters (FEAs) have been successfully employed in the fabrication of field emission displays (1). The promise of a new class of displays featuring low cost, low power consumption and high resolution has driven most of the current research in this area. Many different microfabrication processes have been proposed in the past, ranging from variations to the original Spindt method to more recent methods that utilize the latest developments in microfabrication and micromachining. Numerical, physically based, two-dimensional simulation based on the SSUPREM4 models has been extensively used to simulate the fabrication of devices such as field-effect and bipolar transistors. However, it seldom has been used in the simulation of microfabricated structures such as field emitters.

In this work, we propose the utilization of two-dimensional numerical computer simulation to simulate the fabrication of a self-aligned microemitter using selective etching of silicon. Numerical simulations are based on the SSUPREM4 models, integrated in the Athena process modeling program, available from Silvaco International.

FABRICATON PROCEDURE

The simulated fabrication process(2) for the microemitter is schematically shown in Figures 1 - 8. The following is a description of the simulated process steps.

Fig. 1: An N-type, (100)-oriented wafer, with a background doping of 10 Ω-cm is used as a starting material. A 0.25-μm thick oxide layer is grown on the bare silicon. This layer is subsequently patterned into 4.3μm diameter disks using standard photolithography steps.

Fig. 2: BF_2^+ ions are implanted at 140keV to a dose of 1×10^{17}, this is followed by annealing at 800°C for 30 min. in a N_2 ambient to form a p^+ layer. The patterned SiO_2 disk masks the underlying silicon from the implant. The p^+ layer will act as an etch stop layer for a subsequent anisotropic etch that will define a mold for the emitter. This layer also functions as a conductive gate in the completed structure.

Fig. 3: Oxide is completely removed from the surface of the wafer and an anisotropic etch is performed on the exposed silicon. The etch geometry is the classic, inverted, pyramidal pit, shaped by the crystalline orientation of the substrate and limited in size by the maximum exposed (111) crystal planes.

Fig. 4: A protracted implant drive-in at 1100°C for 60 min. in a N_2 ambient is performed. The drive-in time is crucial, as it will define the height of the emitter tip as well as the relative position of the emitter tip with respect to the gate. This buried p^+ layer will act as an etch stop layer for the backside silicon etch depicted in Fig. 7.

Fig. 5: Oxidation is performed in a wet ambient at 1050°C for 14 min. to form the gate insulator. The oxidation time employed in this step will directly affect the gate to emitter spacing. Wet oxidation is employed to reduce the oxidation time required and thus preventing further diffusion of the p^+ layer.

Fig. 6: WSi_2 is deposited on top of the wafer so it completely fills the mold created by the previous anisotropic etch. The WSi_2 will serve as the conductive material that forms the field emitter tip.

Fig. 7: An anisotropic Si etch is performed on the backside of the wafer. This etch proceeds all the way through the wafer until it encounters the buried p^+ layer which acts a as an etch stop layer.

Fig. 8: The exposed SiO_2 is isotropically etched from the backside of the wafer to expose the emitter tip. This step also determines the size of the emitter gate hole.

SIMULATION MODELS

Two dimensional, physically based computer simulation relies on numerically solving equations at a number of discrete points that represent a cross section of the device being simulated. The collection of points used for this finite element analysis is referred to as the *grid.* Computational nodes are linked to every single point on the grid. Since a computation must be performed at each discrete node, the structure of the grid critically affects the time required to complete a simulation. A coarse grid containing a small number of points is often used to obtain a rough idea of the structure, while a fine grid is used for final results or at critical regions within the structure. The grid used for our field emitter simulation is shown in Fig. 9.

Various numerical models can be utilized in the simulation of the process steps required to complete the device structure. Some of these models trade off computational time for accuracy while others will produce more accurate results depending of the real life processing conditions encountered when fabricating an actual device. The critical processes in simulating the fabrication of the field emitter are Diffusion & Implantation of impurities in Silicon. Other processes used include Wet/Dry oxidation, anisotropic/isotropic etching and deposition of evaporated material.

To form the buried p^+ region that ultimately determines the emitter height, an implantation step is performed. An analytical model is used to simulate the implant profile. The analytical model used in the simulation of implanted profiles relies on the reconstruction of previously measured implant profiles. Spatial-distribution moments are utilized by SSUPREM4 to calculate the profiles of implantation distributions. The method used is based on the theory of "Range Concepts"(3), where an ion-implantation profile is build from a set of previously measured or calculated moments. A Pearson distribution, analytical implant model, is used for the calculation of the implant profile in our simulation.

Diffusion of the implanted profile is the most critical simulation step for this particular field emitter structure. Mathematical equations describing diffusion in silicon have been extensively researched and modeled in the past(4). The concentration and depth of the p^+ diffusion after annealing and/or oxidation steps is determined by analytical calculation of the diffusion equations at each node in the grid.

Analytical oxidation models in SSUPREM4 are limited to planar, bare Silicon surfaces. To obtain accurate solutions in the simulation of arbitrary structures, including structures with steps and trenches, numerical models must be used. In all models, oxidation equations are solved at the silicon/oxide interface to obtain the growth rate at each node in the grid. As oxide grows, its expansion is determined by the chemical reconfiguration of adjacent materials. The expansion of the material promotes a certain flow, which is simulated by the numerical models in SSUPREM4. The compressible model used in our simulation makes use of the reduced hydrodynamic equation to simulate creeping-flow motion in the material(6). Since a slight compressibility is allowed in the numerical model used, solutions using this method are not completely accurate since the internal stress of the material is not taken into account. However, since oxidation is not a critical step in our process, this model is sufficient for our purposes.

The least critical process in our simulation is the deposition of an evaporated metal on the surface of the silicon. A conformal deposition model was utilized to reduce computational time. This simple model deposits a specified thickness of the desired material with unity step coverage. Physical properties of the material such as density and Poisson's ratio are neglected when using this model. More complex models can be utilized to optimize step coverage and deposit material at different angles from the surface normal.

Anisotropic etching capabilities are limited in SSUPREM4 simulations. The etch stop characteristics of an impurity layer can be simulated by noting the junction depth of the impurity in the substrate and then performing a geometrical etch of the material based on the measured doping profile. The exact geometry of the etch is constrained by the number of nodes on the grid that define a line between the specified geometrical coordinates. Isotropic etching was utilized to etch the oxide layer from the tip of the field emitter. This analytical model removes a certain material at a specified rate. Since this model simulates a chemical reaction occurring at an equal rate in all directions, isotropic profiles develop circular cross sections and will simulate undercutting of the mask. All etchants can only be specified for a particular material; residual etching of mask materials is not readily simulated.

SIMULATION RESULTS AND DISCUSSION

Fig. 8 shows the completed field electron emitter structure. The resulting structure was compared to micrographs reported in the literature(2). Dimensions of the simulated structure closely match those of the fabricated device. Since the fabrication process of the device is largely dependent on the etch stop characteristics of the diffused p^+ layer; the depth of this layer is crucial to our results. Boron concentrations of more than 6×10^{19} cm^{-3} dramatically reduce the etch rate of virtually all Silicon anisotropic etchants containing ethylenediamine/pyrocatechol(7). In our simulation, a lengthy drive-in time was required to achieve the required depth as compared to the time reported in the fabrication of the actual device. It was noted that using the reported time would yield a boron concentration of less than 1×10^{16} cm^{-3} at the vertical depth where the etch stop characteristics should significantly slow the etch rate of Silicon in the substrate. This concentration would have a negligible effect in the etch rates of common anisotropic etchants such as EDP and KOH. It is possible that the actual fabricated device was exposed to further, unreported, thermal steps that might account for the discrepancy in our simulation.

The simulated profile in Fig. 1 shows the two-dimensional cross section of the patterned 4.3μm-diameter disk. A wet oxidation at 1100°C for 14 min. is simulated and a geometrical patterning (etch) of the oxide is performed. Fig. 2 shows the profile of the junction depth of the implanted p^+ layer after a simulated implant of BF_2^+ atoms at 140 keV to a dose of 1×10^{17} cm^{-2} and an anneal at 800°C for 30 min. To perform the geometrical anisotropic etch depicted in Fig. 3, measurements were performed on the simulated profile to allow a classic geometrical etch of a triangular shape with sidewalls at 54.7° to the surface normal. The diffused p^+ layer served as the etch mask, limiting the further exposure of (111) planes.

Fig. 4 shows the junction depth profile of the p+ layer after a simulated 1100°C anneal for 60 min. in an N_2 atmosphere; the junction is driven 3.25μm into the substrate. The subsequent wet oxidation at 1050° for 14 min. has only a negligible effect on the diffusion profile; the results of this process step are shown in Fig. 5. Tungsten silicide is deposited to a thickness of 6μm using a conformal deposition model as shown in Fig. 6. The thickness of this deposition is arbitrarily selected for this simulation; however, a significantly greater thickness would be advised in fabricating the device to ensure structural rigidity for handling.

Fig. 7 shows the structure after a simulated backside etch through the entire wafer thickness. This process step would require an estimated 7h to proceed through an entire 500μm-thick silicon wafer. The junction depth of the p^+ layer was used as the etch stop layer for termination of the anisotropic etch. A geometrical etch was performed after carefully measuring the exact position of the junction. The final, simulated process step, was an isotropic oxide etch on the backside of the wafer. This process step exposes the WSi_2 emitter tip; final results are shown on Fig. 8.

The main advantage of two-dimensional, physically based, numerical computer simulation is the fact that results due to changes in process parameters can be readily evaluated. Several simulations were performed while varying some of these parameters. Two of the simulated structures are shown in Figs. 10–11. Variations were made in the thickness of the oxide grown inside the emitter mold and in the p^+ layer drive-in time. Measurements were taken of two important structural properties that affect the performance of the emitter. One of these properties is the gate apex spacing of the emitter, defined as the distance between the gate and the tip of the electron emitter. The second property is the space between the gate and the emitter structure. In our particular device the length of this space is equal to the thickness of the oxide inside the emitter mold separating the conductive p^+ gate from the deposited WSi_2. Plots of these measurements are shown in Figs. 12 – 13.

CONCLUSION

Numerical, two-dimensional simulation of the fabrication of a self-aligned microemitter using selective etching of silicon has been proposed and demonstrated. These techniques can be applied to the simulation of many of the different processes that have been designed for fabricating field electron emitters. Further work in characterizing actual structures predicted by simulation results is needed.

ACKNOWLEDGMENTS

The authors thank Mr. David Horn of SILVACO International Co. for his support and encouragement throughout the course of this work. The authors gratefully acknowledge SILVACO International Co. for providing the software used in our work.

REFERENCES

1. C.A. Spindt, C.E. Holland, A. Rosengreen and I. Brodie: IEEE Trans. Electron Devices ED-38 (1991) 2355.
2. T. Asano and J. Yasuda, Jpn. J. Appl. Phys., 35, 6632 (1996)
3. J. Lindhard, M. Scharff. and H.E. Schiott, Kgl. Dan. Vid. Selsk. Mat.-fys. Medd., 33 (1963)
4. A.S. Grove, Physics and Chemistry of Semiconductor Devices, Wiley, NY, 1967, pp. 38-39.
5. L. Mei and R. W. Dutton, Solid State Technol. 26:139 (June 1983)

6. D. Chin, "Two Dimensional Oxidation, Modeling and Applications". Ph.D Thesis, Department of Electrical Engineering, Standford University, Jun. 1983
7. S. D. Collins, J. Electrochem. Soc., 144, 2242 (1997)

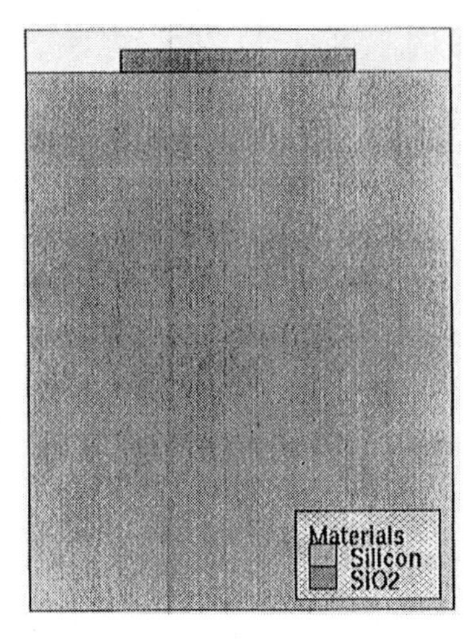

Fig. 1 Structure after oxidation and patterning.

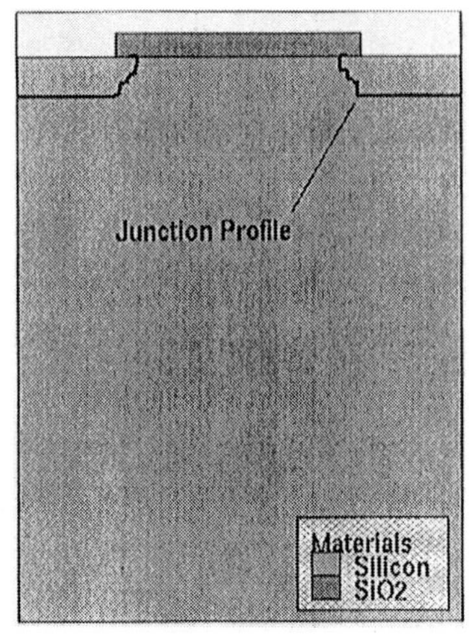

Fig. 2 P+ region junction profile after implantation and annealing.

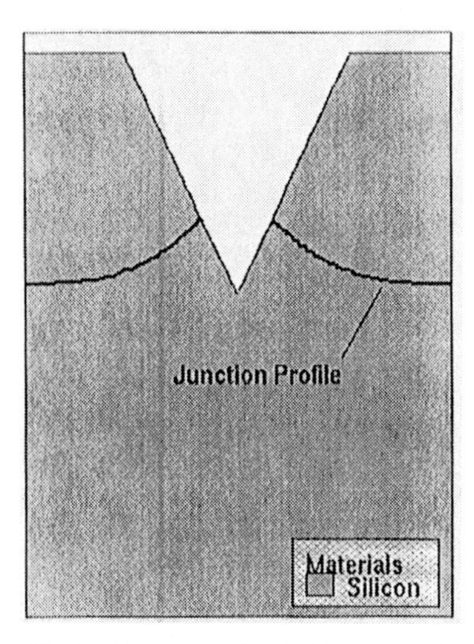

Fig. 3 Structure after oxide removal and anisotropic etching of the substrate.

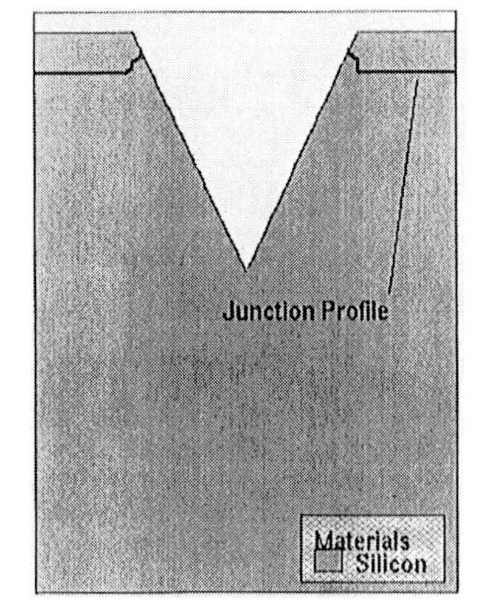

Fig. 4 P+ region junction profile after annealing.

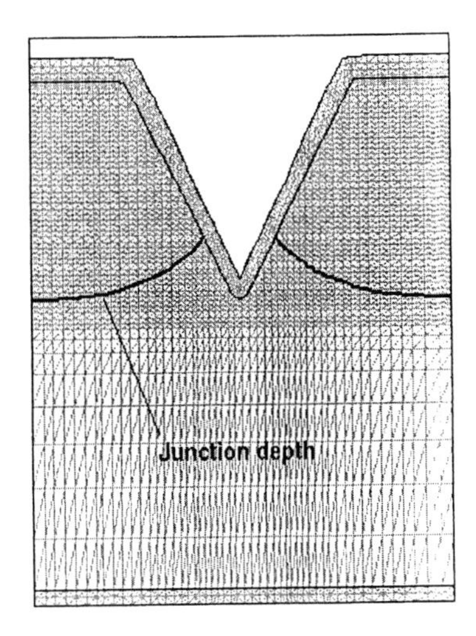

Fig. 9 Grid used in the numerical simulation of the Field emitter.

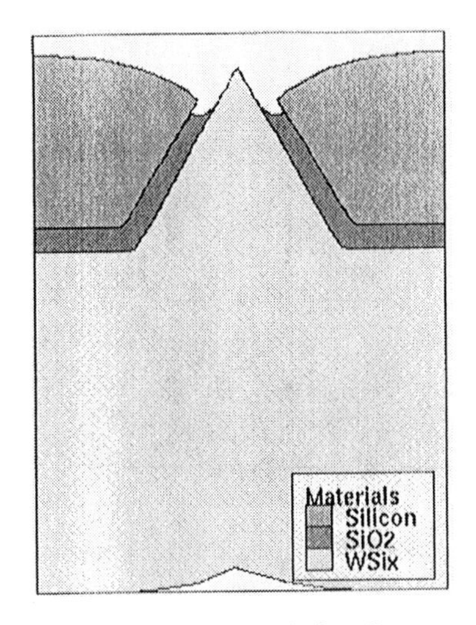

Fig. 10 Sensitivity simulation. Structure after an emitter mold oxidation (wet) at 1050°C for 21 min.

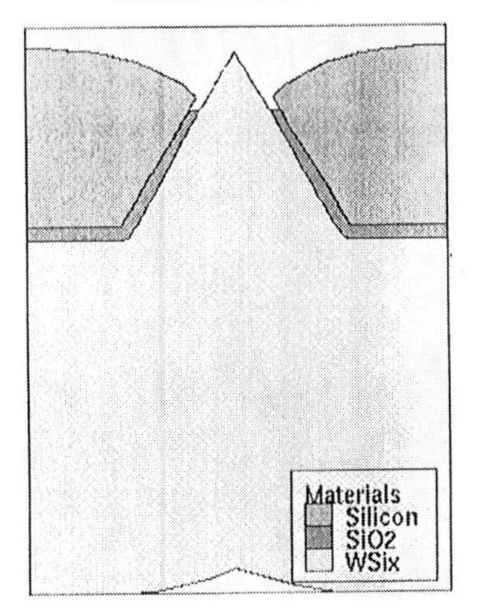

Fig. 11 Sensitivity simulation. Structure after an emitter mold oxidation (wet) at 1050°C for 8 min.

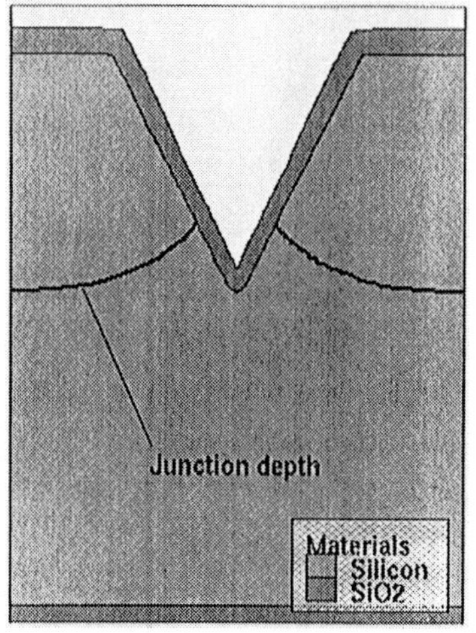

Fig. 5 Structure after thermal oxidation of the emitter mold. Note: wafer thickness not to scale.

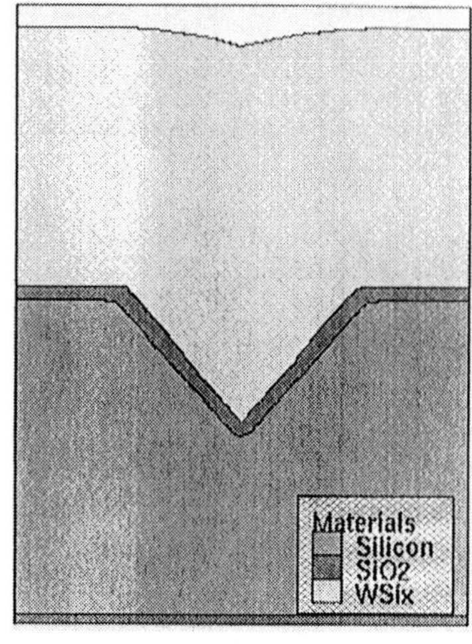

Fig. 6 Structure after WSi_2 deposition. Note: wafer thickness not to scale.

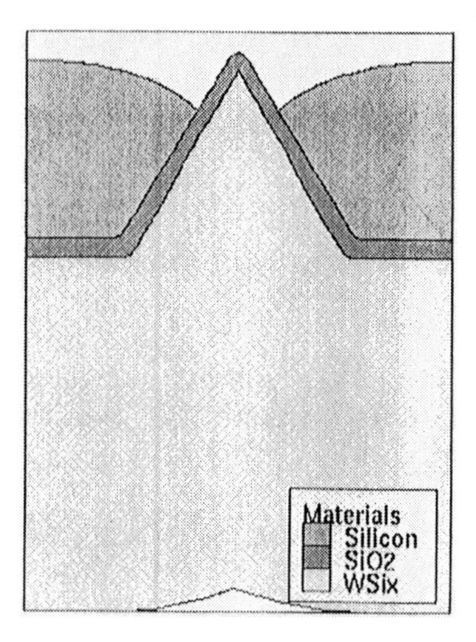

Fig. 7 Structure after anisotropically etching through the entire wafer thickness.

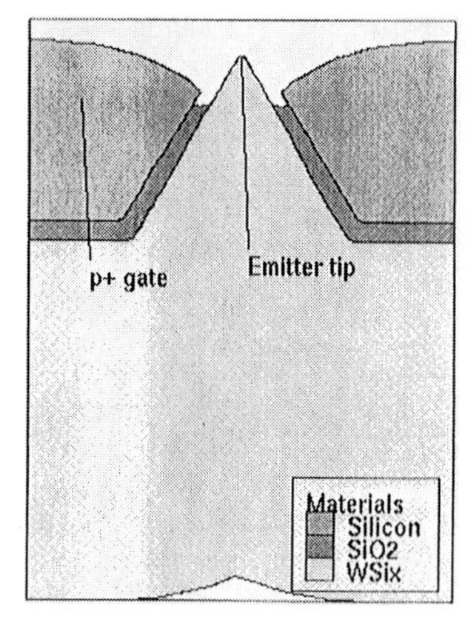

Fig. 8 Completed structure after an isotropic etch of the exposed oxide covering the emitter tip.

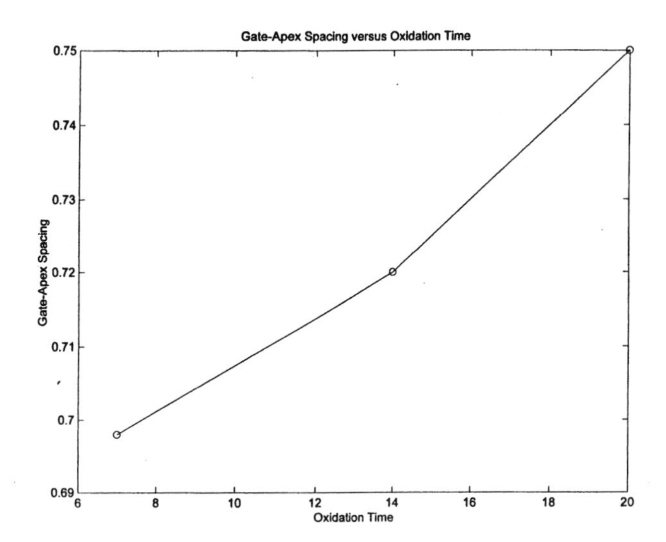

Fig. 12 Gate-apex spacing vs. oxidation time.

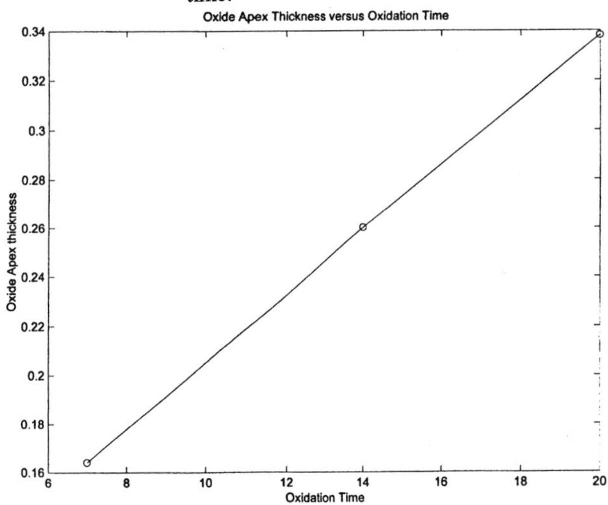

Fig. 13 Oxide apex thickness vs. oxidation time.

MULTICOMPONENT INTERACTIONS OF MOISTURE AND ORGANIC IMPURITIES WITH THE WAFER SURFACE

M. Verghese, E. Shero, and F. Shadman
Department of Chemical & Environmental Engineering
University of Arizona, Tucson, AZ 85712

Adsorption/desorption studies of moisture on the <100> Si wafer surface were conducted using Atmospheric Pressure Ionization Mass Spectrometry (APIMS) and Electron Impact Mass Spectrometry (EIMS) with both regular and fully deuterated water. The resulting outgassing phenomena were modeled using a multilayer model which considers the distribution of chemisorbed and physisorbed moisture molecules and includes an oxidation term to account for the incorporation of moisture into the oxide at elevated temperatures. Moisture can also assist in the adsorption of organic impurities on the substrate surface under certain conditions. Studies were carried out by introducing isopropyl alcohol (IPA) to <100> silicon wafer surfaces challenged with deuterated water. It is proposed that the alcohol's hydroxyl group undergoes both replacement and esterification chemistries with the surface bound, dissociated, water molecules. Experimental results validating the proposed surface kinetics and mechanism is presented.

INTRODUCTION

In modern VLSI manufacturing, the process yield and device performance are sensitive to the presence of homogeneous contaminants from various sources. Allowable contamination levels are becoming exceedingly more stringent as circuit development evolves from the megabit into the gigabit level. By the turn of the century, the maximum permissible defect density for critical process level is projected to be less than $0.0001/cm^2$ [1]. Independent studies by many researchers have shown that even 1 ppb of impurity concentration at critical process steps can severely harm the production of modern solid state devices [2]. Of these contaminants, moisture can be most detrimental because of its high reactivity with silicon. Moisture contamination is hard to control due to its high ambient levels, strong polarity, and its ability to hydrogen bond. The presence of these impurities on the wafer surface may affect the process; for example, such contamination may lead to non-uniformities during oxidation, resulting in uncontrolled variations in the oxide film quality and thickness. Furthermore, the presence of moisture may assist in the chemisorption of other contaminants such as various organics. Organic contaminants, retained on the surface by chemisorption to the surface moisture's hydroxyl groups, can react with the silicon substrate during high temperature treatment to form silicon carbide. This can change the local reaction rate, leading to non-uniformity in the oxide and decline in quality. It has also been found that interfacial carbon contamination and crystal defect has a close relationship [3].

Many questions still remain regarding the interaction of moisture and the oxide surface. There is much debate over the chemical form of moisture on the surface, possible interactions with the surface, retention amounts, and the ease of moisture removal. Fundamental adsorption/desorption studies can give insight the nature and behavior of moisture on the surface. In this paper we present our results from adsorption/desorption studies of moisture and isopropyl alcohol on <100> silicon wafers. A multilayer model was developed to simulate the dynamic interactions of water molecules and the oxide surface. Furthermore, isotopic labeling studies were used to characterize the interaction of surface laden moisture and isopropyl alcohol. Based on the results, a mechanism for moisture assisted organic contamination of the wafer surface will be discussed.

EXPERIMENTAL

Apparatus

Figure 1 shows a schematic of the experimental setup. The main analytical instrument is the VG Trace + Atmospheric Pressure Ionization Mass Spectrometer (APIMS) with a corona discharge source. It is a triple quadrupole mass spectrometer that is able to measure well into the sub-ppb range and hence enables the performance of low loading experiments. Higher concentrations can be measured by an electron impact mass spectrometer (Balzers), and other dedicated analyzers (Meeco). The gas mixing system employs all electropolished stainless steel tubing, all metal mass flow controllers, and is designed to constantly purge the lines to avoid dead volumes. This allows a sharp step input at the test section. The system is also capable of providing deuterated water for studies involving isotopic labeling. The test section is versatile enough to investigate different surfaces in a variety of reactor configurations. The temperature in the test section is controlled by means of a PID controlled furnace capable of a maximum temperature of 1200 °C.

Experimental Procedure

The procedure for a typical adsorption/desorption experiment is shown in Figure 2. First, the surface under investigation is baked while under an ultra-pure N_2 purge. In the case of moisture studies on the wafer surface, this involves a bakeout at 400 °C. This is known to remove most of the chemisorbed water and regenerate the siloxane bridges on the oxide surface [4]. The surface is then exposed to a controlled amount of impurity in the challenge step, and the surface and gas phase concentrations are allowed to equilibrate. The surface is subsequently purged with N_2 to desorb the impurity adsorbed to the surface. A following bakeout can give insight to the total amount of impurity adsorbed.

DISCUSSION

Multilayer Model Formulation

Adsorption/desorption studies conducted in the past have resulted in the formulation of many empirical models [5][6]. However, the goal of this research is to develop a physically realistic model that allows dynamic interactions at the molecular level. Figure 3 illustrates the technique for the formulation of the multilayer adsorption/desorption model.

It has been shown that multilayer coverage occurs even at low concentration challenges. In the case of moisture adsorption on the silicon oxide, the first layer is formed by dissociative chemisorption of water onto the siloxane bridges to form hydroxylated sites. The subsequent layers are physisorbed water molecules that adsorb to form an array of stacks of molecules whose distribution is a function of temperature, flow rate, and other physical properties of the system. The number of water molecules in a given stack can increase by the adsorption of a molecule from the gas phase. Similarly, desorption of a water molecule from the stack to the gas phase leads to a decrease in the size of the stack. Hence, at lower concentration challenges, we expect smaller stacks whereas an exposure to a higher concentration challenge will shift the population to higher stacks as a result of growth of the stacks by adsorption. Consequently, at higher challenges, the number of small stacks approaches zero. At the heart of the model is the inclusion of a variable activation energy of desorption. The first layer consists of dissociated, chemisorbed water and hence has a very high activation energy of desorption. For the physically adsorbed water, the model allows for a variable activation energy which is directly related to the Van Der Waals interactions from the surrounding molecules. Since the molecules on the highest layer have fewer neighboring molecules, they will have the lowest desorption activation energy. The variable activation energy assumption reduces to

$$k_{d,j} = a e^{-b m_j} \qquad [1]$$

The dynamic adsorption/desorption interactions can be followed by means of a site balance. In equation form, the site balances reduce to

$$\frac{\partial n_j}{\partial t} = -(k_a C_g + k_{d,j}) n_j + k_a C_g n_{j-1} + k_{d,j+1} n_{j+1} \qquad [2]$$

The site balance for a stack of molecules (n), j monolayers deep is directly related to the the adsorption constant (k_a), gas phase concentration (C_g), and the desorption constant from a stack one monolayer higher ($k_{d,j+1}$). The negative term in the equation arises from adsorption to or desorption from the stack. The positive term arises from adsorption to a stack one monolayer lower or from desorption from a stack one monolayer higher than the site under investigation. The total loading and the total number of sites is given by the following infinite series:

$$L = \sum_{j=1}^{\infty} \left(j - \frac{1}{2} \right) n_j \qquad [3]$$

$$S = \sum_{j=0}^{\infty} n_j \qquad [4]$$

Hence, the dynamic distribution of water molecules on the oxide surface can be predicted by solving the $j+1$ differential equations simultaneously. The multilayer model can also be linked to an oxidation model by the inclusion of an oxidation term in the site balances.

The multilayer model predicts the adsorption/desorption dynamics of moisture very well. This can be seen from the fits to our experimental data for both the stainless steel and the quartz surface (Figures 4-5).

Moisture Assisted Organic Contamination

The high reactivity of the hydroxylated Si-OH species that arise from the interaction of moisture and the silicon surface can promote further contamination by other compounds. Wafers are exposed to organic contamination in many of the processes they undergo. Contamination by organics has not attracted much concern in the past because they have been thought to be easily controlled. However, the Si-OH species resulting from moisture contamination can strongly hydrogen bond to organic molecules such as alcohols and bind them strongly to the surface. To investigate the propensity of moisture to assist in organic contamination, iso-propyl alcohol (IPA) was used as a model contaminant because of its simple nature and its prevalence in semiconductor processes.

To give insight to the reactions taking place on the oxide surface, deuterated water was used instead of regular moisture. IPA was then introduced to the system and the resulting species formed were monitored. Figure 6 illustrates our proposed mechanism for moisture assisted adsorption of IPA. The high polarity of the surface Si-OD bonds enables the formation of hydrogen bonds to the alcohol hydroxyl group. At higher temperatures, the activation energy for the 'esterification' reaction can be overcome and the iso-propoxy group can be chemisorbed to the surface with the displacement of water (or in the case of the deuterated surface, HDO). The replacement reaction, where the alcohol exchanges a hydrogen with the surface hydroxyl group, is just as viable. Hence, by challenging the surface with deuterated water, we expect to see the production of both HDO (mass 19) and deuterated IPA (mass 46) upon the introduction of IPA. Figure 7 shows that the experimental results validate this proposed mechanism.

Table 1 shows our preliminary findings on the effect of moisture on IPA loading and retention.

Table 1. Effect of Moisture on the Loading of IPA

Moisture Challenge (ppb)	**IPA Adsorbed (moles/cm^2)**	**IPA Desorbed @ 150 °C (moles/cm^2)**	**IPA Desorbed @ 400 °C (moles/cm^2)**	**% Load Retained on Surface**
195.0	3.15 E-11	1.26 E-11	8.65 E-13	57.0%
~ 0.0	1.90 E-11	1.78 E-11	1.09 E-12	0.6%

The surface was first exposed to 195 ppb of H_2O and then to 245 ppb of IPA for a controlled amount of time. The contaminants were then desorbed at 150 °C and 400 °C and the amount loaded, desorbed, and retained were calculated. The experiment was then repeated without the prior exposure to moisture. Comparison of the experimental results shows that the presence of moisture helps adsorb more IPA to the surface oxide and promotes the retention of about 57% of the initial IPA load. With no prior moisture contamination, the IPA is easily removed from the surface oxide.

CONCLUSION

In summary, adsorption/desorption experiments were conducted to investigate the kinetics of moisture contamination and moisture assisted organic contamination of the silicon oxide surface. A multilayer model, which allowed for dynamic interactions, multiple layers, and a variable desorption activation energy was developed and fit to the experimental data. The propensity of moisture to assist in organic contamination was investigated by using IPA as a model contaminant. Isotopic labeling studies were used to give insight to the surface reactions taking place. It is proposed that at higher temperatures, an esterification reaction can take place leading to a surface bound organic species. Initial experimental findings have shown that the moisture contaminated surface oxide has a higher capacity and higher retention characteristics for organics. Hence, moisture acts as both a contaminant and a promoter of contamination by other compounds.

ACKNOWLEDGMENTS

The authors would like to thank the Semiconductor Research Corporation (SRC) for the financial support of this work.

REFERENCES

1. P. Singer, *Semiconductor International*, p46, January 1995.
2. C. M. Osburn, *Microcontamination*, **9**, no.7, p19, July 1991
3. N. Matsauo, K. Kagawa, and T. Miyoshi, J. Appl. Phys. **80** (7), 1 October 1996
4. D.W. Sindorf, G. E. Maciel, *J. Am. Chem. Soc.* **1983**, 105, 1487.
5. A.M. Haider and F. Shadman, *IEEE Trans. Components, Hybrids, and Manufacturing Tech.*, **14**, no. 3, p507, 1991
6. D. G. Coronell, A. D. Johnson, M. S. K. Chen, S. N. Ketkar, D. A. Zatko, J. V. Martinez de Pinillos, *Proceedings Institute of Environmental Sciences*, 1994

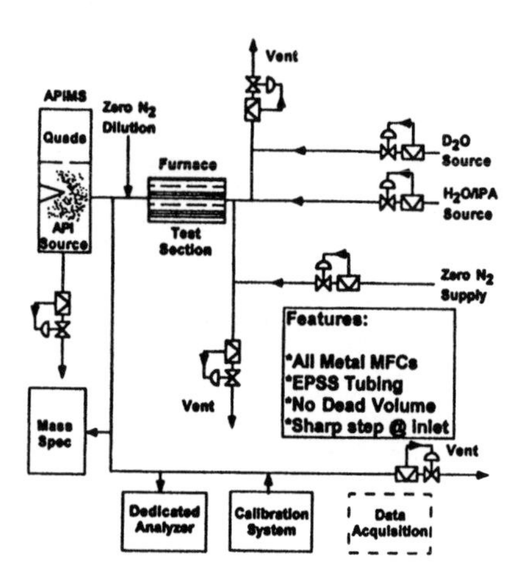

Figure 1. Experimental Setup

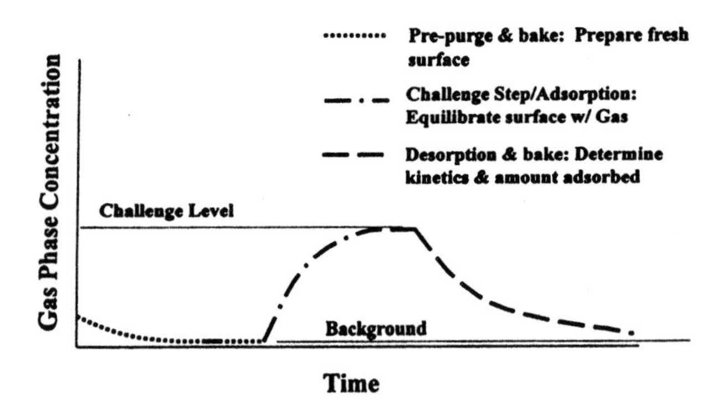

Figure 2. Experimental Procedure

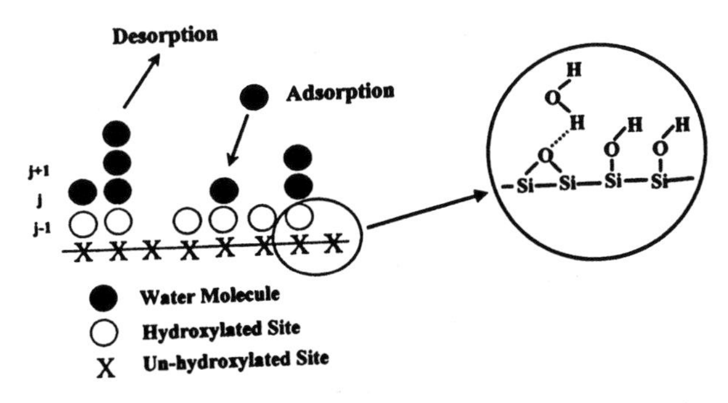

Figure 3. Multilayer Adsorption/Desorption on Silicon Oxide

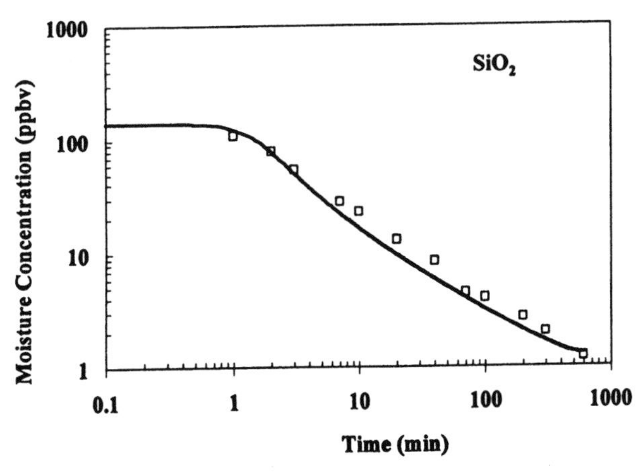

Figure 4. Fit of Multilayer Model to Data (SiO_2)

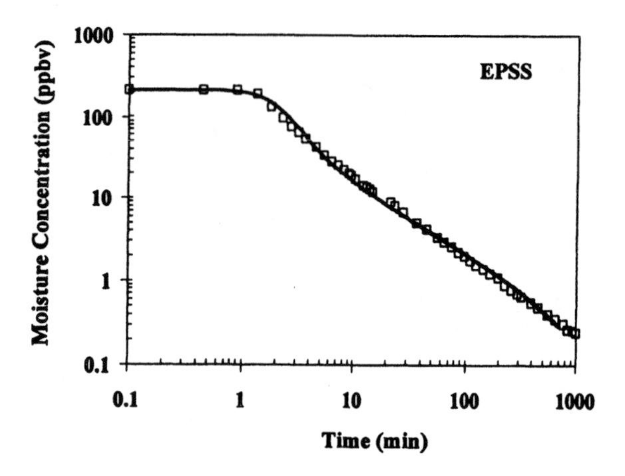

Figure 5. Fit of Multilayer Model to Data (Stainless Steel)

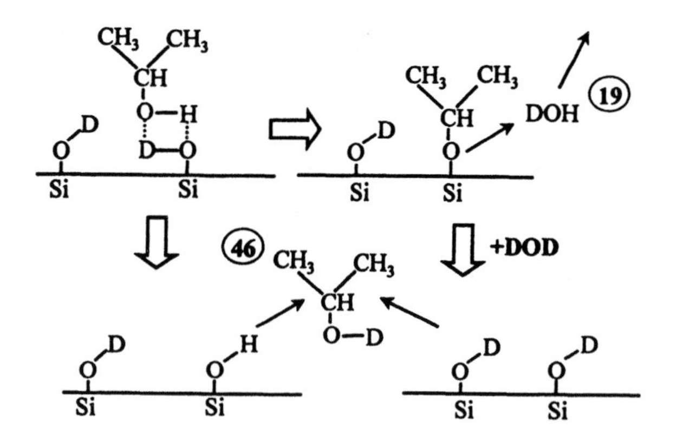

Figure 6. Moisture Assisted Isopropanol Adsorption on SiO_2

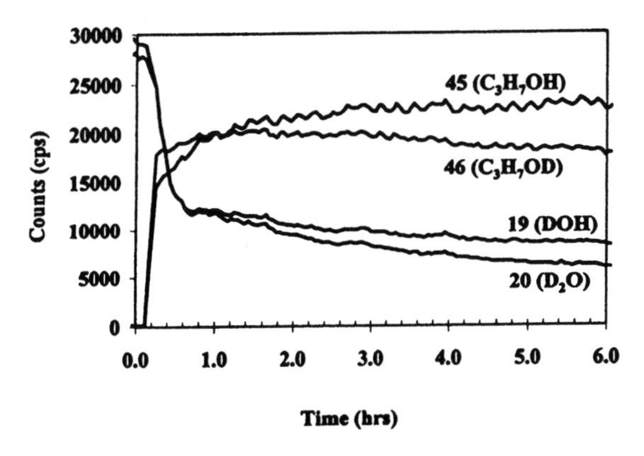

Figure 7. Exposure of D_2O Loaded Surface to IPA Challenge

REPLACEMENT OF Au AND Ag IN RADIOELECTRONICS AND NEW SUBMICRON, LIGA AND ADDITIVE TECHNOLOGIES

T. N. KHOPERIA
Institute of Physics Georgian Academy of Sciences
6 Tamarashvili st., 380077, Tbilisi, Georgia,
E-mail: tekh@physics.iberiapac.ge Fax: (995 32) 954807; Tel: (995 32) 395101

Abstract - The results of systematic investigations of the mechanism and kinetics of the sensitization and activation are described. The new competitive submicron, LIGA and additive technologies for making films and three-dimensional elements for ULSI fabrication are presented. The mechanisms of sensitization and activation are established. As a result implemented of developed technologies in the mass production of articles of piezoengineering, microelectronics, etc. Au and Ag have been adequately replaced by Ni alloys.

The competitive methods of making photomasks with semitransparent submicron size elements based on single, contact, conventional photolithography are elaborated.

The new LIGA method simplifies the formation of adjacent elements of different heights made of various materials on the same substrate. A new, true additive method for ULSI, which excludes the etching of conducting and dielectric layers deposited of different levels as well as cutting in dielectric layers and reactive ion etching is suggested.

Key words: Activation, electroless-deposition, microelectronics, submicron, LIGA, ULSI.

INTRODUCTION

Among disadvantages of the existed methods of metallization are large consume and lose of noble metals, long duration of technological processes, complexity and expensiveness of equipment for vacuum or steam-gas metallization, high energy consumption, the difficulty of obtaining the coatings of uniform thickness on the articles having complex profiles, in some cases, impossibility of plating the inner, hardly accessible surface, especially of small hollow articles, difficulty of continuous metallization of three-dimensional articles, difficulty of alloy deposition of the given chemical and phase compositions and given structures, difficulty of obtaining thin selective, pore-free films or thick films with low internal stress and with high adhesion to the substrate by electroless deposition method on polished dielectrics etc. (1-13).

Many of these disadvantages of the existing methods of metallization are excluded when combining electroless deposition (1-28) and electro plating methods with vacuum-thermal evaporation and deposition from vapour-gas mixtures (1-6, 13, 18, 28).

According to the results of the proposed investigations above mentioned disadvantages are excluded and films with predetermined physical-chemical properties are obtained, in particular, on the basis of nickel alloys with different metalloids and metals (1-6, 13, 18).

The developed methods of metallization of various materials were implemented in the mass production of quartz resonators and filters, monolithic piezoquartz filters, pieziceramic articles for delay lines of colour TV sets and hydroacoustics, etc. As a result, for the first time in the world practice, Au and Ag were adequately replaced by non-noble metal alloys, the processing time for making devices has been reduced 10 times and the manufactured devices are on the level of the world's best standards.

The competitive methods of making photomasks with semitransparent, submicron size elements based on single, contact, conventional photolithography are presented. Thus, a new stage in the development of microelectronics has been achieved (4, 6, 13, 18).

Such advanced methods as LIGA (Lithographie Galvanoformung) method of making the three-dimensional microdevices has some disadvantages. In spite of the fact that LIGA method provides the total revenues of several mld dollars per year and this sum is still growing (29), LIGA method is characterized by significant restrictions. The list of metal materials used as substrate and filling elements is significantly restricted, materials of various chemical nature and different thickness can not be deposited as adjacent elements on the same substrate in the process of one technological cycle (19, 30-34). Accordingly, it is difficult to create gas-removing, heat exchanging channels of efficient design on the same substrate.

The modified, patentable LIGA method proposed by us significantly simplifies the production of devices with adjacent elements of different height made of various materials on the same substrate. Owing to the new technologies and designs of devices elaborated by us, all the above mentioned disadvantages and restrictions have been excluded.

EXPERIMENTAL

The results of investigations of Sn and Pd ion adsorption and desorption were obtained by the methods of radioactive isotopes, X-ray photoelectron spectroscopy

(XPS) and photometry under different experimental conditions (1-4, 13, 15, 18, 25, 35). The samples of glass and quartz plates were immersed into the solution containing 113 Sn and 103 Pd radioactive isotopes introduced as chlorides. Radioactivity of samples relative to b-radiation was measured by gas-flow counter,MCT-17. The measurement precision was Ý 5% and data reproducibility - 30%. Sensitization and subsequent activation of samples, with the exception of specific cases, were carried out in the following solutions: $SnCl_2 \cdot 2H_2O$ - 20 g/l, pH 0.5 for 10min and $PdCl_2 \cdot 2H_2O$ -1.5 g/l, pH=2, respectively. For investigating the adsorbed ion states, serial X-ray photoelectron spectrometer ES-100 was used. The samples were attached to holders and were placed in the spectrometer chamber evacuated to $\sim 6 \cdot 10^{-5}$ Pa at -100^0C .

Results and Discussion

As the kinetic curves of adsorption show in the case of single sinking of glass samples into the solution of sensitization tin ion adsorption is increased during the first 10 min, then remains unchanged (1-4). At single sinking of glass in radioactive tin solution about $5.6 \cdot 10^{-4}$ mg/cm^2 or $3.6 \cdot 10^{15}$ ion/cm^2 tin ions are adsorbed for 15-20 min. The adsorption of tin ions on the quartz under the same conditions is more than 30 times higher than on the glass. In case the glass is not sensitized in advance, but is only activated, the number of the adsorbed palladium ions is several times less under the same conditions (1-4). The number of palladium ions adsorbed on the glass has proved to be greater, than that of tin.

The main results of the XPS experiments (surface concentration and state of ions on the surface) are shown in the Table. Bonding energies of Sn 3d5/2 and Pd 3d5/2 levels, their width E, atomic ratios Pd/Si, Sn/Si, Pd/Sn are given (4,35).

Table Bonding energies of Sn $3d_{5/2}$, Pd $3d_{5/2}$ and ratio of intensity of Sn(II) and Pd(II) ions, adsorbed on the glass

Sample	E, eV		Sn/Si	Pd/Si	Pd/Sn	E',eV	
	Sn $3d_{5/2}$	Pd $3d_{5/2}$				Sn $3d_{5/2}$	Pd $3d_{5/2}$
$SnCl_2$ (salt)	488.2	-	-	-	-	1.8	-
$PdCl_2$ (salt)	-	338.3	-	-	-	-	1.9
Sn(II)/ polished glass	488.2	-	0.018	-	-	1.7	-
Pd(II) polished glass	-	338.0	-	0.023	-	-	2.9
Pd(II)-Sn(II)/ polished glass	487.6	335.6	0.016	0.031	-	1.9	1.4
NaH_2PO_2-Pd (II)/ polished glass	-	335.2	-	0.025	-	-	2.2
NaH_2PO_2-Pd(II)-Sn(II)/polished glass	487.7	335.7	0.025	0.036	1.42	1.8	2.1
Pd(II)-Sn(II)-ground glass	487.4	335.6	0.071	0.099	1.34	1.8	2.7
Pd-Sn-NaH_2PO_2/ ground glass	487.7	335.7	0.059	0.066	1.08	1.8	2.3

Pd-NaH_2PO_2/ ground glass	-	335.0	-	0.042	-	-	2.6

As it was shown above, on the one hand, the surface pretreatment in the $SnCl_2.2H_2O$ solution increases the adsorption of Pd ions, on the other hand tin and palladium ions, as well as reduced palladium atoms exist on the surface after sensitization and activation. Thus we can conclude, that the sensitization stimulates the adsorption of palladium ions and a part of non-reduced palladium ions is reduced by hypophosphite ions (1-4). This is confirmed by the fact that, after surface activation and its hypophosphite treatment, i.e. when the process is carried out without sensitization, palladium atoms are on the surface. A part of palladium ions, not reduced by sensitization, appears to be reduced at the subsequent interaction with hypophosphite according to the reaction :

$$PdCl_4^{2-} + H_2PO_2^- + H_2O = Pd + H_2PO_3^- + 2H^+ + 4Cl^-$$

The proposed mechanism of sensitization and activation is verified also by XPS investigation, showing that reduction of the adsorbed Pd(II) ions to the metallic state at sensitization or without it takes place due to the following treatment of the activated glass in the hypophosphite solution.

Application of the above-mentioned methods enabled us to reveal the mechanism of sensitization and activation, involving the concept of equilibrium shift towards the formation of complex palladium anions, predominance of the number of palladium ions over tin ions on the surface and reduction of palladium ions at interaction with hypophosphite (1-4,25).

The experimental results show that application of sensitization becomes less necessary in case of electroless metal plating of non-metallic materials with greater surface roughness.

The optimal conditions for preliminary treatment of the non-metallic material surfaces depend on its state and nature (1-4,15,18,25,35). The conditions of activation - pH-value of palladium chloride, its concentration, temperature and the surface roughness determine whether sensitization is necessary or not. Sensitization diminishes the induction period of the nickel deposition reaction, promotes complete coverage of the surface and improves the quality of plating.

The study of the influence of a nickel phosphorous layer on the quality of quartz resonator has shown that the best parameters of the device are reached at 0,2-0,4μm thickness of the metallic film. As a result of investigations, the optimal conditions of metallization were established and technological process of electroless nickel plating of piezoelectrical quartz elements with a smooth surface (including polished surface) was developed.

The brief data of main advantages and innovations of our technologies in the field of electroless nickel plating of piezoquartz, lithium niobate and glass, as compared to silver and gold plating are given below:

1). Frequency stability of piezoquartz articles is increased 1.8 times. 2). Absolute size of dynamic resistance of piezoquartz resonators becomes 30% lower and the scattering of the resistance becomes about 40-50% lower as compared to the resonators with silver-plated piezoelements. 3). Good quality, long-term stability of piezoquartz articles are increased. 4). The scattering of dynamic induction and dynamic capacity of piezoquartz articles are decreased providing better conditions for the formation of radio circuits. According to improved above-mentioned parameters, the process of radio circuit optimization is simplified and more narrow-band quartz resonators are obtained. 5). Amplitude-frequency characteristics (spectral characteristics) are improved. 6). Labour-consumption, durability of the process and energy consumption are sharply decreased with simultaneous replacement of Au and Ag by Ni alloys. 7). Yield of end products is increased. 8). Our technology of local chemical quartz etching in combination with photolithography allows us to produce thin, high quality quartz crystal elements by high-productive process. The developed technology is very promising, particularly for deposition on thin piezoelectric layers. As is known, they have such promising properties as low stimulation voltage for making a thin device structure of actuator elements, an optical switch, an optical mirror, a laser diode, etc. (36). 9). The method of connection of holders with piezoelements is elaborated decreasing aging of the resonator 3 times as compared to piezoelements installed by means of soldering. 10). High adhesion reliability (2 times higher than silver films deposited by vacuum sputtering methods). The metallization methods are elaborated allowing to obtain adhesion exceeding the substrate cohesion. 11). High mechanical reliability of resonators with piezoelements, manufactured by the proposed methods. 12). Possibility of realization of assembly by soldering, current conducting glues, micro-welding and thermocompression.

The high productive group technology of electrode and piezoelement formation for making monolithic quartz filters by electroless deposition of nickel-phosphorous alloys with given physical-chemical properties is elaborated.

The method of electroless plating on polished quartz, glass and other materials is developed providing high adhesion and ductility, formation of pore-free films (thin and thick) with given thicknesses. Deposited thick films being the most important for obtaining monolithic quartz filters, in which the effect of energy capture is necessary. In particular, this was achieved by using thin doping of deposited films by "phis.-1" additive in the process of plating, by subsequent treatment of samples providing the obtaining of ductile films with low internal stress and high adhesion to the substrate. This significantly simplifies the obtaining of pads and bumps of IC on ceramics, polymers and on the other materials.

The proposed methods of metallization of different materials are widely used in the enterprises of NIS for production of quartz resonators and filters (several tens of mln.

were produced), monolithic piezoquartz filters, photomasks, piezoceramic articles for hydroacoustics and delay lines of colour TV sets (several hundreds of mln. were produced), casings of integrated circuits and semiconducting apparatus, ceramic microplates, precise microwire resistors and other articles.

Application of the proposed technologies gives a large economic effect.

With this method:

gold and silver are replaced by Ni alloys and the technology is significantly simplified; duration of the technological cycle of metallization is reduced 10 times and labour consuming of the process decreases sharply; the production volume per square meter of the production increases 8 times as compared to the metallization by means of silver paste burning; maintenance, quality and operational characteristics of photomasks increase; the reliability of quartz resonators is increased 1.8 times and dynamic resistance is decreased by 30%, as compared to the resonators with silver plated piezoelements; the accuracy of fixing precise microwire resistors is increased 10 times.

At present, in every developed country many companies produce articles of piezoquartz, ceramics and other piezomaterials with gold and silver plating, only in the field of piezoengineering, whereas, for producing the same articles by means of our technologies, the noble metals are already successfully replaced by non-noble ones in mass production, and the manufactured articles are on the level of the best world standards.

The method of making two-layer photomasks with semitransparent edges of the masking elements in the lower layer

The combination of the vacuum-thermal and electroless methods of metallization gave the possibility to carry out microfabrication (microminiaturization) of selectively semitransparent masking elements of photomasks. Semitransparency (semitransparency in visible and non-transparency in ultraviolet range of the spectrum) of masking edges (with about 3 micron dimensions) of elements in the lower Si layer (deposited by vacuum-thermal method) were obtained under non-transparent masking elements of the upper layer of nickel - phosphorous alloy. This alloy was deposited by electroless method. In the given case a new technology, and a new design for the production of two-layer selectively semitransparent photomask with semitransparent edges (of silicon) were proposed based on application of high-productive, single contact photolithography (1,3,4,6,13).

Semitransparency of such photomasks is reached by the shape identity of the elements of electrolessly deposited upper NiP layer and of the vacuum-deposited lower Si layer. Symmetry of the elements in upper and lower layers coincides. However, the area of upper NiP elements is less than that of corresponding elements of the lower silicon layer (1, 3).

By our technology two-layer film is obtained, the lower semitransparent layer is inert to the etchant, dissolving the upper layer.

The selectively semitransparent double-layer photomasks produced on the basis of the given invention have the following advantages as compared to conventional chromic photomasks:

1) Application of such photomask with semitransparent edges of masking elements significantly simplifying and increasing one of the most important operational characteristics - the precision of photomask alignment. Simplification and increase of alignment precision is induced by the fact that the operator can visually observe the whole IC under the photomask (through the semitransparent edges of masking elements in the visible region of spectrum) in the process of the photomask and IC pictures alignment.

2) Significantly low defectiveness (resulted from pore-free films) as compared to one-layer photomasks since, as a rule, the centers of lower Si layer crystallization do not coincide with the centers of upper Ni-P alloy layer crystallization; transparent defects, pin-holes and holes in the upper layer of Ni-P alloy are not continuation of transparent defects, pin-holes and holes in the lower layer of Si. As a result of mutual lapping of transparent defects in different layers (due to mismatch of transparent defects and of crystallization centers in upper and lower layers of photomask) almost a defectless photomask is obtained.

3) High wear-resistance obtained as a result of baking of Ni-P alloy and formation of hard intermetallic (Ni_3P) substance. In the given case it should be noted, that the edges of the lower layer elements defining the picture (topology) of photomask are not subjected to friction at contact photolithography (as they are protected by the elements of the upper layer), that increases percentage of IC output.

4) The existence of gaps between transparent sections of the photomask substrate and the surface of exposing photoresist, as well as the existence of channels between the upper elements of the photomask, solving the following problems of contact photolithography: a) At the contact photolithography the common problem is the capture of photoresist by a photomask and the swelling of the photoresist. b) At the contact photolithography (in which quinonediazide resists are widely used) unforeseen separation of photomask from IC plate is observed in some cases due to pressure of nitrogen evolved during resist exposure.

By means of the given photomask design the surface of the masking elements being in contact with photoresist is decreased (as in the given case only upper masking elements are connected with photoresist at contact printing), photoresist adhesion to the photomask and photoresist capture by photomask are also decreased. Besides, the existence of gaps and channels between upper masking elements simplifies the removal of gases evolved at the photoresist exposure and eliminates unforeseen separation of photomask from IC plate at contact photolithography. On the basis of our invention practically pore-free, wear-resistant, selectively semitransparent double-layer (Si-NiP) precision photomasks were produced and introduced into radioelectronic industry with large technical-economic effect (1, 3, 4, 6,13).

New, competitive submicron technology on the basis of a single, contact, conventional photolithography

New competitive methods of making photomasks with semitransparent, submicron size elements on the basis of contact, single, conventional photolithography or modified resistless (maskless) technology are proposed (4,6,13,18).

The new, proposed competitive methods solves the main problem of modern microelectronics. The invention allows us to manufacture photomasks with submicron size elements by a high-productive, group method of exposure of the whole substrate. The proposed method is much more advantageous and simple than other expensive and complicated method such as e-beam, X-ray lithography, or the production of photomasks with light phase shift. The new method allows us to avoid the application of e-beam exposure equipment costing more than $4 000 000 and other complicated equipment, as well as X-ray masks with gold masking elements. It also increases the output of production.

On the basis of the new technological principles proposed for manufacturing working copies of submicron photomasks, an inexpensive photomask with elements larger than 1 micron size can be used as a master photomask (4, 6, 13). The above mentioned possibility is due to the fact, that the suggested fabrication method of submicron elements (on working copy of photomask) is not based on transmission of exposing radiation through the similar transparent sections of submicron dimensions on the master photomask. For realization of the invention the transparent sections of photomask are made by selective etching of modified submicron size boundaries between opaque masking elements (on fabricating photomask). The size of both opaque masking elements and transparent sections on the master photomask can be much more than submicron. The invention allows us to obtain more wear-resistant photomask as compared to chromic ones, to increase the alignment precision due to semitransparency of the masking elements (in the visible region of the spectrum), to reduce the reflection coefficient of the masking elements and to provide the sharp contours of the obtained circuit.

On the basis of the proposed invention, a new stage in the progressive trend of microelectronics has been developed. The elaborated design and new technological processes of making photomasks solve problems of contact photolithography and have a number of advantages over the technologies existing so far: 1). Selective semitransparency of submicron masking elements in the visible region of the spectrum that guarantees the high alignment precision and better application conditions (better performance characteristics). 2). High percentage of production output and simplification of the process, significantly cheap price of the manufacture technology.

Scientific basis of the new method of making photomask with semitransparent, submicron size elements consists in that the technological processes are carried out in such a way that the difference between the boundary properties of materials and bulk properties of the same materials are revealed to the utmost.

In order to simplify the removal of gases evolved at photoresist exposure and to eliminate unforeseen separation of photomask from the photoresist (at contact printing) the channels are formed on the photomask from the side of masking elements (to remove gases) by forming the difference in thickness between masking elements of the first and second groups (4,6).

Competitive LIGA and entirely additive methods of making IC multilayers

The elements obtained by existing LIGA method on the same substrate consist of only one metal of the same height. In the modified LIGA method suggested by us the filling elements beside the electroplating are formed by means of electroless deposition or vacuum-thermal evaporation or by deposition from vapor-gas mixture. The modified LIGA, in particular LICHE (Lithography Chemical Forming) and LIEV (Lithography-Evaporation) methods differ from LIGA method existing so far and have the following advantages:

1. The use of dielectric and semiconducting materials (beside metal) as a substrate as well as for filling the regions between the resistive elements is significantly simplified.

2. The solid state, high heat-resistant and plasma-resistant elements with high value of Young's models and rather low gas content are used as intermediate resistive elements. This eliminates the shrinkage of filling elements and the problems of the lift-off process.

3. The elaborated new LIGA and resistless technologies significantly simplify the production of the device having adjacent elements made of various materials with different properties (i.e. they provide an optimal combination of adjacent elements with high wear-resistance and heat conductivity, with different antifriction, lubricate,

optical and other properties) and with different height on the same substrate (4, 6, 13).

The proposed methods allow us to obtain submicron size distances between elements and to regulate the difference between the thickness of the adjacent elements with submicron size precision without using very expensive e-beam and X-ray lithographies. The above mentioned and other advantages significantly increase the possibilities of device design and their functional purposes, provide the simplification of removal of undesirable gases, as well as heat dissipation (4,6).

One object of the invention is to eliminate one of the problems of lift-off lithography connected with polymer resists: gas evolve in evaporating chamber is decreased. The formation of pyramidal, nonisotropic profile of the deposited material is prevented, consequently, favourable conditions are provided for increasing deposition process temperature, i.e. of restrictive characteristics of polymer resists for lift-off lithography are eliminated. Next object of the invention is to expand the range of deviation angle of wall profiles of the resist (as compared to lift-off lithography) providing the absence of merging and breaking of layers deposited on different levels, i.e. on the substrate and the resist. This is provided, in particular, by the fact, that upper layers of unnecessary sections are removed by chemical-mechanical polishing or by selective etching. A further object of the proposed invention is selective electroplating or selective electroless metal deposition for integrated circuit fabrication, in particular, for making single layer or two-layer photomask. This object is achieved by combining the LIGA and the sacrificial layer techniques. In this process, after resist stripping, the sections of contacted underlayer are etched selectively (6).

A new, patentable, entirely additive method of manufacturing of IC multilayers is elaborated by us for the first time (4,6). This method has great advantages: it entirely excludes the etching of conducting and dielectric films deposited on different levels as well as the etching of interconnecting columns (pads).

A novel true additive process of formation of multiple conducting layers, contact filling materials, dielectric layers and pads on Si,GaAs, or on other substrates was developed for VLSI by combination of selective electroless deposition with thin layer LIGA, sacrificial layer and chemical mechanical polishing techniques. This additive process differs from analogues and its prototype in the following: 1) The etching of conducting, dielectric layers and contact filling materials deposited on different levels, the cutting in dielectric layers and patterning processes of dielectric layers, reactive ion etching are excluded in the present invention. 2) The sacrificial layer protecting the conducting underlayer prior to deposition of contact filling materials and preventing the oxidation of catalytic surface of the conducting underlayer at the exposure of the sample to air or at etching process (prior to plating) is not used in the prototype (37,38). The above-mentioned factors increase the adhesion of contact filling materials to conducting layers, as well as decrease the value of transient electrical resistance between them. The above factors are the advantages of the present invention as compared to its prototype (6).

We have elaborated a new method of formation of submicron interconnections for ULSI by combining local electroless deposition or electroplating methods with sacrificial and seed layer techniques and with chemical mechanical polishing (6). The principle of the new selective, local submicron deposition technique developed by us consists in the following: adhesion, seed and sacrificial layers are deposited sequentially onto the substrate (semiconductor or dielectric). The semiconductor mask is formed on the sacrificial layer in order to accomplish the posterior local etching of the sacrificial layer sections and then the local deposition of metal by electroplating or electroless deposition. The sections of sacrificial layer between the elements of semiconductor mask are etched until baring the seed layer sections (between the sections of sacrificial layer and the elements of semiconductor mask). Then the metal or the metallic alloy is deposited locally by electroplating or electroless deposition through the holes of semiconductor plating mask onto naked sections of seed layer. The mechanism of local electroplating of metallic sections onto naked seed layer sections by deposition through the holes of semiconductor plating mask (into semiconductor trenches) consists in the following. For plating the semiconductor electrodes the deposition potential is required that is much higher than for electroplating of metallic seed layer sections located between the elements of semiconductor plating mask. The mechanism of deposition in case of local electroless deposition into semiconductor trenches is as follows. Naked sections of seed layer are characterized by much higher catalitic activity than the sections of semiconductor plating mask, as a result of which local selective deposition of metal through the semiconductor plating mask is accomplished onto the sections of seed layer leaving the semiconductor field free of metal. Further, the semiconductor plating mask is removed, e.g. by chemical-mechanical polishing or selective etching. The rest sections of sacrificial layer and the lateral, adjacent (relative to the metal deposited by electroplating or electroless deposition) sections of seed and adhesion layers are removed by selective etching. As a result, locally deposited sections of adhesion, seed layers and metallic regions locally deposited onto them are formed on the substrate. Such locally deposited metallic regions with submicron width and high-aspect ration can be also obtained, if necessary (6).

ACKNOWLEDGEMENTS

The author is indebted to the Indivisible State Fund of Social Maintenance and Medical Insurance of Georgia for the support of this work.

REFERENCES

1. T.N. Khoperia, Khimicheskoe Nikelirovanie Nemetallicheskih Materialov, Moscow, Izdatelstvo Metallurgia, p. 144 (1982), Monograph.

2. T.N. Khoperia, Proceeding of the 10th International Congress on Metal Finishing (Interfinish'80), Kyoto, pp. 147-151 (1980).
3. T.N. Khoperia, T.J. Tabatadze, T.I. Zedginidze, Electrochim. Acta, Vol. 42, pp. 3049-3055 (1997).
4. T.N. Khoperia, T.J. Tabatadze, T.I. Zedginidze, Proceedings of the International Conference Micro Materials, Berlin, April, pp. 818-823 (1997).
5. T.N. Khoperia and R.G. Kharaty, Plating, V.59, N3, pp. 232-235 (1972).
6. T.N. Khoperia, Replacement of Au and Ag by Ni alloys and New Competitive Submicron,LIGA and Resistless Technologies, Monograph in press.
7. V.V.Sviridov, T.N.Varobeva, T.V.Gaevskaya, L.I. Stepanova, Khimicheskoe Osajdenie Metallov iz Vodnikh Rastvorov, Minsk, p. 270 (1987).
8. K.M. Gorbunova, A.A. Nikiforova, G.A. Sadakov, V.P. Moiseev, M.V. Ivanov, Physiko-Khimicheskie Osnovi Protsessa Khimicheskogo Kobaltirovania, Moscow, Izdatelstvo "Nauka", p.219 (1974).
9. Gavrilov, Chemishe (Stromlose) Vernicklung, p.239, Saulgau, Württenberg (1974).
10. T.N. Khoperia, G.I. Jishkariani, R.G. Kharati, Extended Abstracts, 33th Meeting of the International Society of Electrochemistry, Lyon, France, V.1, pp.401-403 (1982).
11. Kh.B. Petrov, Galvanizirune na Plastmasi, p.247, Technika, Sofia (1982).
12. M. Shalkauskas, A. Vashkialis, Khimicheskaia Metallizatsia Plastmass, Leningrad, Izdatelstvo "Khimia", p.144 (1985).
13. T.N.Khoperia, The 193rd Meeting of the Electrochremical Society, San Diego, Abstract N 261 (1998).
14. C.H. Ting, M. Paunovic, P.L. Pai, G. Chiu, J. Electrochem. Soc, Vol.136, p.462 (1989).
15. T.N. Khoperia and A.V. Ulanova, Extended Abstracts, 40th Meeting of the International Society of Electrochemistry, Kyoto, Vol.2, pp.1297-1298 (1989).
16. L.T. Romankiw, Abstracts, 42nd Meting of the International Society of Electrochemistry, Montreux, Switzerland, Abstract PL 2 (1991).
17. T. Osaka, Abstracts, 42nd Meeting of the International Society of electrochemistry, Montreux, Switzerland, Abstract K.L. 2-1 (1991).
18. T.N. Khoperia, T.J. Tabatadze, T.I. Zedginidze, N.T.Khoperia, Abstracts, Meeting of the Electrochemical Society, Los Angeles, California, May 5-10, p.375 (1996).
19. L.T. Romankiw, Electrochim. Acta, vol.42, pp.2985-3005. (1997).
20. C.J. Sambusetti, E.O. Sullivan, J. Marino, C. Uzoh. Abstracts, The 1997 Joint Meeting of the Electrochemical Society and of the International Society of Electrochemistry, Paris, France, p.535 (1997).
21. T. Osaka, J. Kawaguchi, in Electrochemical Technology: Innovation and New Developments (Edited by N. Masuko, T. Osaka and Y. Ito) pp. 3-17, Kodansha &Gordon and Breach, Tokyo and Amsterdam (1996).
22. C.J.Sambusetti, in Electrochemical Technology: Innovation and New Developments (Edited by N. Masuko, T. Osaka and Y. Ito) pp. 69-91, Kodansha & Gordon and Breach, Tokyo and Amsterdam (1996).

23. Electroless Deposition of Metals and Alloys, Eds M.Paunovic and I.Ohno, PV 88-12, The Electrochemical Society Softbound Proceedings Series, Pennington, NJ, p.306 (1989).
24. T.N. Khoperia, Z.Sh. Glonty, Journal Fisichescoi Chimii, Moscow, V.49, N3, pp.702-705 (1975).
25. T.N. Khoperia, N.A. Balashova, M.I. Kuleznova and B.V. Pailodze, Zashita Metallov, Izdatelstvo "Nauka", V.13, N6, pp.741-744 (1977).
26. T.N. Khoperia, A.V. Ulanova and V.V. Jdanov, Sov. Electrokhimia, Vol.16, pp.1735-1738 (1980).
27. T.N. Khoperia, Abstract, International Conference, Progress in Electrocatalysis, Ferrara, Italy, pp.281-282 (1993).
28. T.N. Khoperia, T.J. Tabatadze, T.I. Zedginidze, Proceeding of the International Symposium Surface Electrochemistry, Alicante, Spain, pp.95-96 (1997).
29. L.T. Romankiw, Abstracts, International Symposium Electrochemical Microsystem technologies, Grevenbroich, Germany, p.13 (1996).
30. W. Ehrfeld, V. Hessel, H. Löwe, Ch. Schulz, L. Weber, Proceedings of the International Conference Micro Materials, Berlin, pp.112-123 (1977).
31. N. Molduvan, J. Göttert, J. Mohr, S. Nedelcu, Ch. OBmann, M. Llie, Proceedings of the International Conference Micro Materials, Berlin, pp.523-526 (1977).
32. G. Engelmann, U. Dietrich, S. Gentzsch, R. Leutenbauer, E. Renger, H. Reichl, Fraunhofer Institut Zuverlässigkeit und Mikrointegraition, Proceedings, Berlin, September, 16-2, 16-7 (1996).
33. A. Thies, G. Schanz, E. Walch, Electrochim. Acta, Vol. 42, pp.3033-3040 (1997).
34. A. Thies et al., Abstracts, International Symposium Electrochemical Microsystem Technologies, Grevenbroich, Germany, p. 55 (1996)
35. T.N. Khoperia, A.V. Ulanova, O.P. Tkachenko, Journal Poverkhnost, Moscow, N7, pp. 96-98 (1987).
36. A. Schroth, M. Tanaka, C. Lee, R. Maeda, S. Matsumoto, Proceedings of the International Conference Micromaterials, Berlin, April, pp. 626-629 (1997).
37. C.H. Ting, M. Paunovic, U.S. Patent 5, 169,680 (1992).
38. C.H. Ting, M. Paunovic, J. Electrochem. Soc., Vol. 136, pp.456-462 (1989).

POLYQUINOLINE/BISMALEIMIDE BLENDS AS LOW-DIELECTRIC CONSTANT MATERIALS

Hari Singh Nalwa, Masahiro Suzuki and Akio Takahashi
Hitachi Research Laboratory, Hitachi Ltd.
7-1-1 Ohmika-cho, Hitachi City, Ibaraki 319-1292, JAPAN

Akira Kageyama
Hitachi Chemical Co. Ltd.
9-25-4 Shibaura, Minato-ku, Tokyo 108, JAPAN

Low-dielectric constant (κ) materials have attracted much attention in the field of microelectronics packaging. Currently a number of low dielectric constant polymers have been reported with κ in the range of 2.2 to 3.0, however, hardly any polymer satisfies all desired material requirements. Polyquinolines are thermoplastic polymers which show low dielectric constant, high thermal stability, good mechanical strength and low moisture absorption. We have blended a polyquinoline designated as PQ-100 with a bismaleimide in order to improve physical properties for microelectronics packaging applications. We have prepared transparent, tough polyquinoline/bismaleimide blend thin films containing 5 to 60 weight % bismaleimide contents as low-dielectric constant interlayer materials for multilevel interconnections. The effect of bismaleimide loading and curing conditions on dielectric, dynamic mechanical and thermal stability properties of polyquinoline/bismaleimide thermoset blends was studied. By the incorporation of the bismaleimide, the blend thin films showed higher glass transition temperature up to 360 °C. The dielectric, thermal and mechanical properties of polyquinoline/bismaleimide blend thin films are discussed.

INTRODUCTION

Recent advances in microelectronics packaging technology have created a great demand for low-dielectric constant (κ) materials. The immediate applications of low-κ dielectrics are in ultra-large scale integration (ULSI), multichip modules (MCM) and printed circuit boards (PCBs) [1-4]. The basic requirements for low-κ materials are many, though some of the most important requirements for these materials are: low dielectric constant and dissipation factor, high thermal stability and glass transition temperature, high tensile strength and modulus, high crack resistance, good adhesion, low mechanical and thermal shrinkage, planarization and low moisture absorption. However, these material requirements may change from one application to another but for an ideal low-κ material, a combination of these properties is needed for applications in microelectronics packaging technology.

The conventional polyimides show dielectric constant of 3.0 and higher [5], however, fluorinated and nanofoamed polyimides offer comparatively lower dielectric constant (κ=2.0) as well as lower moisture absorption [6,7]. The polar carbonyl (C=O) groups in the polyimide backbone are responsible for higher moisture absorption in these polymeric systems which consequentially leads to an increase in dielectric constant. Like polyimides, polyquinolines also show low dielectric constant, high thermal stability and low moisture absorption [8-13]. Though polyquinolines are thermoplastic but they can be modified into thermosetting materials by blending together with bismaleimides which form highly crosslinked polymer network on curing at high temperatures. Our main objectives of blending polyquinoline with bismaleimide are to improve the solvent resistance and glass transition temperature of polyquinoline system as well as to introduce fluidity via blending. We have previously reported that high performance polyquinoline/bismaleimide blend thin films can be obtained without sacrificing the intrinsic physical properties of polyquinoline backbone [14,15]. In this article, dielectric, mechanical and thermal properties of polyquinoline/bismaleimide blend thin films are discussed.

EXPERIMENTAL

Figure 1 shows the chemical structures of a polyquinoline designated as PQ-100 and 2,2-bis(4-maleimide phenoxyphenyl)propane abbreviated as BBMI. The PQ-100/BBMI blends containing 5-60 wt% of BBMI were prepared by dissolving BBMI in cyclopentanone solution of PQ-100 at room temperature. The samples used in this study are referred as follows: A-1 (PQ-100 only), B-1 (5 wt% BBMI), C-1 (10 wt% BBMI), D-1 (15 wt% BBMI), E-1 (20 wt% BBMI), F-1 (30 wt% BBMI), G-1 (40 wt% BBMI), H-1 (50 wt% BBMI) and I-1 (60 wt% BBMI). The blends containing 5-20 wt% and 30-60 wt% of BBMI contents were investigated as the thin films for spin-coating and the adhesive films for laminated thin film multilayer substrate applications, respectively. BBMI exhibits a melting point around 155 °C and a polymerization peak temperature of exothermic reaction around 290 °C. Considering the cross-linking reaction of BBMI, the prepolymer blend thin films were cured at 280 °C for 1 and 3 h , 300 °C for 1, 2 and 3 h as well as at 350 °C for 1 h. Thermomechanical testing was performed with a dynamic mechanical analyzer (DVA-200) at a fixed frequency of 10 Hz between the temperature range of 20-400 °C. DSC of BBMI powder sample was recorded from a 910 Differential Scanning Calorimeter (Du Pont Instruments). Thermal stability of PQ-100 and BBMI samples were recorded under nitrogen atmosphere up to 1000 °C at a heating rate of 10 °C/minute by using Thermogravimetric Analyzer (Hi-Res TGA 2950). The isothermal TGA of cured blend thin films was recorded for 3 hour at 300 and 400 °C under a nitrogen atmosphere.

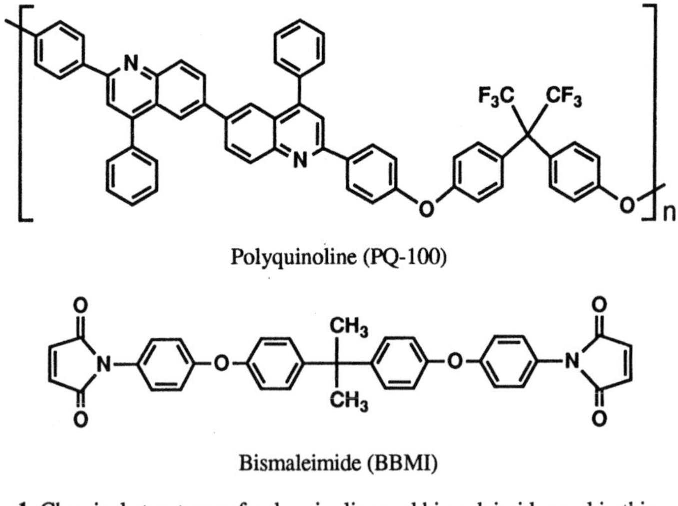

Polyquinoline (PQ-100)

Bismaleimide (BBMI)

Figure 1. Chemical structures of polyquinoline and bismaleimide used in this work.

RESULTS AND DISCUSSION

We have previously reported the effect of BBMI loading and curing conditions on dielectric, mechanical and thermal properties of PQ-100/BBMI blend thin films [14,15]. Figure 2 shows the TGA curves of PQ-100 thin film cured at 280 °C for 1 h and BBMI powder samples. PQ-100 thin films showed no weight loss up to 500 °C however 2, 5 and 10% weight losses were recorded at 530, 538 and 558 °C, respectively. BBMI showed 2 and 5% weight losses at 455 and 471 °C, respectively. Therefore the degradation temperature where a 2% weight loss was observed is about 80 °C higher for PQ-100 thin films than that of BBMI powder. There is a significant difference between the thermal stability of PQ-100 and BBMI. It is apparent that rigid backbone structure of PQ-100 has much higher thermal stability than that of BBMI.

Thermogravimetric analysis of PQ-100/BBMI blend thin films containing 5 wt% (B-1) and 50 wt% (H-1) BBMI contents showed 5% weight loss at 530 and 450 °C, respectively. This indicates a decrease of 80 °C in thermal stability with a ten fold increase in BBMI loading. The 5% weight loss was observed between 450 °C to 535 °C for thin films containing 5 to 60 wt% bismaleimide contents. The thermal stability of PQ-100/BBMI blends decreases with increasing BBMI contents. Figure 3 shows the isothermal TGA of PQ-100 and blend thin films recorded for 3 h at 300 and 400 °C under a nitrogen atmosphere. The PQ-100 thin films showed no weight loss at 400 °C confirming a high thermal stability of polyquinoline backbone. A weight loss of 0.15, 0.40 and 1.0% was observed after 3 hour at 300 °C for D-1, F-1 and G-1 thin films, respectively. These same samples showed a weight loss of 2.0, 5.0 and 7.5% after 3 h at 400 °C, respectively. The H-1 blend thin films showed 1.6, 4.5 and 9.0% weight loss after 3 h at 300, 350 and 400 °C, respectively. These results demonstrate high thermal stability of cured blend thin films which is one of the important requirements.

Figure 4 shows the temperature dependence of storage modulus (Er), loss modulus (Ei) and tan δ of PQ-100 thin films cured at 280 °C for 1 and 3 h and at 300 °C for 1, 2 and 3 h. The DMA scans show Ei peak around 265 °C with a sharp drop in modulus (Er) thereafter, whereas the tan δ peak was centered at 282 °C for all thin films cured at 280 and 300 °C. This indicates that PQ-100 thin films have no effect of curing conditions on mechanical and thermal properties whatsoever [15]. It is due to the fact that PQ-100 is a thermoplastic polymer.

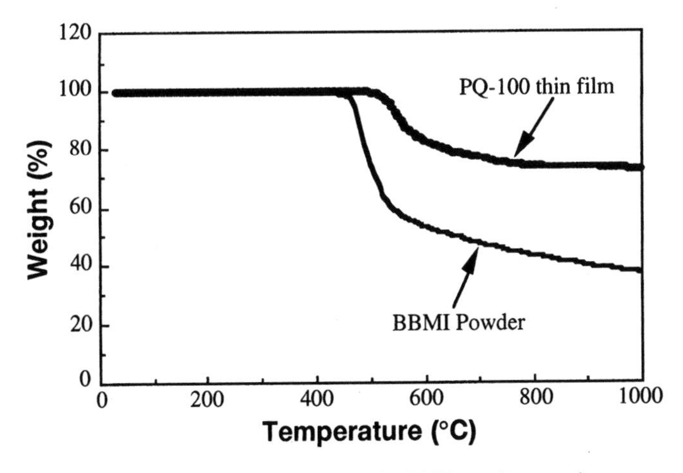

Figure 2. TGA curves of PQ-100 thin film and BBMI powder samples.

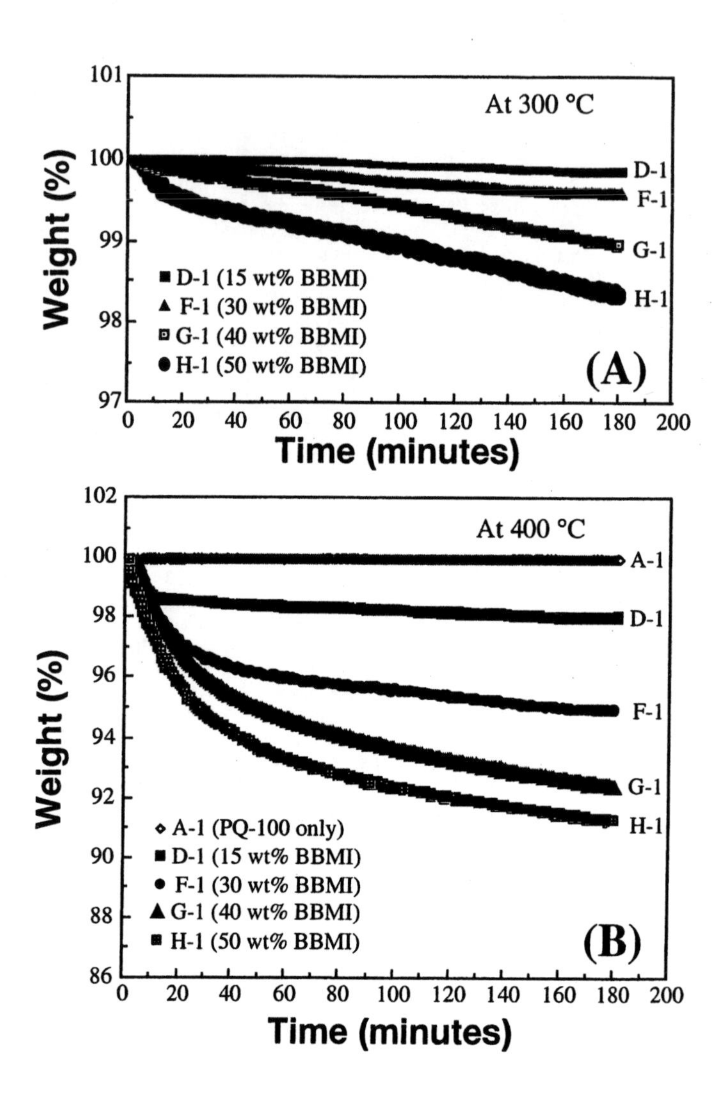

Figure 3. Isothermal TGA of PQ-100 and blend thin films cured at 300 °C for 3 h.

Figure 5 shows the BBMI content dependence of dielectric constant and tan δ of blend thin films cured at 280 °C for 1 h. The dielectric constant of PQ-100 thin films is 2.90 at 1 kHz and it starts increasing with increasing BBMI contents. The dielectric constant reached to 3.10 for blend thin films containing 50 wt% BBMI contents. Likely the tan δ also increased from 0.004 to 0.007 at 1 kHz with increasing BBMI contents.

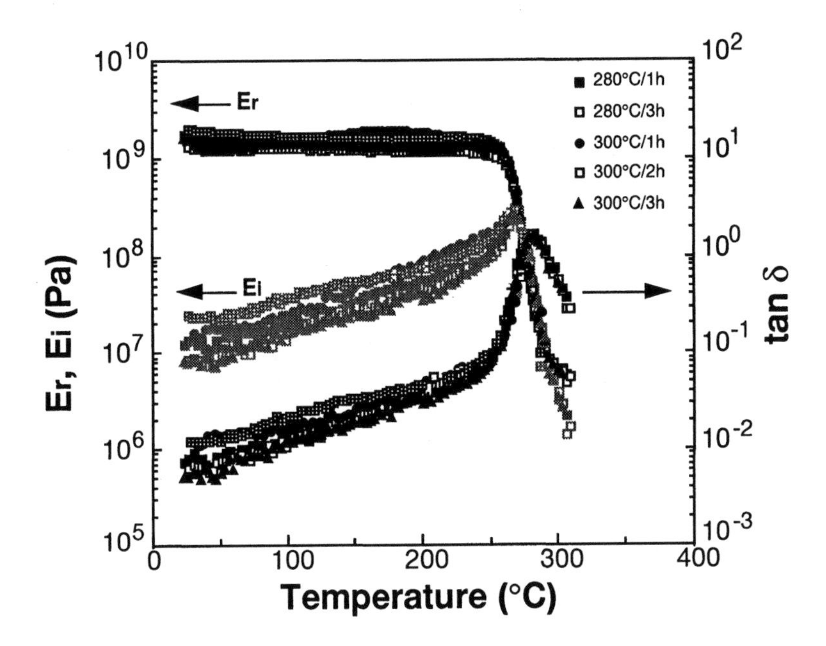

Figure 4. DMA plots of PQ-100 thin films cured under different conditions.

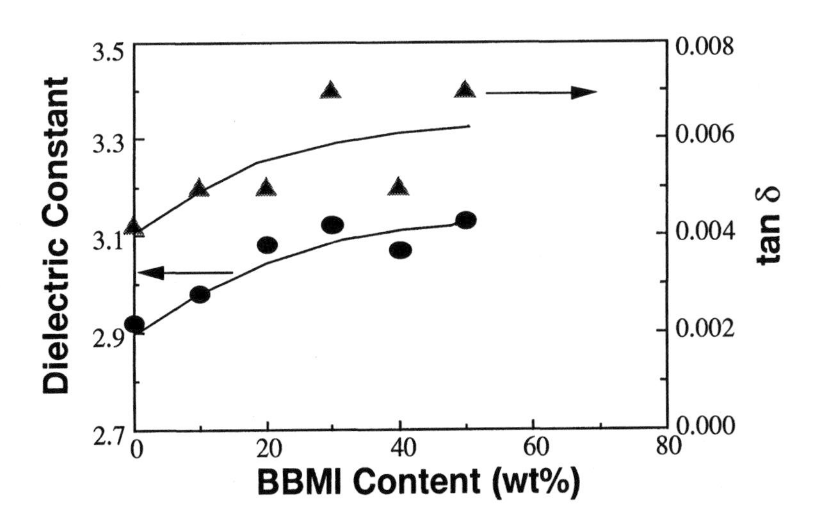

Figure 5. Variation of dielectric constant and tan δ as a function of BBMI content.

Figures 6 shows the DMA plots of C-1 (10 wt% BBMI) blend thin films. The C-1 thin films cured at 280 °C for 1 and 3 h and 350 °C for 1 h showed tan δ peak at 263, 273 and 288 °C, respectively. However when the C-1 films were first cured at 280 °C for 1 h and then subsequently at 350 °C for 1 h, the tan δ peak was observed at 315 °C, showing an increase of 27 °C in Tg. Figure 7 compares the storage modulus,

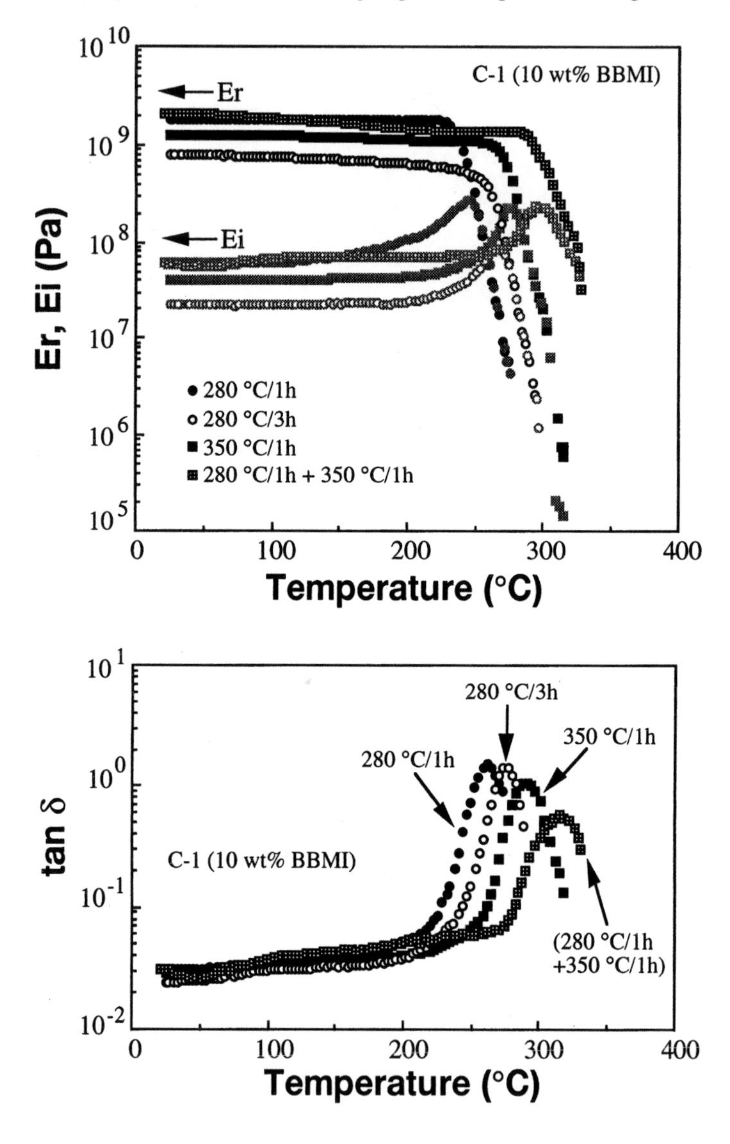

Figure 6. DMA plots of C-1 blend thin films cured under different conditions.

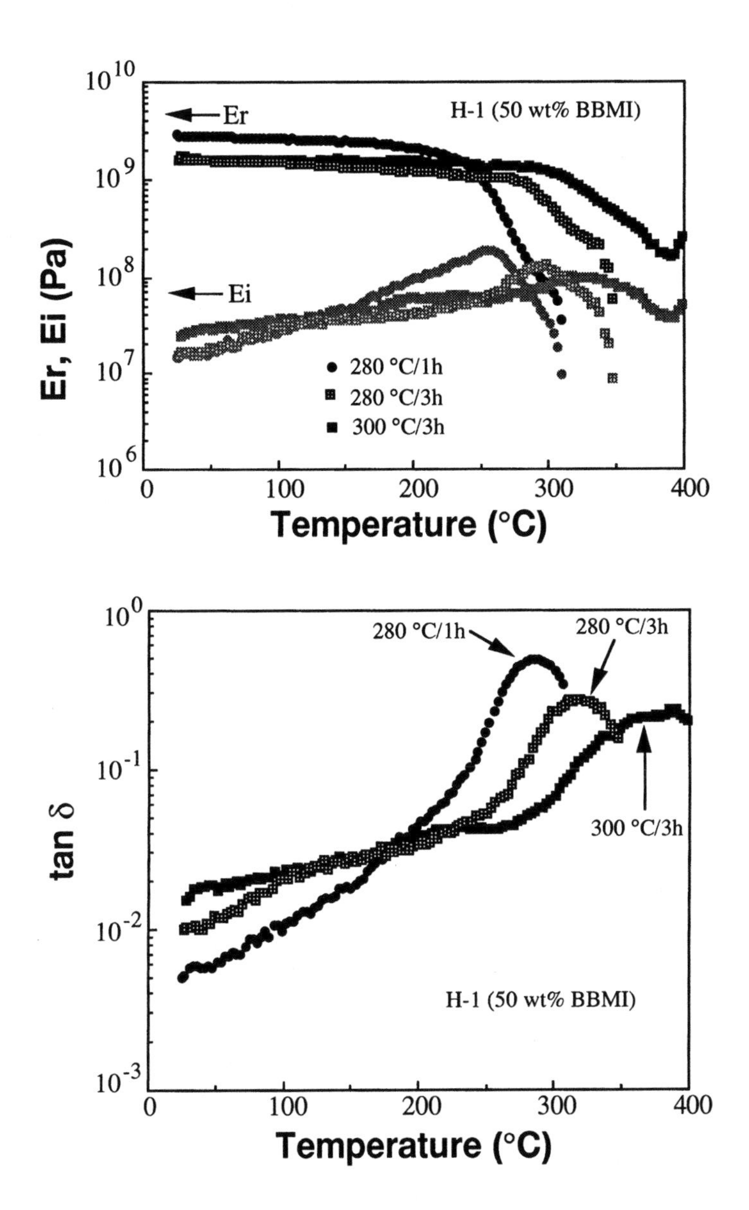

Figure 7. DMA plots of H-1 blend thin films cured under different conditions.

loss modulus and tan δ for H-1 (50 wt% BBMI) blend thin films cured at 280 for 1 and 3 h and at 300 °C for 3 h. As shown in Figure 4 that PQ-100 thin films cured under different conditions showed tan δ peak at 282 °C. In the case of H-1 thin films, the tan δ peak was observed at 286 °C for thin films cured at 280 °C for 1 h. The tan δ peak appeared at 310 and 360 °C for thin films cured at 280 °C for 3 h and 300 °C for 3 h, respectively. The Tg of H-1 thin films was increased by over 70 °C as the curing temperature and time were changed. These two examples indicate that Tg and mechanical properties are significantly improved by changing the curing conditions.

Figure 8 shows the variation of Tg as a function of BBMI content and curing conditions. The Tg increased as the curing time and temperature were increased. The Tg increased by over 20 °C for all D-1 to H-1 thin film samples as the time of curing was raised from 1 h to 3 h at 280 °C. The Tg of C-1 thin films cured at 280 °C for 1 h and then 350 °C for 1 h was found to be 27 °C higher than those cured at 350 °C only. The B-1, C-1, D-1 and E-1 thin films cured at 350 °C for 1 h showed higher Tg than those cured at 300 °C for 3 h. Moreover, the Tg of these blend thin films also increases with increasing BBMI contents. One of the interesting features was that blend thin films cured at 280 °C for 1 h showed a decrease in Tg up to 20 wt% BBMI and then a continuos increase thereafter. However no such feature was observed for samples cured at 280 °C for 3 h and at 300 °C for 1, 2 and 3 h. The Tg as high as 360 °C was recorded for H-1 blend thin films cured at 300 °C for 3 h, whereas the Tg of these H-1 thin films cured at 280 °C for 1 h was 286 °C, therefore, the Tg increased by 80 °C with varying curing conditions. Interestingly, the Tg of blend thin films cured at 280 °C for 3 h and 300 °C for 1 h were found to be exactly in the same temperature range. It seems that BBMI cure reactions were incomplete at 280 °C and further cure occurs at 300 °C which consequently leads to higher Tg. It is apparent that the mechanical and thermal properties are significantly affected by loaded BBMI contents and curing conditions.

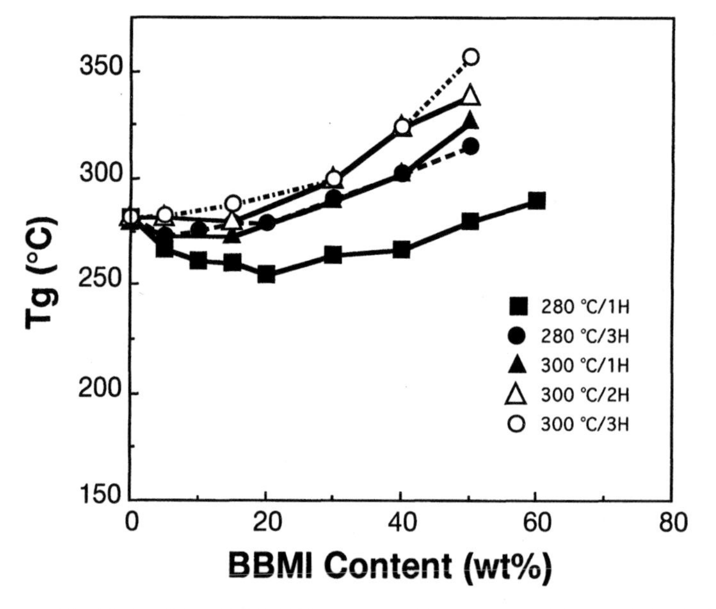

Figure 8. The Tg for PQ-100/BBMI blend thin films cured at 280 and 300 °C. Here Tg corresponds to tan δ peak temperature recorded from DMA data.

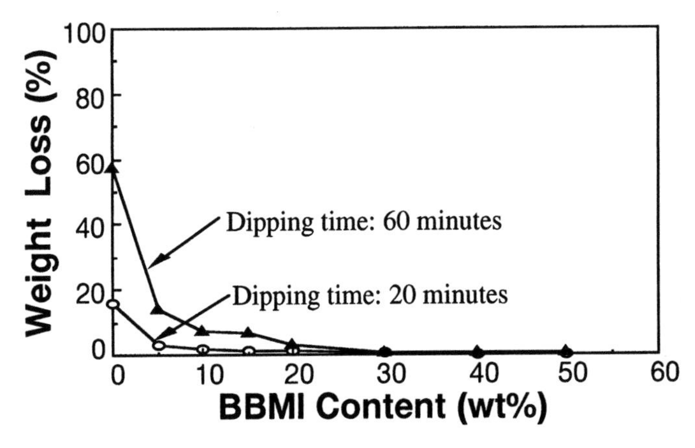

Figure 9. Solvent resistance of cured PQ-100/BBMI blend thin films

The solvent resistance of low-κ materials is one of the necessary requirements for the fabrication of multilayer interconnects. In order to introduce solvent resistance in polyquinoline systems, the effect of bismaleimide blending was studied for the applications of blend thermosets in microelectronics packaging. To examine the solvent resistance of PQ-100/BBMI blends, the thin films cured at 280 °C for 1 h were emmersed in cyclopentanone for 20 and 60 minutes. Figure 9 shows the BBMI content dependent weight loss of blend thin films. It is apparent that the weight loss of cured blend thin films in cyclopentanone was noticeably minimized with increasing BBMI contents. Interestingly, no weight loss was observed for blend thin films containing 30 to 50 wt% BBMI contents. Therefore, the solvent resistance was significantly improved by blending PQ-100 with BBMI.

CONCLUSIONS

In summary, we have shown that PQ-100/BBMI blends having bismaleimide contents between 5-60 wt% have good compatibility, high thermal stability and mechanical strength. Both thermal and mechanical properties of blend thin films were significantly improved by the variation of curing conditions. The highest Tg of 360 °C was recorded for H-1 (50 wt% BBMI) thin films cured at 300 °C for 3 h. We have also demonstrated that the solvent-resistance of PQ-100 films was remarkably improved without adversely affecting intrinsic physical properties of PQ-100. The blending of polyquinoline with bismaleimide provides high temperature-resistant thermoset blends without sacrificing the inherent physical properties of polyquinoline backbone.

REFERENCES

[1] A. Lagendijk, H. Treichel, K. J. Uram and A. C. Jones, eds., Low Dielectric Constant Materials II, Mat. Res. Soc. Symp. Proc., **Vol. 443** (1996).

[2] C. Case, P. Kohl, T. Kikkawa and W. W. Lee, eds., Low Dielectric Constant Materials III, Mat. Res. Soc. Symp. Proc., **Vol. 476** (1997).

[3] M. W. Jawitz, ed., Printed Circuit Board Materials Handbook, McGraw-Hill, New York (1997).

[4] R. R. Tummalla, E. J. Rymaszewski and A. G. Klopfenstein, eds., Microelectronics Packaging Handbook, Vol. 1-3, Chapman & Hall, New York (1997).
[5] M. K. Ghosh and K. L. Mittal, eds., Polyimides: Fundamentals and Applications, Marcel Dekker, New York (1996).
[6] B. C. Auman and S. Trofimenko, Polymer Preprints, **34**, 244 (1992).
[7] B. C. Auman, Proceedings of First International Dielectrics for VLSI/ULSI Multilevel Interconnect Conference (DUMIC), Santa Clara, CA, p.295 (1995).
[8] J. F. Wolfe and J. K. Stille, Macromolecules, **9**, 489 (1976).
[9] W. Wrasidlo, S. O. Norris, J. F. Wolfe, T. Katto and J. K. Stille, Macromolecules, **9**, 512 (1976).
[10] W. H. Beever and J. K. Stille, J. Polym. Sci. Polym. Symp. **65**, 41 (1978).
[11] J. K. Stille, Macromolecules, **14**, 870 (1981).
[12] N. H. Hendricks, M. L. Morrocco, D. M. Stoakley, and A. K. St. Clair, Proceedings of the 4th International SAMPE Electronics Conference, SAMPE Pub., p.544 (1990).
[13] N. H. Hendricks, L. C. Hsu, C. Taran, N. M. Rutherford and J. Y. Chee, Proceedings of the 5th International SAMPE Electronics Conference, SAMPE Pub., p.365 (1991).
[14] H. S. Nalwa, M. Suzuki, A. Takahashi and A. Kageyama, Appl. Phys. Lett., **72**, 1311 (1998).
[15] H. S. Nalwa, M. Suzuki, A. Takahashi, A. Kageyama, Y. Nomura and Y. Honda, Chem. Mater., **10**, 2462 (1998).

DEPOSITION OF LOW-K DIELECTRIC FILMS USING TRIMETHYLSILANE

M.J. Loboda, J.A. Seifferly, R.F. Schneider, and C.M. Grove
Dow Corning Corporation, Midland, MI 48686

ABSTRACT

A simple process for the deposition of a low-k dielectric film has been developed using the organosilicon gas trimethylsilane and nitrous oxide. The process can be implemented in standard equipment designed for plasma enhanced chemical vapor deposition. The resultant film is a random network of C-Si-C and O-Si-O bonds. The desirable properties of the film include k<3.0, low stress, low leakage current density and high thermal & oxidative stability. These properties can meet the requirements imposed on new low-k materials intended for state-of-the-art integrated circuit manufacturing technology.

INTRODUCTION

In advanced semiconductor integrated circuit technology there is now a well-defined need for thin film materials with dielectric constant significantly lower than plasma deposited SiO_2 (k~4). Films deposited by spin-on processes and plasma enhanced chemical vapor deposition (PECVD) have been explored as candidate "low-k" materials and include polyamides, fluorocarbon polymers and porous silicon-based oxides. While many of these materials have been shown to provide the desired permittivity reduction, difficulty can be encountered at the process integration stage, including poor thermal and oxidative stability, incompatibility with standard etching techniques, and low breakdown strength.

Low-k materials based on silicon oxides will likely be the easiest to integrate as a replacement for SiO_2 in current integrated circuit (IC) manufacturing processes. Over twenty years ago it was demonstrated that plasma polymerization of simple organosilicon molecules could produce dielectrics with permittivity less than SiO_2 [1,2]. The siloxane materials used in these experiments have relatively low vapor pressure compared to the gases commonly used for PECVD, and would pose challenges to implement in high volume manufacturing. Recently, a new CVD process based on methylsilane, $(CH_3)SiH_3$, has been reported for deposition of low-k dielectrics [3]. While methylsilane has vapor pressure suitable for CVD applications, like SiH_4 it is a dangerous material to handle. Also, this particular application requires a unique CVD tool and process.

We have previously shown that the non-pyrophoric, organosilicon gas trimethylsilane, $(CH_3)_3SiH$, is easily integrated into standard PECVD processes for traditional silicon-based dielectrics [4]. Using a generic SiH_4-based plasma oxide process as a starting point, we have developed a low-k dielectric thin film process by replacing SiH_4 with trimethylsilane. This simple process produces a material which can meet the integration requirements associated with semiconductor device manufacturing.

This paper will report on the deposition process and standard film characteristics of low-k dielectrics deposited from trimethylsilane. It will be shown these dielectric films can be produced with permittivity values that parallel those films deposited by plasma polymerization of organosilicon molecules [1,2]. Properties will be reviewed in the context of specifications for dielectrics commonly required in IC manufacturing processes.

EXPERIMENTS

The low-k dielectric films discussed in this paper were deposited using a gas mixture of trimethylsilane (Dow Corning® 9-5170 Trimethylsilane (3MS), Semiconductor Grade) and 99.99% nitrous oxide. Film deposition was performed in standard, capacitively coupled parallel plate PECVD systems (PlasmaTherm, I.P. System 790 or when noted, an Applied Materials P5000 PECVD system). The deposition equipment was configured identically to that used to deposit oxide films using SiH_4/N_2O gas mixtures. Films were deposited onto bare silicon substrates or oxidized silicon substrates coated with 0.5 um of blanket aluminum. Film thickness was measured using a spectroscopic ellipsometer. Film stress was calculated from wafer curvature change as measured using a profilometer or optical cantilever. Dielectric properties were measured by sputtering gate electrodes of 1-4 mm nominal diameter through a shadow mask onto the low-k films to form capacitors. Capacitance measurements were performed at 1 MHz using a Keithley 590 CV measurement unit. Current-voltage measurements on the capacitor structures were performed using a Hewlett-Packard 4140A pA/DC Voltage Source.

PECVD Process Results

Table 1 shows the typical ranges of gas flow, pressure, RF power and temperature explored to date in the low-k process. Within this process space the relative dielectric constant of the films was in the range $2.6<k<3.0$. Table 1 also highlights the major impact each parameter had on the film process. While the k values did not change significantly, the film uniformity degraded as the pressure was increased. Typical thickness variation in the PlasmaTherm 790 system ranged 10-20%, while the values in the Applied Materials

P5000 system ranged 3-10% as measured on 150 mm and 200 mm silicon wafers. At higher values of gas flow and RF power deposition rates of 6000 A/min in the PlasmaTherm 790 system and 20000 A/min in the Applied Materials P5000 system.

The effect of the plasma chemistry on the film process was studied by substituting helium for nitrous oxide while maintaining constant total flow. Figure 1 shows the growth rate as a function of N_2O concentration. Over the range investigated, the dielectric constant of the films (k~3.0) and the composition of the films did not vary. Growth without nitrous oxide under the identical conditions, i.e., a-SiC:H film deposition, resulted in very slow growth rates, on the order of 200 A/min.

The composition and molecular structure of the film have been studied by Rutherford backscattering spectrometry (RBS), hydrogen forward scattering spectroscopy (HFS) and Fourier transform infrared spectroscopy. The composition of the films described in Figure 1 is nominally $C_3H_3Si_2O_2$, with density ranging 1.36-1.40 g/cm^3. Throughout the range of gas mixtures shown in Figure 1 these properties did not vary significantly. Figure 2 shows an infrared spectrum obtained on a 3MS-based low-k film. The key features of the spectrum are the C-H stretching vibration near 2960 cm^{-1}; Si-H stretching vibration at 2180 cm^{-1}; the antisymmetric and symmetric deformation of C-H in a Si-$(CH_2)_n$-Si at 1410 cm^{-1} and 1360 cm^{-1}, respectively; the bend in Si-CH_3 at 1260 cm^{-1}; Si-O-Si at 1030 cm^{-1}; and the Si-C stretch vibration near 790 cm^{-1}. The RBS and infrared spectroscopy data suggest the structure of this material to be a network of -O-Si-O-,-C-Si-C- and O-Si-C bonds, throughout which are randomly distributed hydrogen or methyl group terminations.

Table 1 - PECVD process conditions used in the deposition of low-k dielectric films from trimethylsilane. Specific impact on the process is also noted.

Process Parameter	Range	Process Impact
Temperature	300-400 °C	• no impact
Pressure	2.8-3.5 torr	• uniformity
Power	100-350 W	• k and uniformity
N_2O Flow	100-400 sccm	• growth rate
Trimethylsilane Flow	50-100 sccm	• growth rate

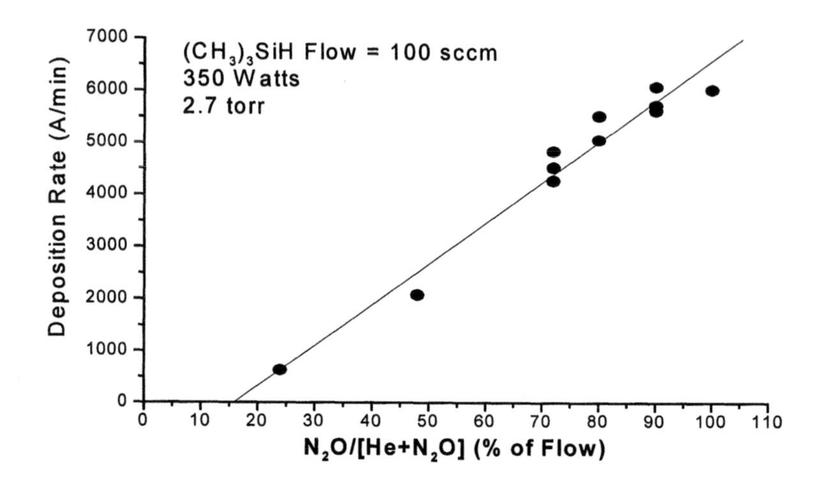

Figure 1. Variation of film deposition rate as a function of nitrous oxide gas concentration in an $(CH_3)_3SiH-N_2O-He$ plasma. Total gas flow=600 sccm.

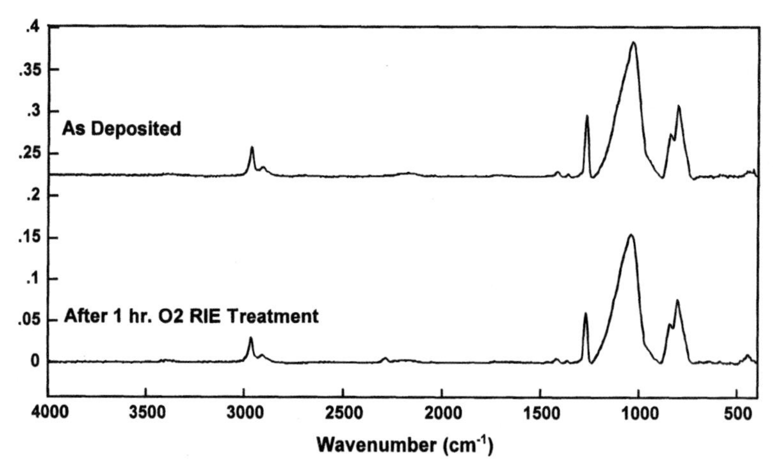

Figure 2. Infrared spectra measured on 3MS-based low-k films as deposited and after exposure to an O_2 reactive ion etch for one hour.

Table 2 - Properties of low-k films deposited from trimethylsilane and nitrous oxide gas mixtures by PECVD corresponding to the conditions listed in Table 1.

Property	Value	Measurement
Relative Dielectric Constant	$2.6<k<3.0$	• metal insulator semiconductor and metal insulator metal capacitors
Leakage Current Density	$J\ (A/cm^2)<1x10^{-10}$ @ 0.5-1.0 MV/cm	• metal insulator semiconductor capacitors
Breakdown Field	> 4 MV/cm	• metal insulator semiconductor capacitors
Basic Thermal Stability	$\Delta k<2$ % $\Delta t_f<1$ %	• Diffusion furnace anneal for 8 hours @450 °C in N_2
Film Stress	$-30<\sigma<40$ MPa	• wafer curvature
Film Stress Thermal Stability	$\Delta\sigma<+20$ MPa (tensile increase)	• wafer curvature measurements vs. temperature to 450 °C
Coefficient of Expansion	CTE>Si	• wafer curvature measurements vs. temperature
Oxidative Stability I	$\Delta k<1$ % $\Delta t_f<1$ %	• one hour oxygen plasma at 800 mT, RF=500 W and T=100 °C
Oxidative Stability II	$3<\Delta k<6$ % $\Delta t_f<1$ %	• one hour oxygen reactive ion etch at 800 mT, RF=500 W and T=100 °C

TRIMETHYLSILANE LOW-k FILM PROPERTIES

Applications of low-k thin film materials as interlevel dielectrics in integrated circuit wiring structures require that properties such as film stress, thermal stability, and ease of integration into CMOS process technology are comparable to that associated with plasma deposited SiO_2 films. Table 2 summarizes the properties measured on 3MS-based low-k films which are relevant to these applications.

Capacitance values typically drop ~1% following anneal cycles. This is likely due to improved contact resistance and not a change in the dielectric material. The leakage currents developed in these materials are comparable to standard plasma oxides. Therefore, it is not believed that an SiO_2 cap or liner would be needed to integrate the 3MS-based low-k film in an interlevel metal dielectric process. Figure 2 shows the change in the infrared spectra following one hour of oxygen reactive ion etching. There was no significant change in thickness observed. Integration of the carbon hydrogen stretch peak would indicate approximately 6% of the C-H bonds were removed by the exposure to the O_2 plasma. The splitting of the Si-H stretch peak may be indicative of separation of H-SiO_x and H-SiC_x bond environments. The integrated Si-H peak area did not change significantly. Finally the infrared analysis shows no formation of -OH bonds as a result of the oxygen reactive ion etch plasma, which has been observed on other Si-O materials containing CH_3 groups [5]. The lack of formation of -OH bonds suggests there would be no process failures related to -OH outgassing during via etch and resist strip processes, resulting in successful integration with via fill processes.

The film stress and wet etch rates of 3MS-based low-k materials have reduced values

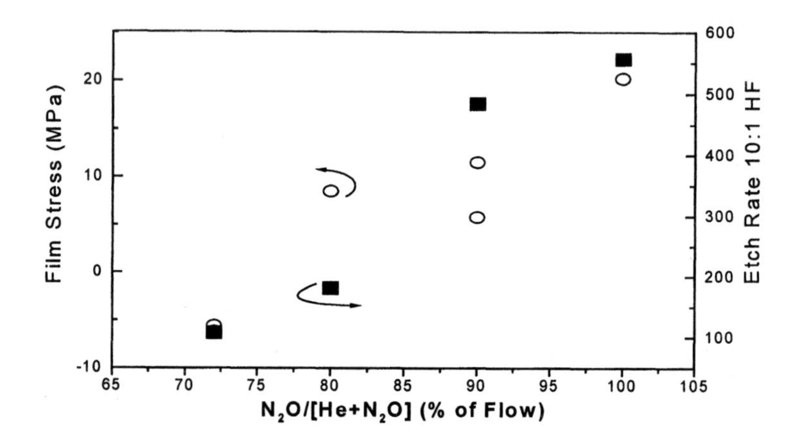

Figure 3. Variation of 3MS-based low-K film stress and wet etch rate as a function of N_2O concentration corresponding to Figure 1.

compared to SiH_4-based oxide and nitride films. Figure 3 shows the variation of these properties as a function of N_2O gas concentration for the films shown in Figure 1. In addition, Table 2 shows that for these materials, the as-deposited film stress does not change significantly during even severe temperature cycling. Since the films in Figure 1 and 2 did not show any variation in composition or density, it appears that the build up of the small amounts of tensile stress can lead to a correlated increase in the wet etch rate.

The step coverage was evaluated using an Applied Materials P5000 deposition system configured for SiH_4/N_2O plasma oxide processes. Films were deposited onto metal structures formed on 150 mm wafers at a wafer temperature of about 350 °C. Figure 4 shows the results obtained from X-SEM analysis. The step coverage and gap fill is better than that observed for SiH_4-based oxide processes traditionally run in this type of PECVD system.

ORIGINS OF LOW-k PROPERTY

The observation of the reduced permittivity in this trimethylsilane/nitrous oxide process can be explained by considering film density and the electronic polarizability of an O-Si-C material and how these properties interact via the Clausius-Mossoitti relation

$$(k-1)/(k+2)=N\alpha/\varepsilon_0$$

where N is the density of molecules and α is the polarizability. The equivalent bulk density of these films, nominally 1.4 g/cm^3 is very low compared to plasma deposited SiO_2 films (2.2 g/cm^3), so there is less polarizable entities per unit volume. Review of the Pauling estimates of the electronic polarizabilities of C, Si, and O [6] shows that carbon

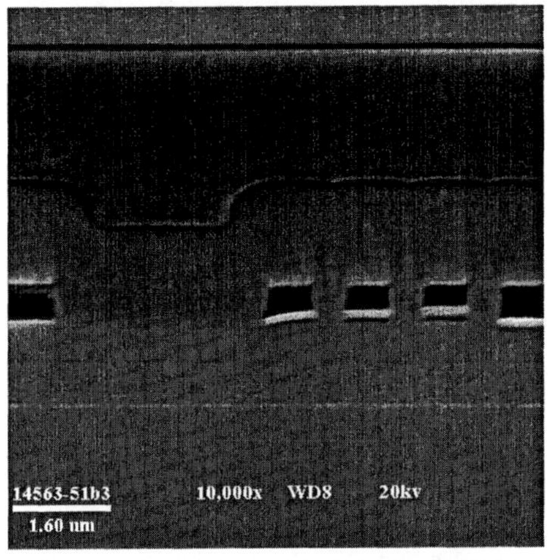

Figure 4. Step coverage observed with the 3MS-based low-k process in an Applied Materials P5000 Oxide PECVD system.

has significantly lower polarizability than both oxygen and silicon. This implies that for tetrahedral bond arrangements such as SiO_2, replacement of some of the silicon and oxygen atoms in the with carbon atoms would result in a reduced net polarizability. Inspection of the Clausius-Mossoitti relation shows that the combined effects of reduced density and polarizability would result in permittivity lower than SiO_2.

SUMMARY

The plasma assisted deposition of dielectric films from trimethylsilane and nitrous oxide gas mixtures can be controlled to produce a highly stable low-k dielectric film. The process can be performed in standard PECVD equipment. The benefits of improved process safety and tool performance from the use of a non-pyrophoric gas are realized. The properties of these dielectric films deposited from trimethylsilane and nitrous oxide gas mixtures, such as low permittivity, low stress, high thermal stability and high dielectric isolation are compatible with the requirements needed for the fabrication of the high performance interconnect structures needed for increased processing speed in modern integrated circuit designs.

ACKNOWLEDGEMENTS

The authors would like to thank W. Ken Weidner of Dow Corning and the Broadway Services Engineering Team (Chandler, AZ) for their assistance in this project.

REFERENCES

[1] G.W. Hill, Microelectronics and Reliability, **4**, 109 (1965)

[2] M. Maisonneuve, Y. Segui, and A. Bui, Thin Solid Films, **33(1)**, 35 (1976)

[3] S. McClatchie, K. Beekmann, and A. Kiermasz, Proc. 1998 Dielectrics for ULSI Multilevel Integrated Circuits Conference, 311 (1998)

[4] M.J. Loboda, J.A. Seifferly, C.M. Grove, and R.F. Schneider, in Silicon Nitride and Silicon Dioxide Thin Insulating Films, M.J. Deen, W.D. Brown, K.B. Sundaram, S.I. Raider, eds., PV97-10, p. 443, Electrochem. Soc. Proc. Series, Pennington, NJ, (1997)

[5] A.D. Butherus, T.W. Hou, C.J. Mogab, H. Schonhorn, J. Vac. Sci. Technol., B-**3**, 1352(1985)

[6] C. Kittel, Introduction to Solid State Physics, Ch. 13, 5th edition, John Wiley & Sons, New York (1976)

THERMAL DECOMPOSITION OF LOW-K PULSED PLASMA FLUOROCARBON FILMS

Brett Cruden, Karen Chu, Karen Gleason, Herbert Sawin
Massachusetts Institute of Technology
Cambridge, MA 02139, USA

Low-k fluorocarbon films have been deposited by a pulsed plasma chemical vapor deposition (CVD) with a variety of different precursors. Deposition rates and resulting film composition have been characterized as a function of pulsing parameters. To examine the thermal decomposition process, we have constructed a novel apparatus for observation of decomposition, utilizing laser interferometry to examine changes in film thickness/properties during the heating process. With this technique, we have observed at least two mechanisms for thermal decomposition and will suggest possible mechanisms for the decomposition based on correlation with XPS data and other observations in the literature. We have been able to eliminate the most labile decomposition route by increasing substrate temperature during deposition. Enhanced decomposition due to ex-situ oxygen exposure has been observed. Mechanisms associated with loss related to oxygen uptake will be presented.

INTRODUCTION

Over the past few years, the development of a viable low-k thin film has become a topic of intense interest to the microelectronics industry.[1] Continually decreasing device dimensions have ensured that intermetal capacitive effects coupled with line resistivity are quickly becoming the bottleneck in device speed. As a result, a replacement dielectric for silicon dioxide is desired. Many materials have been and are being examined as possible candidates. The search process has been further complicated by the many properties required for integration into present processing schemes.

In many respects, an ideal new low k dielectric would be polytetrafluoroethylene (PTFE). It has the lowest dielectric constant of any non-porous material (k = 2.0). The thermal stability is adequate for integration (~400 °C). Possible drawbacks, however, include its generally low adhesion strength and weak mechanical properties typical of polymers (i.e. soft). Another concern is the difficulty in coating a thin film of this material on devices.

In our lab, we have deposited PTFE-like films by pulsed plasma enhanced Chemical Vapor Deposition (PPCVD). It has been shown that some degree of control over film structure and properties can be obtained by using a pulsed plasma.[2] Different chemical species (e.g. CF_2) are relatively more abundant in the afterglow than during the plasma on time. Thus, by turning the plasma on and off, one can control relative species fluxes during the deposition process, which in turn allows for film composition control. Pulsing the plasma also reduces the effects of gas-phase fragmentation by electron impact and film damage by ion bombardment. Work has also been performed with different gas phase precursors,[3] which undergo different reaction pathways in the plasma, forming different species and ultimately films with different properties.

While these films generally have optical and dielectric properties similar to bulk PTFE ($n_f \sim 1.38$, $k \sim 2.0$), their structure is quite different. Because of the different reaction pathways available in the plasma and ion bombardment to the surface, the films tend to have varying degrees of branching and cross-linking. Unlike PTFE, the carbon to fluorine ratio is always less than 2. While the adhesion appears adequate (passes a standard scotch-tape test), the thermal stability of these films is typically less than that of bulk PTFE.

In this work, we have looked at the thermal stability of fluorocarbon films formed by PPCVD from various precursors, including hexafluoropropylene (HFP), tetrafluoroethylene (TFE), and two hydrogen containing fluorocarbons which will be referred to as HFC-A and HFC-B. Changes in film composition are measured and allow us to propose pathways for film loss. A novel new method for studying thermal decomposition of thin films is also presented.

EXPERIMENTAL

Films were deposited in one of two parallel plate reactors. Both reactors have a showerhead gas feed through the top electrode, and the wafer rests on the bottom electrode. In the first reactor (A), the top electrode is grounded, and the bottom is powered with a gap spacing of 1". The bottom electrode is also built with electrostatic clamping abilities and a He backside feed. Lower electrode temperature is controlled via a circulation bath with temperature capabilities in the range of –30 °C to 150 °C, depending on the bath fluid used. In reactor (B), the top electrode is powered, and the bottom is grounded with a gap spacing of 0.5". Cartridge heaters can heat the bottom electrode up to 400 °C. In both systems, a pulse generator/function generator/RF amplifier train is used to power the system at 13.56 MHz with on time powers of 100 W except where noted. Both are equipped with a laser interferometry system for in-situ monitoring of deposition rates. Typically, a flow rate of 50 sccm of precursor gas is used, with a pressure set point of 1 torr.

Film compositions are analyzed by X-ray photoelectron spectroscopy (XPS). Mg $K\alpha$ X-rays are used for all measurements. The carbon 1s spectra obtained are fit with Gaussian-Lorentzian (70/30) curves with FWHM of 2.0 ± 0.1. In all cases, the CFx peaks are readily identified. The spectra are corrected for charging by setting the CF_2 peaks to 292.0 eV. After this correction, positions of the other fit fluorocarbon peaks agree well with literature values.[4] Additional lower energy C peaks are sometimes observed. As these peaks do not appear charge shifted, they are due to carbon on the metallic sample holder and are ignored for this analysis.

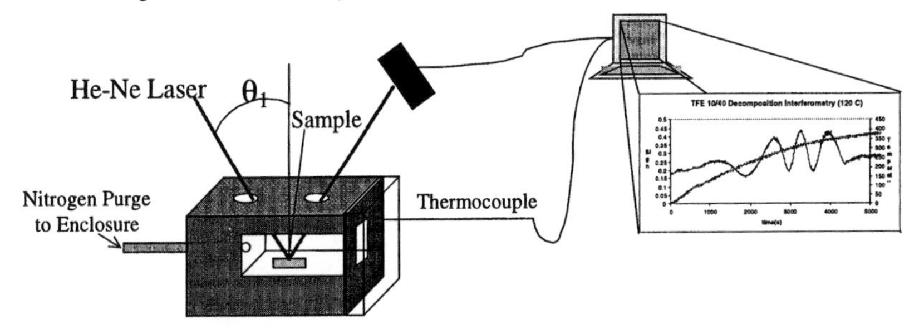

Figure 1. Interferometry for Thermal Stability (ITS) apparatus. The Al block can be heated to over 400 °C.

Interferometry for Thermal Stability (ITS)

A simple, yet novel, apparatus has been constructed for thermal stability measurement of thin films. In this device, depicted in Figure 1, reflectivity of the thin film on a reflective substrate (e.g. silicon) is monitored as the sample is heated. As in a typical interferometry setup, a HeNe laser is directed at the sample, and the reflected beam is monitored by a photodetector. The intensity of the signal will be a function of the substrate reflectivity, film roughness (light scattering), film absorption, film thickness and refractive index. Neglecting signal changes due to absorption or scattering, the intensity can be described by the following equation:

$$I = I_0 \sqrt{\frac{1}{2} + \frac{1}{2}\cos(2\pi \frac{t}{\Delta d_{opt}})} \tag{1}$$

where I is the signal intensity, I_0 is the maximum intensity, t is the film thickness and Δd_{opt} is given by:

$$\Delta d_{opt} = \frac{\lambda}{2n_f \cos(\theta_2)} \tag{2}$$

where λ is the laser wavelength (6328 Å), n_f is the film refractive index and θ_2 is the off-normal angle inside the film, or

$$\cos(\theta_2) = \sqrt{\left(1 - \left[\frac{\sin(\theta_1)}{n_f}\right]^2\right)}. \tag{3}$$

For a film that is somewhat absorbant or reflective, but still has sufficient degree of transmission, the signal will still be modulated as in equation (1), so thickness change can be estimated utilizing the fact that peak-trough separation will correspond to a thickness change $\Delta t = \Delta d_{opt} / 2$.

The sample is placed inside a heated aluminum block as shown in Figure 1. The laser enters and exits through two holes on the top of the block. The holes are covered by a transparent glass slide, and a lid blocks off the entrance port, so the ambient surrounding the sample can be controlled. A continuous nitrogen purge at low flow rate runs to the system to keep it under a positive pressure of inert gas. The block temperature is monitored along with laser signal throughout the course of an experiment to obtain loss versus temperature data.

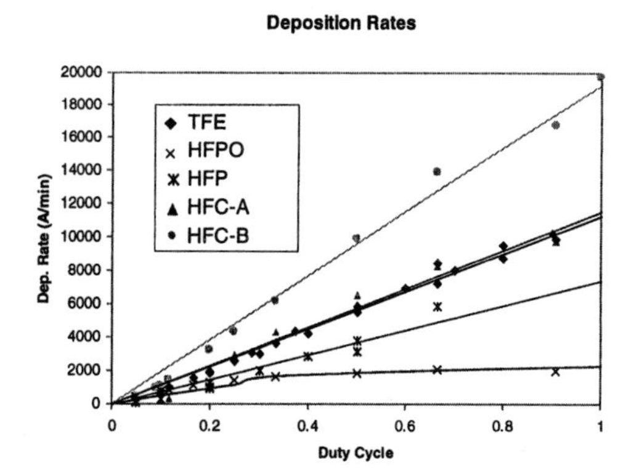

Figure 2. Deposition Rates for different precursor gases as a function of duty cycle.

RESULTS

Deposition

Room Temperature Deposition. Deposition rates in reactor A are shown as a function of pulse duty cycle ($t_{on}/(t_{on}+t_{off})$) in Figure 2. For the precursors of interest, uninhibited TFE, HFP, HFC-A and HFC-B, the deposition rate is approximately linear with duty cycle. This implies that deposition is dominated by processes during the plasma on time.

The deposition rate for the HFC-B reached as high as ~2 micron/minute while the deposition rates for the HFC-A and TFE species were about half of that. The HFP shows a slightly lower maximum deposition.

The HFPO films exhibit a different behavior entirely, where deposition levels off, or even decreases, at higher duty cycles. Oxygen present in this precursor contributes significant etching during the on time period. Hence, the off time becomes an important period for deposition.

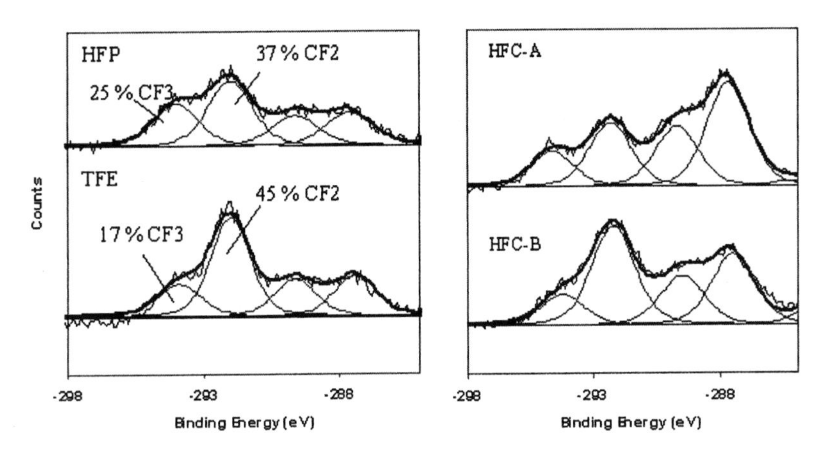

Figure 3. C 1s XPS for (left) HFP, TFE, (right) HFC-A, HFC-B deposited films. All conditions were 10 ms on/90 ms off, 100 W peak pulse power.

The chemistry and structure of these films is shown in Figure 3. Previous work has demonstrated that pulsing conditions can affect film composition.[2] Here, we demonstrate the effect of precursor on film composition. For these conditions, some of the original molecule structure is maintained. TFE shows a large CF_2 component, while HFP shows a slightly larger CF_3 peak. On the simplest level, this difference can be related to the precursor structure, as HFP has a CF_3 group, and TFE is simply linear CF_2. The hydrogen containing species show a greater amount of quaternary carbon. The energy of this peak, however, indicates that the carbon is still in a fluorine environment. FTIR data (not presented here) show minimal amounts of hydrocarbon in the film. These results imply that hydrogen is effective in removing fluorine from the films (either from gas phase elimination or extraction from the film) but does not incorporate to any significant extent. The HFC-B shows both a high CF_2 peak and greater amounts of quaternary carbon, apparently somewhere between TFE and HFC-A films in structure. A relationship between refractive index and film structure is apparent. The HFP, TFE, and

HFC-B films all show a refractive index in the range of 1.38 ± 0.02. The HFC-A films, however, show a refractive index of 1.45 ± 0.01. The difference can be attributed to the low degree of fluorination on the latter film.

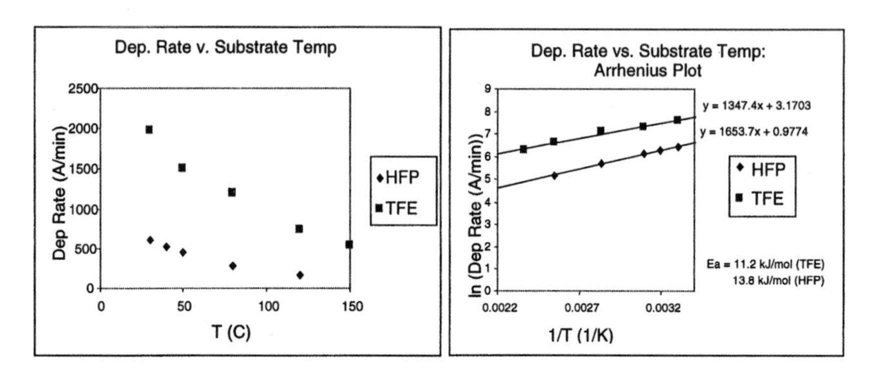

Figure 4. (left) Deposition rates for elevated Temperature Deposition of TFE and HFP. (right) Arrhenius plot.

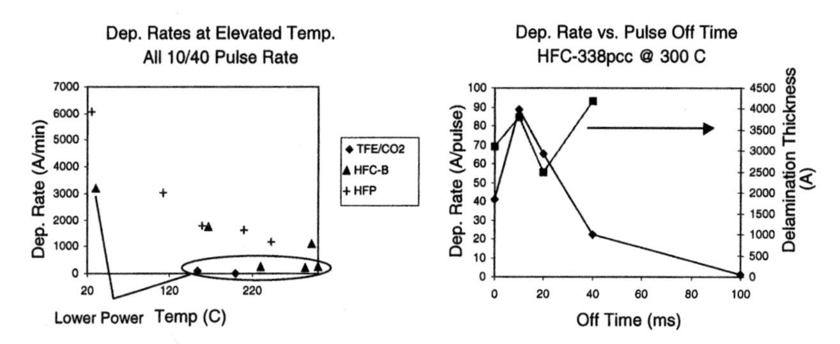

Figure 5. (left) Deposition Rate v. Temperature for a variety of precursors. All are 10 ms on/40 ms off at either 100 or 200 W on time power. (right) Normalized deposition rate (dep./pulse) for HFC-B at a variety of off times, with 10 ms on @ 200 W. The squares indicate at what thickness the film delaminates.

Elevated Temperature Depositions. Films were also deposited at elevated temperatures. Figure 4 shows deposition rates versus substrate temperature up to 150 °C for TFE and HFP. The rate shows an Arrhenius dependence with negative activation energies. The activation energies are –11.2 kJ/mol for TFE and –13.8 kJ/mol for HFP. These values are typical of absorption limited processes, and agree with activation energies for a similar process with HFPO (-13.8 kJ/mol.)[5] Very little change in film structure is detected by XPS at these temperatures. However, a slight reduction in the CF_3 peak is observed.

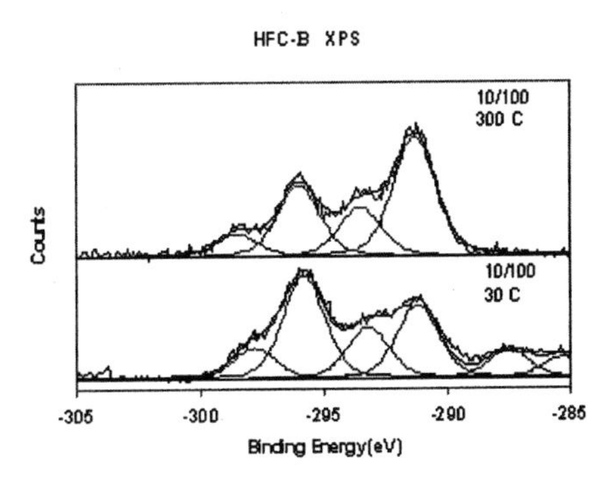

Figure 6. C 1s XPS for HFC-B pulsed plasma deposited films at different substrate temperatures. Note the higher deposition temperature favors more quaternary C over CF_2.

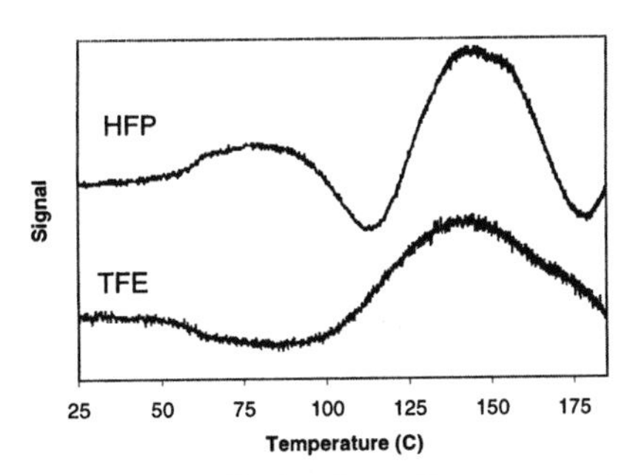

Figure 7. ITS signal vs. Temperature for HFP and TFE films. Temperature is ramped to 200 C over the course of 1 hour. The higher frequency in signal for HFP indicates a faster decomposition.

Reactor B was used to deposit films at temperatures up to 300 °C. The plot of deposition rate versus temperature for a variety of precursors is shown in Figure 5 (left). Of the precursors used, HFC-B and HFP both showed appreciable deposition rates. The deposition of HFC-B as a function of pulse off time is shown in Figure 5 (right). The HFC-B films deposited at 300 °C temperatures showed a tendency to delaminate near 0.5 μm of thickness. It was found that the delamination thickness was not affected greatly by pulsing, but was reduced at lower powers. This is consistent with a reduction of compressive stress by reduced ion bombardment. Figure 6 shows the effect of temperature on film composition. The increase in quaternary carbon, a cross-linking species, explains the increased stress within the film and the higher refractive indices observed at these temperatures (~1.5). Though not shown here, the HFP films show similar structure, with large amounts of quaternary carbon in the film.

Thermal Stability

Interferograms from the ITS apparatus for decomposition of one micron thick films deposited from HFP and TFE at room temperature are shown in Figure 7. The initial change in signal is due to thermal expansion. Where the signal first changes direction, near 80 °C, is the onset of decomposition. The slower cycle period for the TFE films indicates better thermal stability. Figure 11 shows minimal HFC-A decomposition up to 350 °C. Though not shown here, the HFC-B shows decomposition rates between that of HFP and TFE films. The difference between the HFP and TFE films is explained by comparison to NMR results of Lau, et. al.[6] They have shown that low temperature decomposition in cross-linked plasma polymerized fluorocarbon films is dominated by the loss of CF_3 end group species. The underlying mechanism can be explained by an adaptation of gas phase perfluorocarbon cracking observed by Tortelli, et. al.[7]

$$\begin{array}{c} CF_3 \\ | \\ (CF_2)_{i,\ i\ small} \\ | \\ (CF_x)_mCF_y\text{-}C\text{-}(CF_x)_n \\ | \\ (CF_x)_l \end{array} \rightleftharpoons \begin{array}{c} CF_3 \\ | \\ (CF_2)_i \\ | \\ (CF_x)_mCF_y\text{-}C\bullet + \bullet(CF_x)_n \\ | \\ (CF_x)_l \\ \downarrow \\ (CF_x)_mCF_{y-1}\text{=}\underset{\displaystyle (CF_x)_l}{\underset{|}{CF}} + \bullet(CF_2)_iCF_3 \end{array} \qquad \text{(R1)}$$

Tortelli observed this reaction to occur in gas phase perfluorocarbons as low as 150 °C. The lowest temperature decomposition occurred in more highly substituted carbons, due to larger steric strains. For these cross-linked films, steric strains could be very high and decomposition may occur more readily.

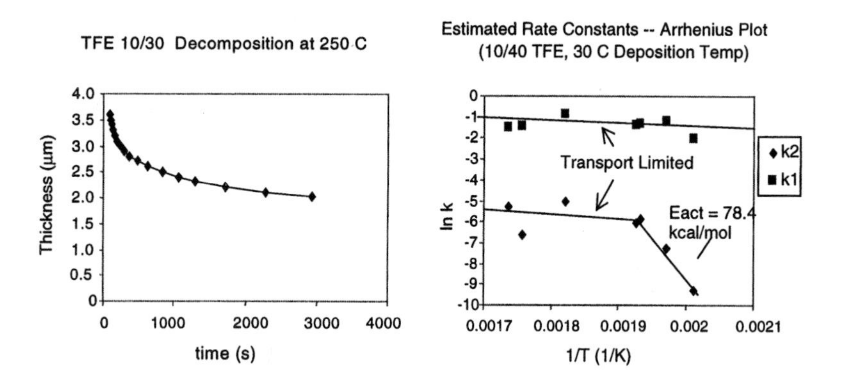

Figure 8. (left) Decomposition of a TFE film at constant temperature, fit by two exponential decays. The time constants ($\tau = 1/k$) for decomposition at 250 °C are 75 s and 20 min. (right) Arrhenius plot of rate constants for decomposition process.

The decomposition of TFE films was examined at constant temperatures as in Figure 8. The decomposition here can be fit by the sum of two exponentials, indicating that decomposition can be described by two separate first-order processes:

$$\frac{dX}{dt} = \frac{dX_1}{dt} + \frac{dX_2}{dt} = -k_1X_1 - k_2X_2 \tag{4}$$

$$X_i = X_i^{\,0} \exp(-k_i t)$$
$$k_i = A_i \exp(-\frac{E_i}{RT}) \tag{5-6}$$

This decomposition was performed at various temperatures to obtain activation energies as shown in Figure 8. Above a certain decomposition rate, the rate constant was found to only change very slightly with temperature. It appears as if the process is not activated, and the limiting step in decomposition is the transport of product out of the film. At very low rate constants, a dependence on temperature is apparent, giving an activation energy of 78.4 kcal/mol. This energy is comparable to a typical C-C bond strength and is similar to that observed for bulk Teflon decomposition (83 kcal/mol):[8]

$$-(CF_2-CF_2)_n- \longrightarrow -(CF_2-CF_2)_m^{\bullet} + {}^{\bullet}(CF_2-CF_2)_{n-m}- \quad \text{(initiation)} \quad \text{(R2a)}$$

$$^{\bullet}(CF_2-CF_2)_m- \longrightarrow CF_2{=}CF_2 + {}^{\bullet}(CF_2-CF_2)_{m-2}- \quad \text{(propagation)} \quad \text{(R2b)}$$

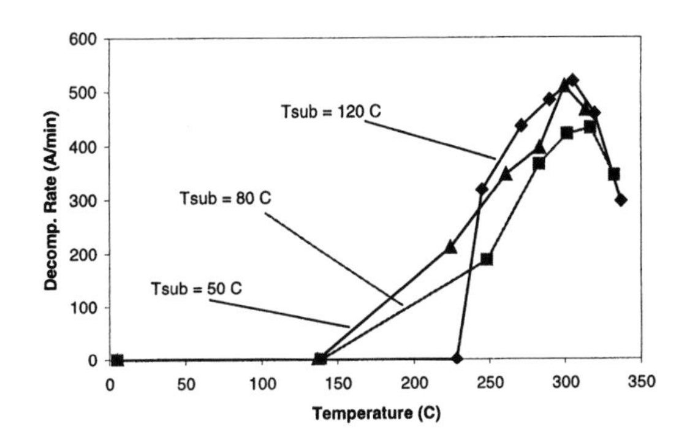

Figure 9. Decomposition Rate for TFE films deposited at different substrate temperatures (10 ms on/40 ms off).

The effect of higher temperature deposition on the TFE film stability is shown in Figure 9. The initial decomposition near 150 °C is due to the more labile decomposition mechanism. The film deposited at a higher substrate temperature does not show this low temperature decomposition. The peak observed between 250 and 300 °C still occurs, however. Thus, raising temperature during deposition is mainly effective in reducing the loss component due to the more labile elimination route.

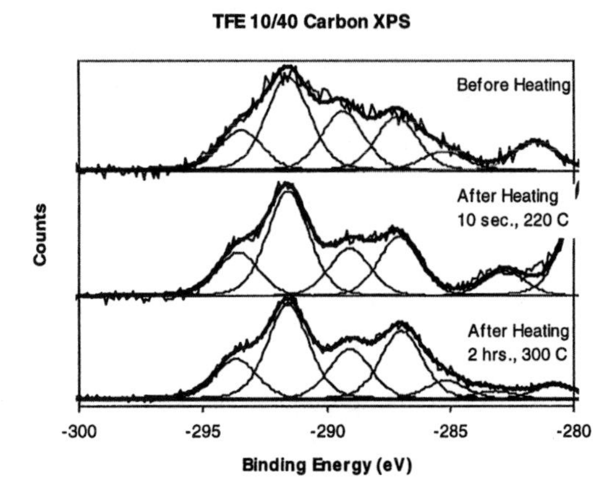

Figure 10. C XPS of TFE 10/40 Film before and after decomposition

XPS of these films before and after decomposition is presented in Figure 10. All of the fluorocarbon peaks are reduced relative to the quaternary peak upon heating. The quaternary peak resembles a fluorinated graphite and appears more stable than the fluorocarbon species. This would explain the higher thermal stability of HFC-A, with its large quaternary C peak. The decomposition thus involves bulk film loss with possible defluorination.

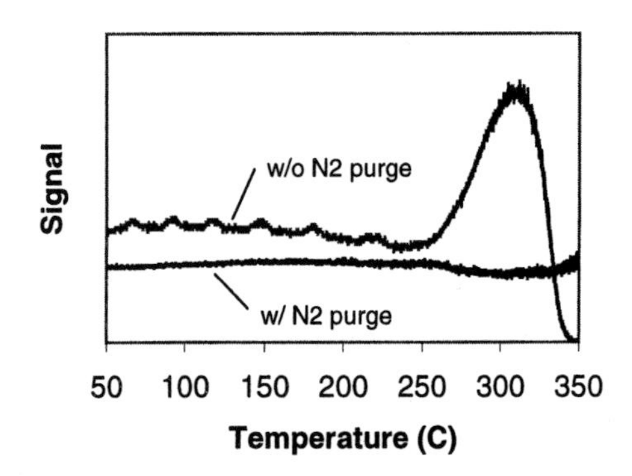

Figure 11. ITS Data with and without nitrogen purge. Ambient oxygen in the latter case appears to enhance decomposition

Some work was done to examine the effect of ambients on film loss. Figure 11 shows the interferograms of HFC-A with and without a nitrogen purge. Complementary to this

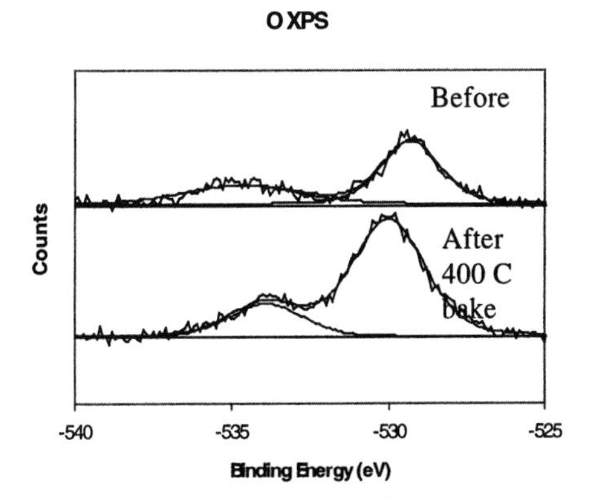

Figure 12. O XPS of HFC-A film before and after heating to 400 C in an uncontrolled ambient. The rightmost peak is attributed to the sample holder. The left peak represents oxygen within the film. Oxygen incorporation during the heating cycle is observed.

data is before and after XPS (Figure 12.) There is a significant degree of oxygen uptake during the process. Oxygen is also incorporated into the film due to ambient exposure. The free radicals in the film readily bind with oxygen radicals in the air, forming an unstable carbonyl moiety, degrading the thermal stability. Decomposition mechanisms of this nature have been observed in the decomposition of irradiated PTFE.[9]

$$(CF_x)_nCF_2\bullet \xrightarrow{O_2} (CF_x)_nCF_2O\text{-}O\bullet \xrightarrow{\Delta} F_2 + CO_2 + (CF_x)_{n-1}CF_2\bullet$$

$$(CF_x)_nCF_2\bullet \xrightarrow{\Delta} (CF_x)_{n-2}CF_2\bullet + C_2F_4 \qquad \text{(R2,R3)}$$

Although the films deposited at very high temperatures give too high a refractive index to be worthy of further examination as a low k dielectric material, the effect on their thermal stability was examined. As expected, high temperature deposition improves the thermal stability of the film. The films are more graphitic in nature, having significantly higher quaternary carbon concentrations (Figure 6.)

CONCLUSIONS

Deposition of fluorocarbon films by Pulsed Plasma enhanced CVD has been studied with four different precursors. The film compositions and properties are found to be a function of both the pulse conditions and gas phase precursors. Variation of substrate temperature during deposition, while reducing deposition rate, is found to improve thermal stability.

The decomposition mechanisms are also studied. Films with significant amounts of quaternary carbon (graphitic) show high thermal stabilities, but also have higher refractive indices, and thus higher dielectric constants. In films deposited from TFE, two decomposition mechanisms were identified. A loss mechanism associated with CF_3 end groups is proposed. Bulk film loss by a mechanism similar to bulk PTFE decomposition, is also observed, with the remaining film being rich in quaternary carbon.

Films are also observed to incorporate oxygen, due to radical sites left behind during the plasma deposition process. These sites form unstable species in the film, which must be eliminated if good thermal stability is to be achieved.

ACKNOWLEDGEMENTS

Thanks to DuPont's Zyron® Electronic Gases Group for funding of this project. I would also like to thank Andrew Campbell for his contribution to initial stages of this work.

REFERENCES

[1] L. Peters, *Semiconductor International*, **21**(10), 64 (1998).
[2] S. J. Limb, D. J. Edell, E. F. Gleason and K. K. Gleason, *J. Applied Polymer Science*, **67**,1489 (1998).
[3] C. B. Labelle and K. K. Gleason, *JVST A*, In Press), (1998).
[4] A. Dilks and E. Kay, *Macromolecules*, **14**,855 (1981).
[5] S. Limb, Ph.D. Thesis, Mass. Inst. Techn., Cambridge (1997).
[6] K. K. S. Lau and K. K. Gleason, *J. Phys Chem. B.*, **102**,5977 (1998).
[7] V. Tortelli, C. Tonelli and C. Corvaja, *Journal of Fluorine Chemistry*, **60**,165 (1993).
[8] J. C. Siegle, L. T. Muus, T.-P. Lin and H. A. Larsen, *Journal of Polymer Science: Part A*, **2**,391 (1964).
[9] U. Lappan, L. Haubler, G. Pompe and K. Lunkwitz, *Journal of Applied Polymer Science*, **66**,2287 (1997).

THERMAL STABILITY OF SPUTTERED TA/PECVD SIOF FILMS FOR INTERMETAL DIELECTRIC APPLICATION

Kihong Kim, Sangjoon Park*, and G. S. Lee
Solid State Laboratory
Electrical and Computer Engineering
Louisiana State University
USA

The characteristics of Ta films that were sputtered on fluorinated silicon (SiOF) substrate were investigated and compared with those for Si and thermal oxide substrates. The deposition rate of the Ta film increased almost linearly with increasing target power. At a target power of 150W, bcc-Ta and β-Ta were observed for Si and thermal oxide substrates, respectively. Meanwhile, the transition between bcc and β was observed for SiOF substrates. To investigate Ta as a diffusion barrier between Cu and different substrates, Cu was overcoated on 20nm-Ta and 60nm-Ta, then annealed in the temperature range 400 - 800 °C. It was observed that no diffusion occurred up to 600 °C regardless of the substrates and Ta thickness. However, 20nm-Ta and 60nm-Ta started to fail at 700 and 800 °C, respectively.

INTRODUCTION

As the number of metal layers in integrated circuit increases, the role of intermetal dielectric (IMD) material has become more important to prevent the degradation of signal transmission caused by parasitic capacitance. The delay constant, RC (where R = the resistance of the interconnect metal and C = the capacitance of IMD material) that is related to the rate of signal transmission can be reduced effectively by using Cu as an interconnect metal and SiOF film as an IMD material. Cu as an interconnect metal also provides many advantages such as the ability to reduce the number of metal layers, minimizing power dissipation and improving reliability [1]. Meanwhile, SiOF film by plasma enhanced chemical vapor deposition (PECVD) has attracted interest for its easy use and inexpensive precursors as well as its low dielectric constant. In this study, the characteristics of Ta as a diffusion barrier between Cu and SiOF were discussed for the possible use of Cu/SiOF in integrated circuit. In further, the moisture absorption problem in SiOF film was mentioned in brief.

*Permanent address: APEX Co., 128 Chukbuk Ri Namyi Myun Chongwon Kun, Korea.

EXPERIMENT

The SiOF film was prepared in a conventional parallel plate PECVD system. Boron doped Si wafers with (100) orientation were used as substrates. The precursors used were 40 sccm of Si_2H_6 (5% in He), 100 sccm of N_2O, and 30 sccm of CF_4, which gave dielectric constant of 3.5 [2]. The process pressure, rf power, and substrate temperature were maintained at 700 mTorr, 50W, and 180 °C, respectively. The Fourier transform infrared (FTIR) spectra of the SiOF films were observed for moisture absorption problem in SiOF film. Ta was sputtered at room temperature by using dc magnetron sputter. The process pressure was maintained at 10 mTorr with flowing 56 sccm of Ar. The structure of the as-deposited Ta films on different substrates were investigated by x-ray diffraction (XRD). Cu was sputtered onto Ta without breaking vacuum, then sheet resistance of Cu/Ta films were measured by four point probe after these films were annealed in the temperature range of 400 - 800 °C for 30 min in N_2 ambient. All measurements were carried out at room temperature.

DISCUSSION AND ANALYSIS

Figure 1 displays the typical FTIR spectra for the SiOF film. The spectrum for the film after 10 month aging test was compared with the spectra for the film as-deposited and with post-deposition anneal [2]. The film with post-deposition anneal was exposed in the air (≈25 °C and 60% humidity) for 10 month for this aging test. It has been concerned that the SiOF films are known to have a stability problem with respect to moisture absorption [3]. It was observed that Si-OH peak at ≈920 cm^{-1}, that seemed to

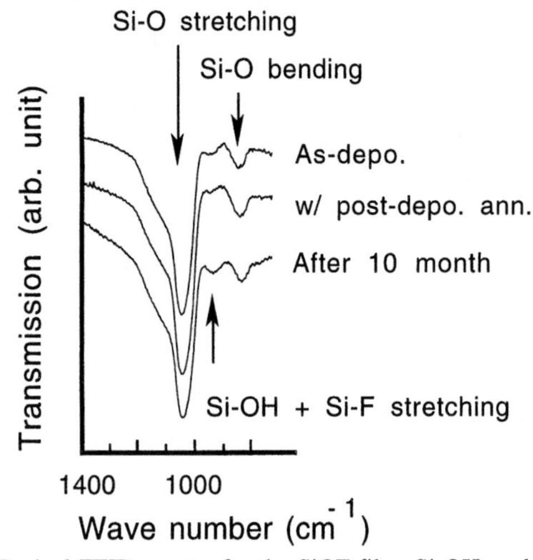

Fig. 1 Typical FTIR spectra for the SiOF film. Si-OH peak appeared again after aging test, but the positions for Si-O peaks were undisturbed.

have disappeared with post-deposition anneal at 400 °C for 30 min, appeared again after aging test. However, the peak positions for Si-O stretching and bending vibration modes were undisturbed.

Figure 2 shows the typical I-V characteristic for the SiOF films. The film after aging test was compared with our previous result [2]. The ramp voltage at the rate of -1 V/s was applied to the gate area introducing electron injection. Because the following reaction may have occurred [4]

$$SiF + H_2O \rightarrow SiOH + HF,$$

it is supposed that the outgasing HF degraded the film, thus causing larger amount of leakage current compared with the one before aging test. It was suggested that capping the SiOF film with a thin layer of undoped oxide [4] or adding Ar gas during deposition [5] can improve the stability of the SiOF film in terms of moisture absorption .

Ta was sputtered onto substrates at room temperature by dc magnetron sputter. Figure 3 shows the deposition rate as a function of target power. For the target power of 50W, the deposition rate was $\approx$20 nm/min, meanwhile, it increased almost linearly to $\approx$90 nm/min for the target power of 200W. It was observed that the structure of Ta had dependence on target power. For all substrates, Ta had both β and bcc structures at 50W. At 150W, however, bcc-Ta and β-Ta were dominant for Si and thermal oxide substrates, respectively. On the other hand, Ta for SiOF substrate showed mixture of two structures that was closer to bcc. These results are shown in Fig. 4.

Figure 5 shows the sheet resistance of Cu/Ta on different substrates (thickness of SiO_2 and SiOF $\approx$100 nm, respectively) as a function of annealing temperature. Cu with

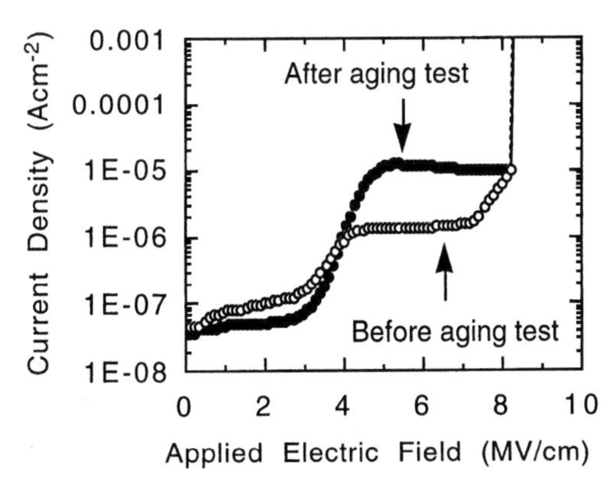

Fig. 2 Typical I-V characteristic for the SiOF films before and after aging test. It was observed that the film after aging test was degraded because of moisture absorption.

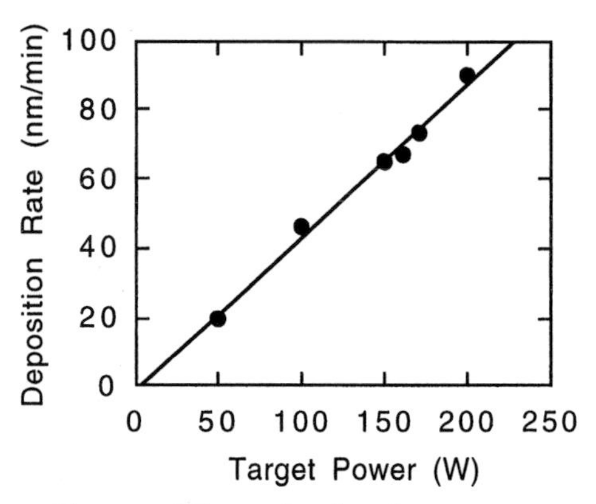

Fig. 3 Deposition rate of Ta as a function of target power.

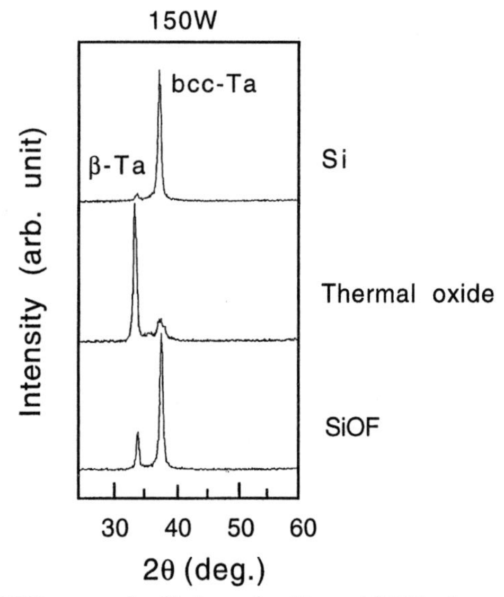

Fig. 4 XRD pattern for Si, thermal oxide, and SiOF substrates at target power of 150W.

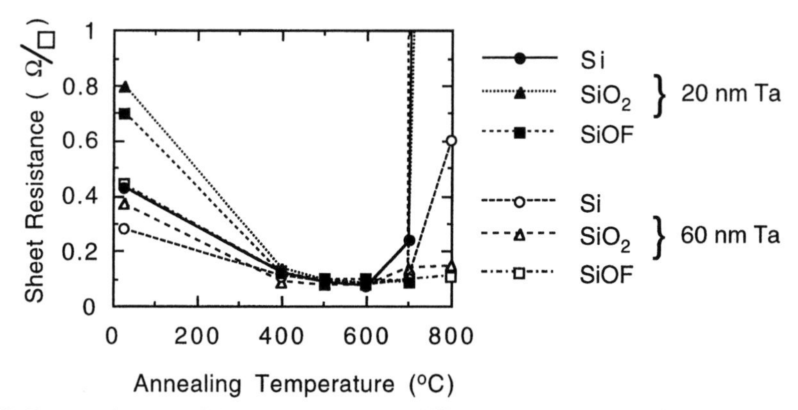

Fig. 5 Sheet resistance of Cu (150 nm)/Ta on different substrates as a function of annealing temperature.

thickness ≈150 nm was sputtered on 20 nm and 60 nm Ta, respectively, that exhibited the structures as shown in Fig. 4. Before annealing, thicker Ta showed lower sheet resistance for all substrates, and the value of sheet resistance was different for different substrates. After annealing at 400 °C, however, the sheet resistance decreased and was almost independent of the thickness of Ta and substrates. The same result was observed for the annealing up to 600 °C. After annealing at 700 °C, the sheet resistance for Cu/20 nm Ta/Si increased abruptly indicating the reaction between Cu and Si. On the other hand, the sheet resistance for 60 nm Ta still did not change. Finally, after annealing at 800 °C the sheet resistance for all 20 nm Ta increased dramatically and the color of Cu changed to dark brown indicating significant reaction between Cu and all substrates. For 60 nm Ta, however, only Si substrate seemed to fail.

CONCLUSION

In this study, we investigated the characteristics of Ta for the possible use of Cu/SiOF in integrated circuit. It was identified that Ta was stable as a diffusion barrier between Cu and SiOF for annealing up to 800 °C. Thus, it is promising that this SiOF film with moderately low dielectric constant of 3.5 could be used with copper [6].

REFERENCES

[1] P. Singer, Semicond. Int., **11**, 67 (1997).
[2] K. Kim, D. Kwon, and G. S. Lee, Thin Solid Films (to be published).
[3] V. L. Shannon and M. Z. Karim, Thin Solid Films, **270**, 498 (1995).
[4] M. T. Weise, S. C. Selbrede, L. J. Arias, and D. Carl, J. Vac. Sci. Technol. A **15** 1399 (1997).
[5] K. Kim, S. Park, and G. S. Lee (unpublished).
[6] P. Singer, Semicond. Int., **6**, 91 (1998).

A HIGH SELECTIVITY TO LOW K MATERIALS PLASMA ASHING PROCESS.

Mohamed Boumerzoug, Han Xu, Richard L. Bersin
Ulvac Technologies Inc., 401 Griffin Brook Drive, Methuen, MA 01844

The stripping of resist and post-etch residue cleaning in structures containing low-k materials have been investigated. Low temperature plasma processing with oxygen and fluorine chemistry allowed us to obtain clean samples with high selectivity to organic interlayer dielectric (ILD). We have successfully developed a process to strip the resist and to clean post-etch residue for via containing low-k as ILD. SEM characterization shows residue free samples with no evident attack of the low-k layer.

The experiments were conducted with the ENVIRO™ post-etch residue cleaning tool from Ulvac Technologies Inc.. For the low-k material we used benzocyclo-butene (BCB), Flare and Silk. All three materials are reported to be potential candidate for application as ILD in multilevel metallization. The resist and the low-k etch rates were measured using a nanospec and an ellipsometer respectively. For the low-k the refractive index was measured before and after processing.

INTRODUCTION

Low dielectric constant materials are being introduced as ILD to replace conventional oxides and nitrides. They have the advantage of allowing higher performance in integrated circuit devices [1]. The most promising candidates show good stability and compatibility with processes involved in device fabrication [2]. In our investigation we have considered three materials BCB, Silk and Flare. BCB is reported to absorb very low amount of moisture and good adhesion to oxide and metal layers [3]. Flare and Silk are reported to have low outgassing and high thermal and mechanical stability [4].

In wafer processing, the stripping of resist is needed for every lithographic step. One of the challenges to integrate low-k materials is to obtain a high selectivity to them during the stripping process [5]. Conventional resist stripping combines high temperature and plasma oxygen to obtain high rates. However, high temperature processing with oxygen plasma will not only remove the resist but also attack the low-k layer due to its organic nature. Therefore a low temperature stripping process is needed. It is well known that adding a small amount of fluorine to the oxygen plasma increase the stripping rate of the resist at low temperature. However in this particular application the content of the fluorine must be controlled to obtain high ash rate, high selectivity and not alter the properties of the low-k layer.

In a previous publications we demonstrated that an ion-assisted process combined with microwave downstream plasma could be successfully applied to clean post etch residues from via holes and metal lines with conventional SiO_2/TiN/AlCu structure [6,7]. However, replacing conventional oxide with low-k material introduce new challenge due the organic nature of low-k.

The ENVIRO™ tool with its combination of microwave downstream and RIE capabilities is very suitable for resist stripping after via etching where the organic component is exposed in the via sidewall. Considering that conventional oxygen stripping damages the organic layer making it porous and hygroscope and leading to via poisoning in subsequent metallization steps, the ENVIRO™ tool can strip polymers from the via sidewall at low temperature without damaging the exposed organic layer.

To demonstrate these capabilities a baseline process was developed for structure containing resist/low-k. The objective is to develop a process that can strip resist with good selectivity to low-k. The process was also applied to remove etch residue from via that combine low-k with AlCu metal layer (SiO_2/low-k/TiN/AlCu). The process was then used to clean via that contains low-k as ILD and Cu as metal layer.

EXPERIMENTAL DETAILS

All the investigated structures were fabricated on 8-inch wafers. After the etching step the wafers were processed with the ENVIRO™ tool developed by Ulvac Technologies Inc. followed by DI rinse. The ENVIRO™ tool incorporates both microwave downstream and non-damage RIE plasma for conventional resist stripping and selective layer removal according to the chemical composition of each layer [8]. The RIE and microwave plasma parameters of pressure, gas mixture, temperature and power are tunable trough a software user interface. The system can process up to six different gases including O_2, NF_3 and CF_4. The wafer is placed on an insulated hotplate that is located in the downstream area. Resist removal is monitored by optical emission. During the resist stripping with the oxygen/fluorine plasma effective detection of the stripping endpoint characteristic was performed by monitoring the intensity of the OH radicals using their optical emission at 307nm.

Microwave energy with frequency of 2.45 GHz propagates from the microwave power supply to the plasma source through a wave-guide. The tool is equipped with a load-lock. This configuration will ensure process reproducibility, excellent wafer temperature control; and minimize the contamination due to the exposition to atmosphere.

Blanket wafers of resist, oxide, Flare, BCB and Silk were processed and thickness and refractive index were measured with a nanospec and an ellipsometer respectively. After, the dry processing with the ENVIRO™ and DI rinse the wafers were broken for SEM cross section inspection. SEM was used for checking the cleanliness of the investigated structure and for any undercut of low-k and other layers due to our dry cleaning process.

RESULTS AND DISCUSSION

Resist:low-k Selectivity

Figure 1 shows resist and BCB etch rate vs. CF_4 flow rate in oxygen/fluorine mixture ($O_2/CF_4/N_2/H_2$) microwave downstream plasma at room temperature. As shown the etch rate of BCB increases with fluorine content. Adding a moderate amount of fluorine to the oxygen increases etching rates while maintaining a good selectivity (resist:BCB >10). For such amount of fluorine the refractive index of the BCB was not changed. The refractive index of Flare and Silk were not also affected by similar processing conditions and a selectivity of 5:1 (resist:Flare and resist:Silk) was obtained. However, if the fluorine content is higher than 30 sccm the properties of the BCB are altered. Because of the presence of silicon in the BCB the ashing process must be adjusted to keep the Si content unchanged. For example if the ashing is performed with fluorine/oxygen mixture, very low fluorine will led to a selective removal of polymer and leaving a film that is Si rich while high fluorine content could selectively remove the silicon. Figure1 also shows the resist ash rate at the same conditions. The etching rates of the resist are higher than BCB rates and increase with the fluorine content. Our investigation on blanket wafers shows that high temperature oxygen plasma modifies the low-k properties as shown in the refractive index changes. The changes could be attributed to a selective depletion of carbon from the film [9].

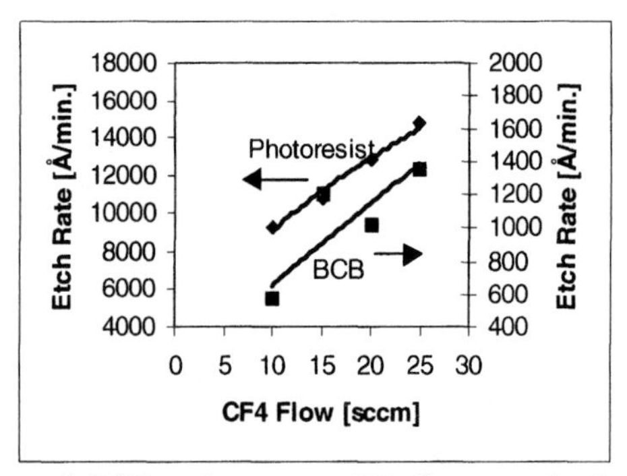

Figure 1: Resist and BCB etch rate vs. CF_4 flow rate in O_2/CF_4 microwave downstream plasma.

Figure 2a shows a resist/SiO_2/BCB structure that was etched down to the BCB layer. This structure was made to check the selectivity of the oxygen/fluorine mixture microwave downstream plasma process at room temperature with the SEM. Figure 2b shows the same structure after the removal of the resist. It shows clearly that while the resist is completely removed little etching occurs to the BCB.

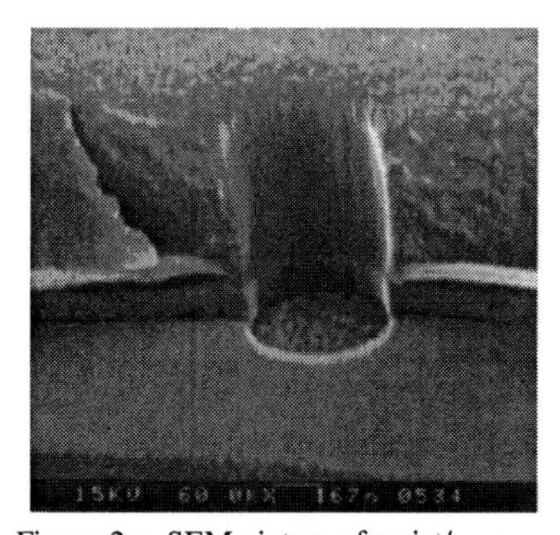

Figure 2a: SEM picture of resist/ BCB/SiO_2 before the resist stripping step.

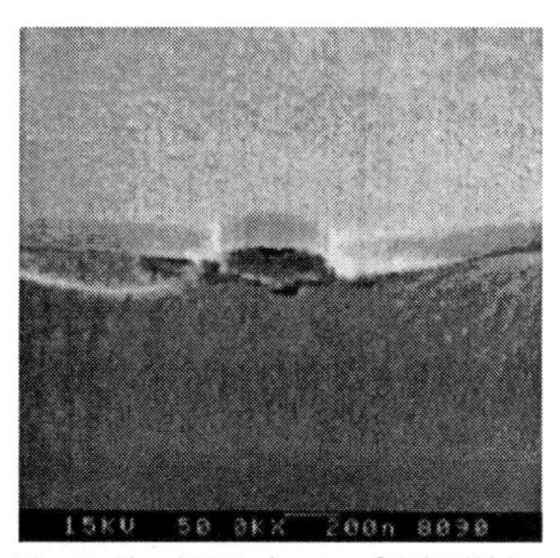

Figure 2b: SEM picture of BCB/SiO_2 after the resist stripping step.

Application to SiO_2/low-k/SiO_2 Structure

Figure 3 shows a typical SiO_2/low-k/SiO_2 structure etched down to the SiO_2 . The top SiO_2 layer serves as a hard mask that does not erode during the etching of the low-k layer. Resist layer, not shown in the picture was used to pattern the oxide.

The residue on the top oxide layer is a mixture of oxide and organic compounds that

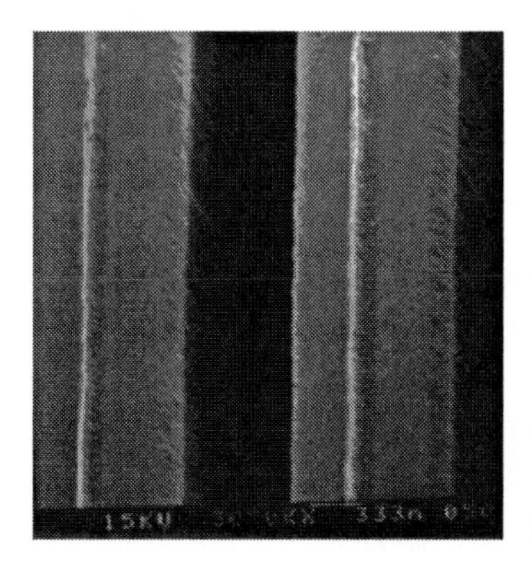

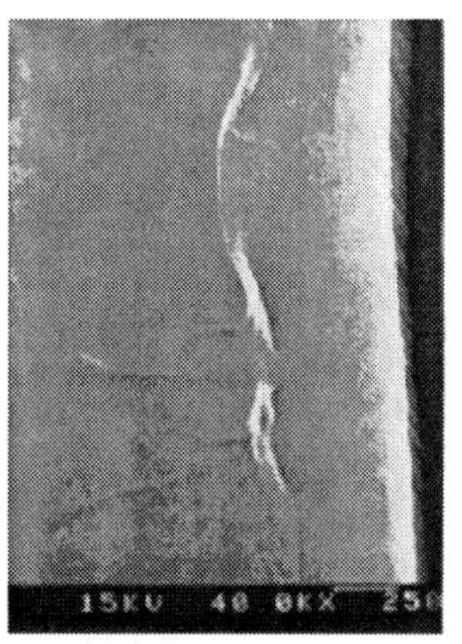

Figure 3: SEM of SiO_2/low-k/SiO_2 structure after etching process.

were sputtered from the vertical wall and lower oxide layer during the etching. Experiments performed at high temperature lead to a significant undercut of the low-k layer. However, microwave downstream oxygen/fluorine plasma at low temperature was successfully used for the cleaning of the residues without undercutting the organic layer. The oxide loss was also kept at an acceptable level. Figure 4 shows a SEM picture of the Low-k pattern after the DI rinse following the dry cleaning process. As shown the top surface is very clean and no residue is visible on the sidewall. Therefore oxygen/fluorine downstream plasma at low temperature followed by DI rinse is effective in removing post etch residue from Low-k containing structure.

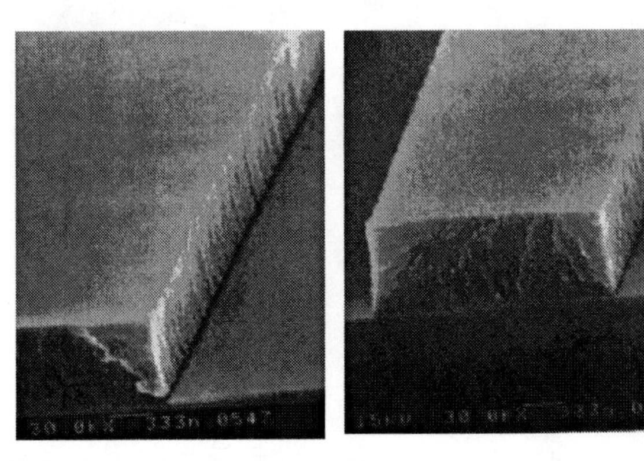

Figure 4: SEM of SiO_2/low-k/SiO_2 structure after dry cleaning process.

Post-via Etch Residue Cleaning

The process was also applied to clean via that are made of low-k as ILD and AlCu as metal layer with TiN as antireflective layer. TiN has the advantage of integrating well as barrier and glue layer for tungsten deposition. Figure 5 shows a top view and a cross section of a via after the etching step. During the etching a sidewall polymer is intentionally formed to control the via's vertical profile. Material sputtered from the side and bottom of the via during the etching is also incorporated into the sidewall polymer. As shown, a thick polymer is formed on the sidewall of the via. Some residues are also visible at the bottom. The residue on the sidewall must be removed to ensure good adhesion of the subsequent filling. Any residue on the bottom of the via is also undesirable and will lead to high via contact resistance if not removed.

The residue could be removed using wet cleaning technique, hydroxylamine based solvent has been found to be effective solution for conventional oxide/TiN/AlCu via structure. However, beside the concerns associated with the safety and solvent disposal, the replacement of oxide with low-k has lead to an increasing concerns of the attack of the low-k by hydroxylamine based solvent. Therefore a dry process will be desirable. The process should be selective to both TiN and low-k layers.

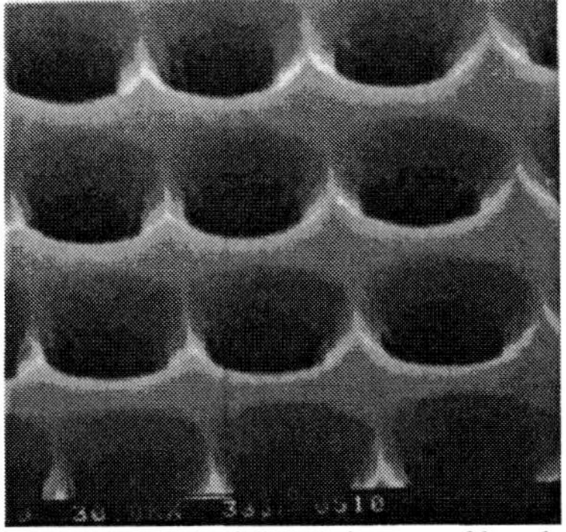

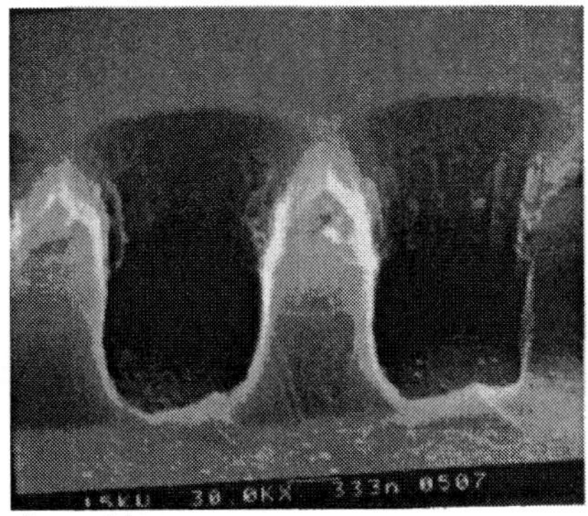

Figure 5: Typical SEM pictures of low-k via after etching process.

Our investigation shows that with oxygen fluorine chemistry TiN and low-k losses are very sensitive to the processing temperature. Both show severe undercut when processed at high temperature. Figure 6 shows a cross section of via after dry cleaning process at low temperature followed by DI rinse. There is no residue left and there no undercut of the low-k and TiN layers.

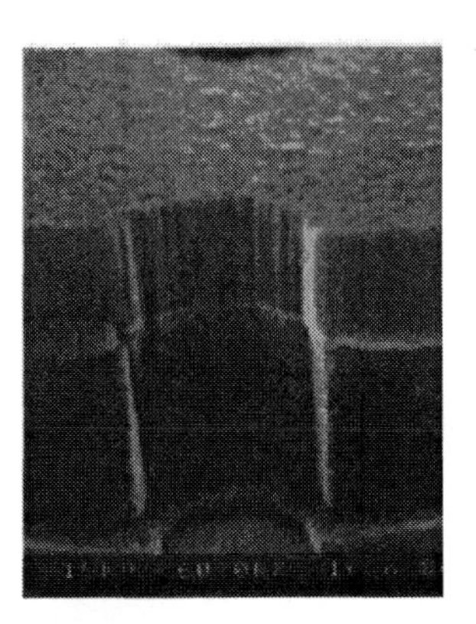

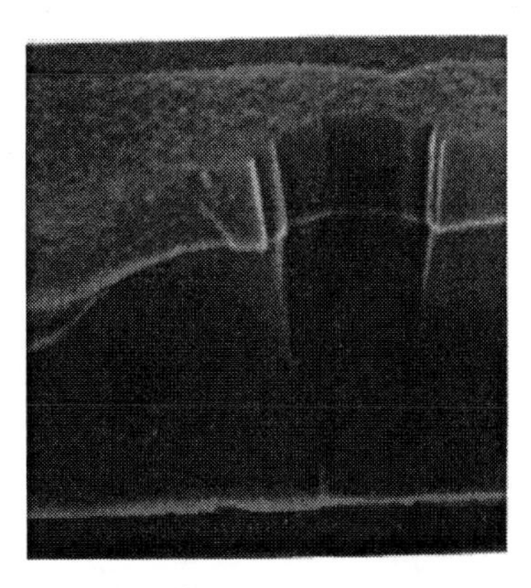

Figure 6: SEM cross section of via after dry cleaning process.

In a third tep the process was tested on SiO_2/low-k/SiN_x/Cu via that are etched down to the Cu. Figure 7 shows a cross section of a typical via. The top layer of oxide is used as mask to protect the low-k layer during the etching. A layer of SiN_x is deposited on top of Cu and is used to enhance the adhesion to low-k and avoid the oxidation of the Cu layer during the etching. The oxide opening is achieved with fluorine based plasma. The

Figure 7: SEM picture of SiO_2/low-k/SiN_x/Cu via after etching process.

low-k is anisotropically etched with oxygen or oxygen/fluorine mixture plasma at low pressure. The via is then completed by etching the SiN_x layer down to the Cu in a separate step using dry or wet step. As a result of the RIE via etching, polymer film is formed on the bottom and sidewall of the via. The nature of etch residues in Cu via are reported to be CH, CO, CF_x CuO, and CuF_x . Figure 8 shows a cross section of a via after the dry cleaning process followed by DI rinse. In this case too, oxygen/fluorine plasma at low temperature was found to be very effective in removing the post etch residue without undercutting the low-k layer.

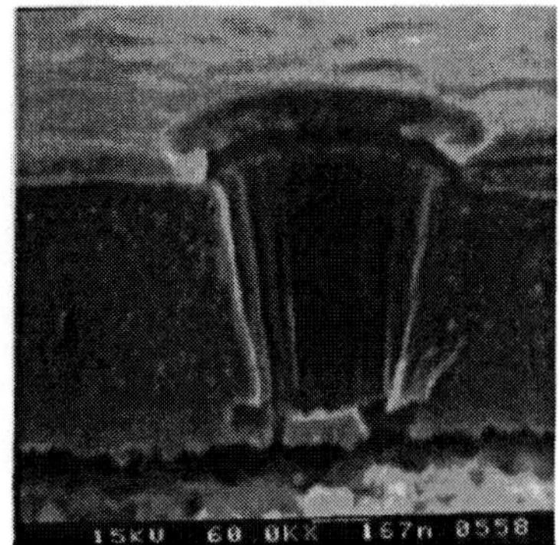

Figure 8: Cross section of a SiO_2/low-k/Cu via after dry cleaning.

CONCLUSION

Low dielectric constant materials are promising candidates as interlayer dielectric (ILD) to replace conventional oxides and nitrides. The removal of resist and post etch residues in structure containing low-k materials is very challenging. Consequently new process chemistry must be developed to remove this new type of residues and keep a high selectivity to low-k.

A resist stripping and post-etch residue cleaning process suitable for the integration of low-k materials is demonstrated. Oxygen and fluorine plasma processing at low temperature was found to be necessary to obtain high selectivity to low-k films without altering their properties.

REFERENCES

[1] W. W. Lee, P. S. Ho, MRS Bulletin,**22,** 19 (1997).
[2] N. P. Hacker, MRS Bulletin,**22,** 33 (1997).
[3] S. Bothra, M. Kellam, P. Garrou, J. Electron. Mater. **23**, 819(1994).
[4] L. Peters, Semicon. International, September 1998, pp. 64.
[5] R. S. List, A. Singh, A. Ralston, G. Dixit, MRS Bulletin **22**, 61 (1997)
[6] M. Boumerzoug, H. Xu, R. L. Bersin, Mat. Res. Soc. Symp. Proc. **495**, 345 (1998).
[7] H. Xu, M. Boumerzoug, R. L. Bersin, SEMI Symposium on Contamination-Free Manufacturing for Semiconductor Processing, F-1 (1998).
[8] R. L. Bersin, M. Boumerzoug, Q. Geng, I. Nakayama, H. Xu, Semiconductor Fabtech , **6**, 341(1996).
[9]. M. Z. Karim, D. R. Evans, Future Fab, **1**, 213 (1996).
[10] Kazuyoshi Ueno, Vincent M. Donnelly, and Takamara Kikkawa, J. Electrocehm. Soc. 144, 2565 (1997).

LOW K INTEGRATION ISSUES FOR 0.18 μM DEVICES

Rao V. Annapragada and Subhas Bothra
VLSI Technology, Inc.
1109, McKay Dr., MS 02, San Jose, CA 95131

The microelectronics industry has accumulated considerable experience using spin on polymers such as methyl silsesquioxane (MSQ) and hydrogen silsesquioxane (HSQ). The requirements for low k materials for 0.18 μm and beyond devices has resulted in renewed interest in these materials. These materials will be the most likely candidates for integration into the 0.18 μm devices considering the large accumulated knowledge in using these materials in the previous generations and also due to the fact that other purely organic materials and aerogels have far more formidable challenges to surmount before they can be successfully integrated. Vapor deposited low k material based on methyl silane and hydrogen peroxide chemistry can be considered similar to MSQ. In this paper process integration challenges involved in incorporating these spin on materials and the vapor deposited material into 0.18 μm devices in non etch back mode are presented.

INTRODUCTION

Enhancement of the cracking resistance of silicate SOGs is obtained by incorporating organic side groups such as -CH3. The resulting structure consists of single, double or triple chains with -O-Si-O- backbone rather than a large three dimensional network. These chains are loosely stacked together through weak Van der Waal's bonds providing the flexibility to glide and twist. Besides an improved cracking resistance, the loosely stacked chain structure results in a low density. Lower density combined with CH_3 group incorporation results in a low dielectric constant. The approximate empirical formula of MSQ is $CH_3SiO_{1.5}$.

Cracking resistance has been improved in hydrogen silsesquioxanes by incorporation of H to form a terminating -Si-H bond in a -O-Si-O backbone. This results in the molecules forming a chain or a two-dimensional layered structure. The density of the resulting material is lowered and dielectric constant of about 3.0 is obtained. The approximate empirical formula of HSQ is $HSiO_{1.5}$.

Low k vapor deposited material based on the methyl silane and hydrogen peroxide chemistry can be considered similar to the MSQ material. In the vapor deposited process methyl silane (CH_3SiH_3) and H_2O_2 are used as precursors to form CH_3SiO_x (x = 1.5 - 2.0). The dielectric constant of the material (k=2.8) is reduced due to the inclusion of the methyl groups in the oxide lattice as in SOG MSQ. In this process, the gaseous precursors react at the substrate surface to form a liquid which then polymerizes into the doped silicon dioxide [1]. The process has close resemblance to the sol gel process by which SOGs are deposited and hence may be termed vapor gel process. One of the advantages of the process is that the thickness of the film deposited can be quite high due to the solvent free operation, as much as 1.5 KÅ compared to about 4000 or 5000 Å for MSQ and HSQ. The larger thickness has the advantage of having thicker low k material over the metal lines after CMP of the cap oxide.

OXIDATION AND SHRINKAGE

In MSQ materials there is considerable shrinkage of the material (up to 25 %) resulting from the loss of mass when exposed to an oxygen plasma to remove the resist material after the via etch. The amount of shrinkage depends on the extent of diffusion of oxygen plasma species into the SOG exposed in the via. The amount of shrinkage increases with the amount of carbon incorporated. Both MSQ and vapor deposited low k material suffer from this problem. In the case of HSQ the shrinkage is considerably small (2-5%) as hydrogen is oxidized to water (Fig.1). The larger the shrinkage, the more severe is the via bowing and the step created in the SOG which results in poor barrier metal coverage in the via. One solution was to densify the exposed SOG in the via using RIE treatment [2]. The energetic bombardment from Ar or oxygen species densifies the initial 200 Å or so of the SOG preventing further diffusion of the plasma species into the interior of the exposed via which causes porous oxidation and moisture uptake. Outgassing of the moisture during the metal deposition step may prevent metal deposition(Figs.2 and 3). E-beam treatment of the exposed SOG in the via side wall also serve the same purpose.

However, the applicability of these techniques may be problematic in the sub 0.18 µm devices as the via aspect ratios become large. Unlanded vias are even more difficult

Fig.1. W filled via with good outgassing of HSQ prior to metal deposition. Via bowing in HSQ can also be seen.

to be plasma treated by this method. The aspect ratio will be even higher if the resist thickness increases as the treatment is done prior to stripping the resist and as the ions have to impinge on the nearly vertical surfaces to densify the film. The process window available is also rather small as the pressure has to be low and the ions should have sufficient energy to densify the film on the via side walls. If the conditions are not met a very substantial amount of the material will be oxidized. The problem is less severe in the HSQ films. Nevertheless, to improve the yield and reliability some kind of treatment of the SOG exposed in the via is preferred. If no via treatment is done there should at least be a degassing step of about 1 to 2 minutes at the temperature of the subsequent metal deposition step to drive out the moisture from the via resulting from the oxidation of the HSQ SOG in the via. However, there have been concerns about the thermal stability of the film at these temperatures and may result in some loss of the dielectric constant as the film loses hydrogen.

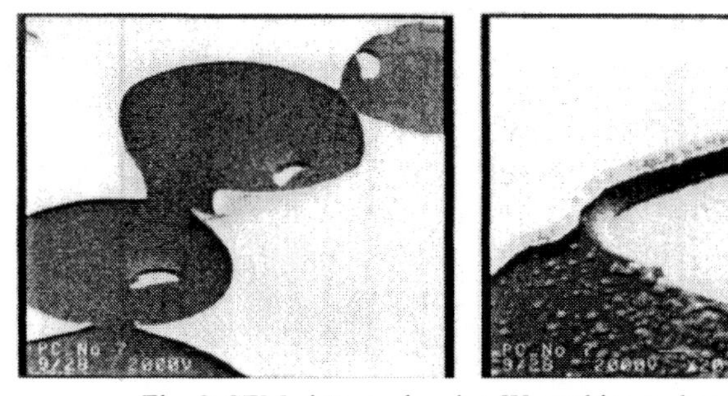

Fig. 2. SEM pictures showing W cracking and peeling due to moisture outgassing from HSQ via.

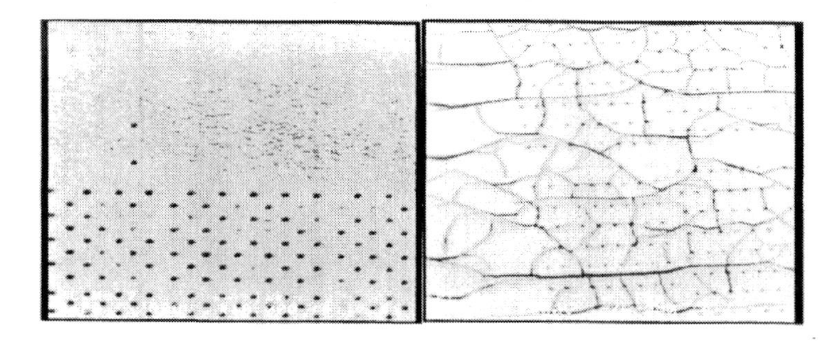

Fig 3. Cracks in the metal and powdery metal due to moisture outgassing from the vapor deposited low k material.

FORMING GAS BASED RESIST STRIPPING

Forming gas based resist stripping has been suggested as a method to alleviate the problem of oxidation of the organic SOGs [3]. As the resist is removed in a reducing ambient the organic groups are prevented from oxidation and resulting moisture uptake, alleviating the problem of via poisoning. The process seems to be quite promising. However, one of the challenges is to keep out oxygen as oxygen leaking into the system may result in considerable amount of oxidation of the low k material. In addition, the resist removal rate is low at about 0.25 um/min resulting in a low throughput process.

WET CLEAN TO REMOVE THE POST ETCH POLYMER RESIDUE

Another challenge for the HSQ films is that the hydroxylamine based wet clean to remove post etch plasma residue attacks the material oxidizing and hydroxylating the material. From our experiments on blank wafers coated with HSQ material dipped in the hydroxylamine bath at 70 C for 20 min resulted in complete loss of the hydrogen and large increase in the OH groups (Fig. 4). It will be a considerable challenge to develop a suitable wet chemistry that would effectively remove the etch residue without attacking the HSQ material. Significant levels of Al and Ti incorporated in the polymer residue makes the removal of these residues very difficult. Fortunately, the clean solvent does not attack the MSQ and vapor deposited low k materials probably due to the poor wetting characteristics of these carbon containing materials. Recent developments involving NF_3 and O_2/NF_3 plasmas to remove the etch residue may be suitable for the HSQ materials [8]. However, there have been no reports on the interaction of these dry cleans with HSQ.

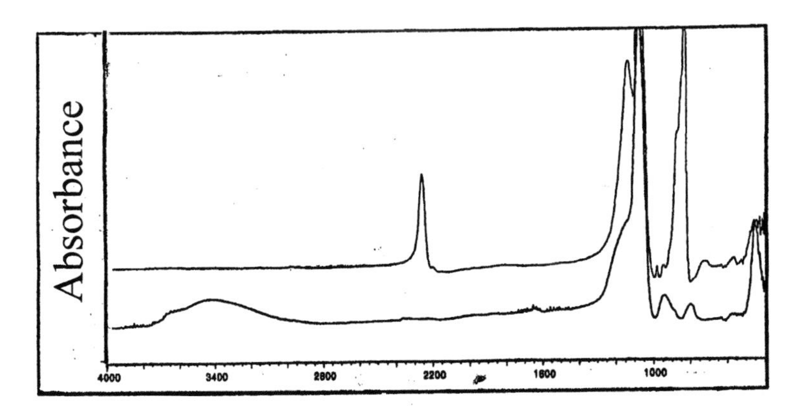

Fig. 4. HSQ material showing complete loss of hydrogen when dipped in hydroxylamine clean bath.

ADHESION OF THE CAP OXIDE TO THE LOW K MATERIAL

The presence of methyl groups in MSQ and low k vapor deposited films makes these films hydrophobic as it is difficult to form SiOH bonds by breaking the Si-CH_3 bonds. While it is advantageous to have a surface that is hydrophobic, it may cause adhesion problems while depositing a PECVD cap layer for the same reason that it is difficult to form bridging bonds between the two layers due to the presence of methyl groups. It may be necessary to modify the surface of these materials prior to depositing the cap layer. In the plasma treatments it is important to convert the top thin layer to a dense oxide using oxygen ion bombardment. The top dense layer would then prevent the further diffusion of the oxygen species to the bulk of the material. Without densifying the top layer the entire thickness of the material would be oxidized. Parallel plate reactors are suitable for this densification, as the bias voltages generated across the wafer causes the ions to accelerate in the sheath imparting it the kinetic energy necessary for densification. However, it is important to select the process parameters carefully to avoid the complete oxidation of the film. The advantage of doing plasma treatments is that the treatment can be done in situ in a parallel plate reactor prior to deposition of the cap oxide.

Some of the other methods to improve the adhesion include e-beam curing and ion implantation both of which effectively modify the surface of the film. However, both these methods increase the process complexity due to the additional step involved.

In the case of HSQ, it is relatively easy to break the hydrogen bonds and during the cap oxide deposition some of the Si-H bonds are decomposed forming a transition oxide layer [4]. The formation of a transition oxide layer is beneficial from the point of view of the adhesion. The properties of the transition oxide layer, however, depend on the temperature of deposition and the source gases used. It was shown that lower temperatures and SiH_4-N_2O chemistry are favorable to the high temperature TEOS cap, as the later would consume more hydrogen from the HSQ material increasing its dielectric constant. However, it is not clear what the optimum conditions are from the point of view of adhesion. The adhesion of the cap layer to the HSQ is still a cause for concern. We have noticed quite a few instances when the cap layer was found to be peeling during CMP of the cap layer.

In the case of MSQ, the adhesion of the material to the liner oxide is also of concern. During ashing in oxygen plasma, the force generated due to the shrinkage of the material in the via may be sufficient to delaminate the material from the liner oxide. Attempts to improve adhesion by using adhesion promoters did not seem to have significant effect.

DIRECT DEPOSITION ON METAL

As the interconnect line and space measurements shrink it is important to have as much of low k material as possible between the metal spaces. However, a base layer is deposited prior to the low k layer as the base layer would offer protection as a barrier layer and as a protection layer to the underlying metal during plasma etch back. However, in non etch back mode the primary purpose of the base layer would be to act as a moisture diffusion barrier. Putting a thick PECVD oxide as a base layer would reduce the effective dielectric constant. There have been reports of successfully integrating the low k materials with base layers as thin as 500 Å. There have also been reports of putting the HSQ directly on the metals [7]. However, to ensure that the devices are protected against moisture diffusion it is preferred to put at least 500 Å thick base oxide.

E-BEAM TREATMENT AND ION IMPLANTATION

To integrate organic SOG materials in non etch back mode, e-beam treatment [5] and ion implantation [6] have been implemented to avoid the problem of via poisoning. In both these methods the entire thickness above the metal layers were converted and densified, so that there are no organic groups remaining above the metal layer. Lack of organic groups results in the prevention of via poisoning. However, as the low k material is completely converted, the intermetal dielectric constant is close to that of SiO2 and results in reduction of only the intra metal dielectric constant. Moreover, as the material converted and densified over large metal areas and over widely spaced metal layers is same, the necessity to convert the higher thickness over the large metal areas results in complete conversion of the material in wide spaces. This results in the loss of low dielectric constant in wide metal spaces and some reduction in the narrow metal spaces. It is also important to carefully optimize the e-beam curing process to avoid any possible gate oxide damage and Vt shifts.

HARD MASK TO PREVENT OXYGEN PLASMA ASHING DAMAGE

As the low k materials exposed in the via are damaged due to oxygen plasma ashing, a hard mask approach can be used. The vias are initially etched into a nitride or other suitable hard mask initially and the resist stripped using standard oxygen plasma ashing. The hard mask then can be used to etch vias into the low k material. In this approach, the low k material is not exposed directly to the oxygen plasma environment and the oxidative damage is minimized. This approach is particularly well suited for organic materials, as the oxidative damage is more significant in these materials. As the hydroxylamine solvents are not an issue, there is no need for using new chemistries for post etch polymer removal. As the possible delamination of the MSQ material is a consequence of the shrinkage resulting from the oxidation, the problem is also addressed by this approach. In case of HSQ, non hydroxylamine solvents are still required to prevent oxidative damage.

CONCLUSIONS

MSQ SOG material has the advantage of higher thermal stability compared to HSQ but has poor oxygen plasma resistance and may cause serious via poisoning problems without proper via treatment. The vapor deposited low k material has the advantages of solvent free operation, deposition in a vapor deposition equipment, integrated processing capability to deposit the base oxide, low k material and the cap layer in one cluster tool. The solvent free operation helps in the deposition of thick (up to 1.5 μm) low k material. However, the poor oxygen plasma resistance due to the methyl groups as in MSQ SOG (though the organic content is smaller) may cause via poisoning problems. HSQ material on the other hand has lower thermal stability and better oxygen plasma resistance. Resistance to hydroxylamine post etch residue clean is poor. Considering these factors it can be concluded that all the three materials considered need to be integrated carefully by optimizing the process steps involved.

REFERENCES

1. S. McClatchie et al., 3rd International DUMIC, Feb. 10-11, 1997.
2. Masafumi Miyamoto et al., 14th International VMIC, June 10-12, 1997.
3. Duong Nguyen et al., 13th International VMIC, June 18-20, 1996.
4. Ishita Goswami et al., 14th International VMIC, June 10-12, 1997.
5. L. Forester et al., 13th International VMIC, June 18-20, 1996.
6. Ching-Hsing Hsieh et al., 14th International VMIC, June 10-12, 1997.
7. V. Mcgahay et al., 13th International VMIC, June 18-20, 1996.
8. Jimmy Martin et al., 14th International VMIC, June 10-12, 1997.

PULSED ELECTRODEPOSITION OF COPPER METALLIZATION FROM AN ALKALINE BATH

A. Krishnamoorthy, C.Y. Lee, D.J. Duquette, and S.P. Murarka,
Materials Research Center, Rensselaer Polytechnic Institute, Troy, NY 12180

ABSTRACT

This paper describes the role of pulse parameters and bath composition on the electrochemical deposition of copper as a metallization technique for the fabrication of multilevel interconnects. The electrochemical behavior of an alkaline solution of varying composition was studied using potentiodynamic polarization experiments. A potential range corresponding to the desired current density was thus selected. Pulse plating experiments were performed at various forward and forward/reverse pulse cycles. Cross sections of the patterned wafers were examined using SEM. Electrodeposition using forward/reverse pulse cycles on the order of 1 ms provided excellent feature filling. These results provide a basis for further work for plating on wafers with smaller feature size and higher aspect ratios.

INTRODUCTION

Continued advances in ULSI technology have necessitated the search for new interconnection materials to replace aluminum. Recently, copper metallization in multilevel interconnect systems has been successfully demonstrated (1,2). Copper's lower resistivity, together with a switch to low-κ dielectrics, will reduce the RC delay that is of rising concern with smaller, more densely packed metal lines. These high aspect ratio trenches require a deposition method whereby they can be filled. Electroplating has been used in the manufacturing of printed circuit boards (3) and is now considered to be the leading method for copper metallization. Electroplating offers the advantages of high copper film quality and excellent feature filling capability, as well as low cost and high throughput. However, there are numerous practical issues that accompany the use of acidic copper baths, such as the need for brighteners and levellers, and safety and health concerns. As an alternative to acid solutions, the present work considers the electrodeposition of copper from a simple alkaline bath.

Pulse plating involves the periodic on/off switching of a potential (or current). Fig. 1 shows the application of short and long pulses at various potentials and on/off times (henceforth denoted t_{on}/t_{off}). Switching off the pulse allows the metal-solution interface to trace back to its steady-state potential. This in turn relaxes the concentration gradient. A reverse pulse of duration t_{rev} may be applied either in addition to or instead of t_{off}. During a combination forward//reverse pulse (denoted $t_{on}//t_{rev}$), the plating period is followed by a period in which metal ions are dissolved into solution. Some of the benefits reported from pulse plating include increased plating rate, higher purity, reduced porosity, and fine-grained deposits (4).

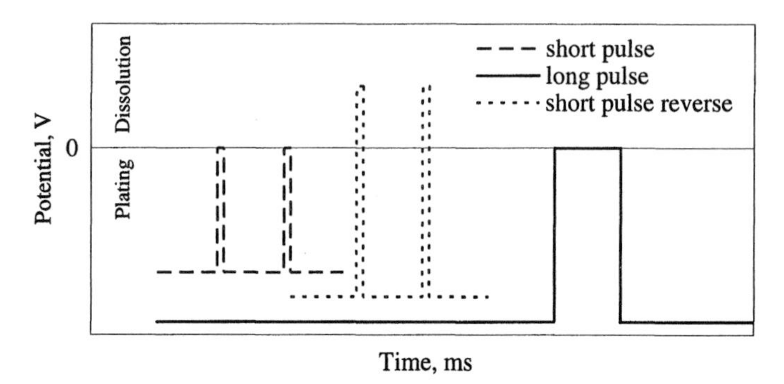

Fig. 1 : Schematic representation of pulse cycles

EXPERIMENTAL PROCEDURE

The electrochemical response of the copper in alkaline plating solution was observed by use of potentiodynamic experiments. The experimental apparatus for the polarization experiments included a saturated calomel reference electrode, a platinum counter electrode, and a Pine™ rotating disk electrode. A cylindrical copper rod (99.99% purity) of diameter 1.2 cm^2 functioned as the working electrode. The rod was covered with epoxy, polished on its exposed face, and fixed to the rotating disk electrode. The concentration of copper in the test solution was varied from 0.04M to 0.12M to determine the effect of bath composition on the polarization response. The solution also contained ammonium sulfate as a supporting electrolyte and ammonium hydroxide and Ethylene Diamine (ED) as complexing agents. The polarization experiments were initiated after a 1 minute soak at the open circuit potential (OCP) and the polarization curves were recorded utilizing a Gamry™ CMS 100 potentiostat at a scan rate of 2 mV/s. From these measurements, the desired potential range for optimum plating was selected.

Potentiostatic plating was performed on sections of both blanket and patterned silicon wafers. The Si wafers were p-type device quality wafers subjected to wet oxidation at 1050°C for 1.25 hours to develop 0.7 μm of oxide. A copper seed layer was sputtered at a base pressure of 10^{-7} Torr and an argon pressure of 5 mTorr. The sputtered layers were 30 nm thick and exhibited a resistivity of 2.1 μΩ·cm. The patterned wafers had a 0.5 μm minimum feature size with a 2:1 aspect ratio. The electroplating experiments were performed with a Dynatronix™ power supply. The wafer sections were clipped onto the rotating disk electrode and plated in solution containing a copper-phosphorous anode.

Plating in only the forward direction was performed at various pulse cycles ranging from 10 to 1000 Hz at an applied bias of -750 mV vs SCE. For plating with both forward and reverse pulses, a reverse potential of -100 mV vs SCE was established. On the blanket wafer specimens, the resistivity of the deposited film was computed from the sheet resistance and thickness. To investigate the effect of annealing on resistivity, the wafers were subjected to 30 min. anneal at 250°C. The patterned specimens were observed using a scanning electron microscope to determine step coverage and deposit quality.

RESULTS AND DISCUSSION

Selection of Bath Composition from Polarization Curves

The polarization curves in Fig. 2 show the effect of the addition of ammonium hydroxide on a solution containing 0.08M copper and 0.08M ammonium sulfate. With no ammonium hydroxide in solution, no plating was observed. When $CuSO_4$ is exposed to ammonia, cupric hydroxide, forms and precipitates. With an excess of ammonia, the hydroxide redissolves, forming an amino complex, e.g., $[Cu(NH_3)_4]^{2+}$ (4) . This complex formation led to a decrease in the open circuit potential of 300mV (SCE) and plating occurred upon cathodic polarization. Presence of small amounts of ethylene diamine (ED) resulted in smoother and brighter deposits.

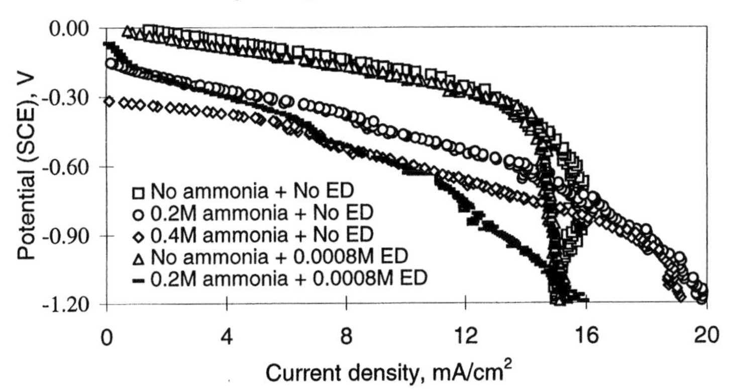

Fig. 2 : The influence of ammonia and ethylene diamine (ED) on electrochemical behavior of copper

Fig. 3 shows the polarization response of the copper rod in solutions containing 0.04, 0.08, and 0.12 M copper. A concentration of 0.04M copper showed a low current density not conducive to effective plating. The solution containing 0.12M copper yielded a higher current density and thus a higher plating rate, but resulted in a rough and powdery deposit. A concentration of 0.08M copper was found to provide a better quality deposit and a reasonable current density over a range of potentials; thus, this solution was selected for subsequent plating experiments.

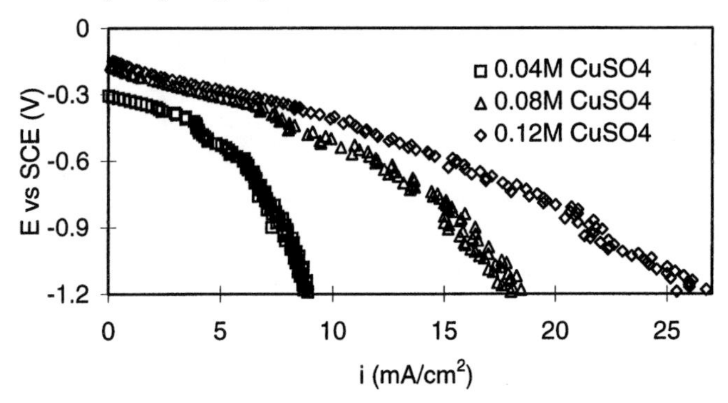

Fig. 3 : The influence of copper concentration on cathodic polarization.

Hydrogen Evolution

Copper electroplating involves the reduction of copper from the +2 state to the 0 state. While this reduction is occurring, it is also possible for other species present in the test solution to become reduced, thereby lowering the plating efficiency. To examine plating efficiencies, cathodic polarization experiments were performed in test solutions without copper for three different base materials: stainless steel, copper, and platinum. By using different working electrodes, each with a different affinity for hydrogen evolution, it is possible to determine whether or not hydrogen is evolved at the copper surface. In this way, the potential range where hydrogen evolution becomes possible can be avoided.

The polarization curves shown in Fig. 4 were obtained in a test solution containing no copper. 'Region A' corresponds to hydrogen evolution arising from the reduction of water and 'Region B' corresponds to the reduction of oxygen (the test solution was not deaerated). The current due to oxygen reduction was on the order of 10^{-5} mA/cm^2 while that of hydrogen evolution was on the order of 10 mA/cm^2. Water is stable above a potential of approximately -1.05V (SCE) and hydrogen cannot be liberated on the copper rod above this potential. Point C is the steady state potential of copper in test solution (-0.16V vs SCE); above this potential, copper dissolves in ammoniacal solutions and forms $[Cu(NH_3)_2]^+$ complex (5).

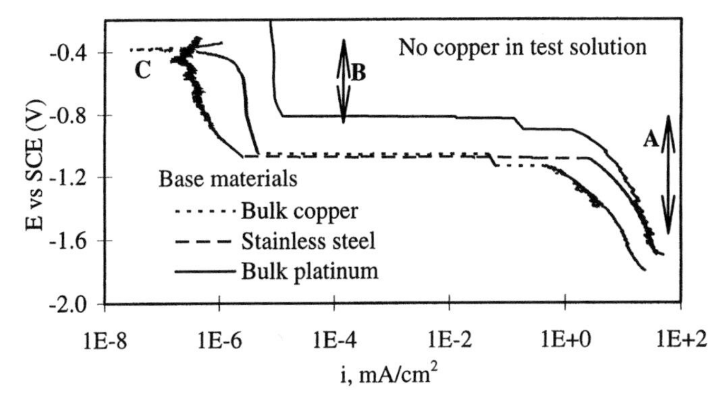

Fig. 4 : Cathodic polarization of S.S, Cu, and Pt electrodes in a solution containing 0.15M $(NH_4)_2SO_4$, 0.2M NH_3, 0.004M $C_2H_8N_2$ and no $CuSO_4$

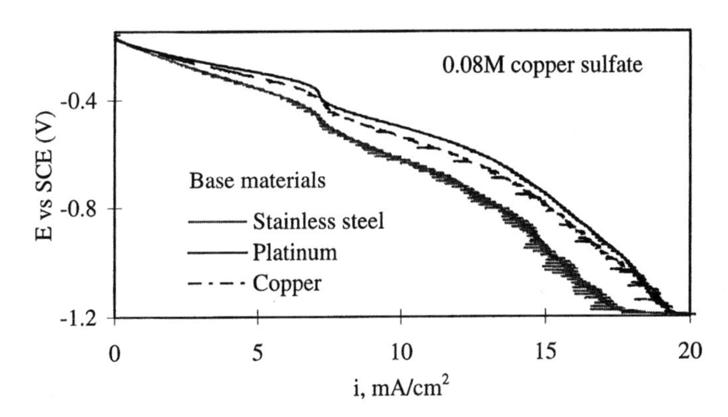

Fig. 5 : Cathodic polarization of stainless steel, copper, and platinum electrodes in a solution containing 0.15M $(NH_4)_2SO_4$, 0.2M NH_3, 0.004M $C_2H_8N_2$ and 0.08M $CuSO_4$. pH = 9.5.

In a solution containing 0.08M copper in addition to the supporting electrolyte and complexing agent, all three substrates showed similar behavior, represented in Fig. 5. Although hydrogen evolution was possible at potentials less noble than -1.05V (SCE), the reduction from copper complex ions to copper was predominant and most of the current was due to this reaction. By carrying out cathodic deposition at fixed potentials in the range -350 to -750 mV (SCE), the charge associated with reduction/deposition was determined. The weight values calculated on the basis of charge and weight gain were almost equal, indicating that the current due to hydrogen evolution was insignificant.

Safe Range of Potential for Electrodeposition

Fig. 6 shows different potentials of interest to electrodeposition. The range of potential over which water is stable was taken from the Pourbaix diagram of copper in water (6). This range was wider for copper-ammonia system because of a shift in hydrogen evolution potential to a higher negative value. The potential at which hydrogen evolution began in the ammoniacal solution was 0.27 mV more negative than in water. The potential region in which plating without hydrogen evolution occurred was from -0.2 to -1V (SCE).

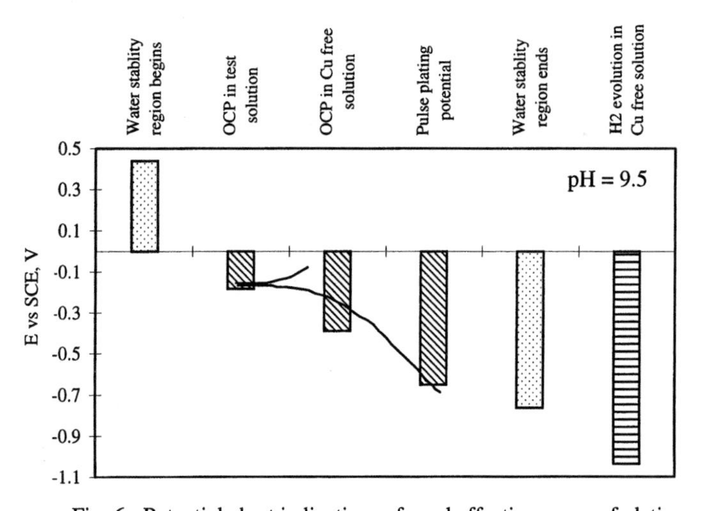

Fig. 6 : Potential chart indicating safe and effective range of plating

Plating Rate Determination

On the basis of the cathodic polarization response of the copper rod in various test solutions, a plating bath composition [0.08M $CuSO_4$, 0.15M $(NH_4)_2SO_4$, 0.2M NH_3, and 0.004M $C_2H_8N_2$; pH = 9.5] for pulse plating experiments was selected. The plating rate was determined at various potentials in the DC mode and in the potential pulse mode (using a 90/10 ms on/off cycle), see Fig. 7. In the potential range of -400 to -700 mV (SCE), plating rates were higher during pulse plating than during the DC condition. However, the curves merge at more negative voltages, indicating that both then proceeded at approximately the same rate.

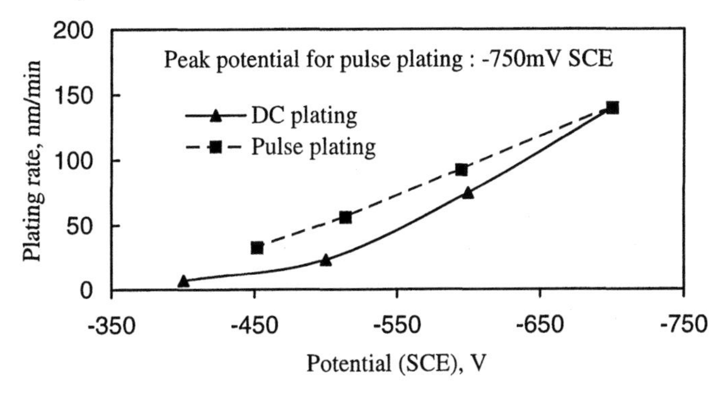

Fig. 7 : Plating rate as a function of potential in DC and pulse plating

Plating Results

The resistivity values of the deposits before annealing ranged from 2.2-2.5 μΩ·cm as shown in Figure 8. A 5% reduction in resistivity was observed upon annealing.

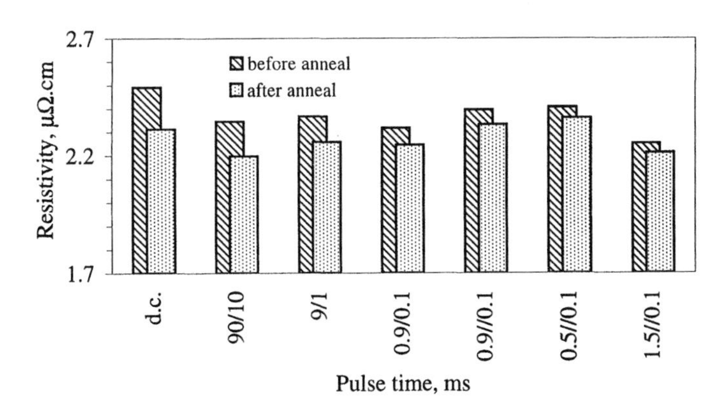

Fig. 8 : Resistivity chart of copper deposits

Figure 9a shows typical SEM cross-sections of the pulse-plated specimens. The complete void-free filling of features required a thorough optimization of the pulse parameters. By operating in forward pulse mode with various t_{on} / t_{off}, the concentration gradient that resulted due to deposition could be relaxed. However, this relaxation was not sufficient to yield a void-free fill of trenches. At pulse on/off cycles of 90/10, 9/1, 0.9/0.1 ms at -750 mV (SCE), the trenches were not completely filled and the voids extended upwards. However, the size of the voids was observed to diminish when the pulse cycle was on the order of 1 ms. This suggested the need for dissolution that could supply cations at the trench bottom and hence pulse reverse plating was attempted.

It is well known that raised features are dissolved in electrolytes at a higher rate than recessed regions such as inside a trench (7-10). By applying pulse potential waveforms with reverse cycling, the excess deposition at the mouth of the trench was dissolved while that at the bottom was relatively unaffected. This also allowed the replenishing of copper ions from solution to the depleted interface, and successive pulses allowed the trenches to be filled. Figure 9b shows the cross-sections of specimens plated under forward/reverse pulse conditions 0.5//0.1, 1.5//0.1 and 0.9//0.1 ms at -750//-100mV(SCE). The 0.5μm features were completely filled under the condition 0.9//0.1 ms. By suitably fixing the forward and reverse potential and pulse times, it was possible to achieve void-free feature filling.

On the basis of these results, pulse reverse plating of trenches of smaller feature size and higher aspect ratios is being examined.

CONCLUSIONS

The use of pulse plating in the electrodeposition of copper from an alkaline solution has been examined. Cathodic polarization curves demonstrate the effect of solution chemistry (e.g. Cu^{2+} concentration, pH) and potential ranges on plating. Pulse reverse plating provided successful gapfill capability for 0.5 μm, 2:1 aspect ratio trenches when pulse cycle was on the order of 1 ms. Further work is being pursued to study the effectiveness of the deposition parameters on features of smaller size and higher aspect ratio.

ACKNOWLEDGMENTS

The authors wish to thank Semitool, Inc. (Kalispell, Montana), especially Dr. Henry Stevens.

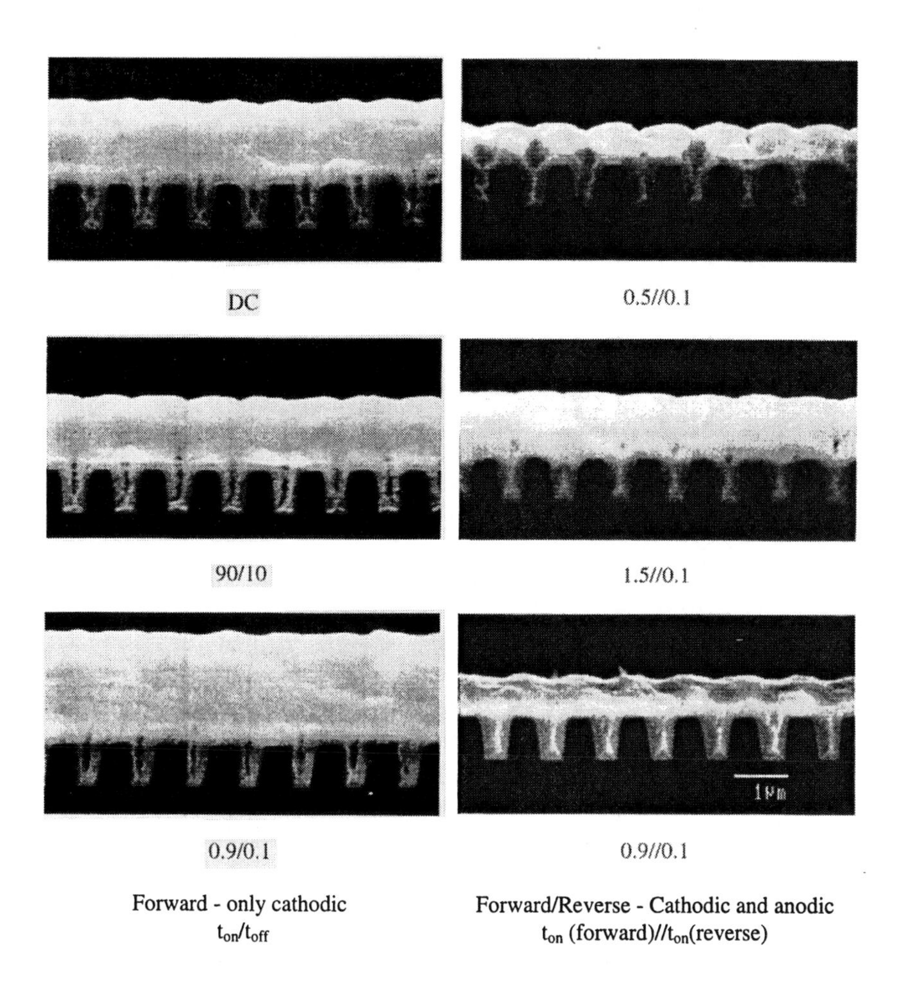

Fig. 9 : Trench filling by pulse plating as a function of pulse cycle time

REFERENCES

1. D. Edelstein, et al., *Proc. IEEE IEDM*, 773 (1997).
2. V.M. Dubin, C.H. Ting, and R. Cheung, in Proc. International VLSI Multilevel Interconnect Conference, VMIC Catalog No. 97 IMSIC-107, p.69 (1997).
3. E. Yung and L. Romankiw, *J. Electrochem. Soc.,* **136**, 756 (1989).
4. G. Devaraj, S.K. Seshadri, *Plating & Surface Finishing*, 72-78 (1992).
5. G.D. Parkes, *Mellor's Modern Inorganic Chemistry*, John Wiley & Sons, NY (1967) p. 650.
6. M. Pourbaix, *Atlas of electrochemical equilibria in aqueous solutions*, p. 384, National Association of Corrosion Engineers, Houston, Texas (1974).
7. R.J. Contolini, A.F. Bernhardt and S.T. Mayer, *J. Electrochem. Soc.,* **141**, 2503 (1994).
8. S.H. Glarum, J.H. Marshall, *J. Electrochem. Soc.*, **132**, 2872 (1985).
9. A.C. West, C.C. Cheng and B.C. Baker, *J. Electrochem. Soc.,* **145**, (9) 3070 - 3074 (1998).
10. R.J. Contolini, S.T. Mayer, R.T. Graff, L. Tarte, and A.F. Barnhardt, *Solid State Technology*, **40**, 155 (1997).

Chemical Mechanical Polishing of Copper Using Silica Slurry

Seiichi Kondo, Noriyuki Sakuma, Yoshio Homma and *Naofumi Ohashi

Central Research Laboratory. Hitachi, Ltd.
1-280 Higashi-Koigakubo, Kokubunji-shi, Tokyo 185-8601, Japan
* Device Development Center, Hitachi, Ltd.
6-16-3 Shin-machi Oume-shi, Tokyo 198-8512, Japan
e-mail; condor@crl.hitachi.co.jp

ABSTRACT

A new Cu-CMP (copper chemical mechanical polishing) process is proposed by clarifying the mechanism of the surface reaction between Cu film and the slurry. The proposed CMP mechanism consists of Cu surface oxidation, oxidized layer protection by an inhibitor, polishing of the protection layer on top of protrusions by fine abrasive, and etching of the oxide by an acidic media. A silica-abrasive-based slurry is proposed according to the model. The slurry contains silica abrasive, citric acid, hydrogen peroxide and benzotriazole. It is shown that a large rate-ratio of the CMP removal to etching is required for reducing the Cu dishing, and the silica abrasive is effective to reduce scratches. Practically large CMP rate of more than 150 nm/min is obtained by the slurry. This new slurry is very promising for improving both yields and performance of the damascene Cu interconnect process.

INTRODUCTION

Copper (Cu) metallization has been investigated for a multilevel interconnection of the next generation ultralarge scale integrated circuits (ULSIs) because of its low resistivity and high electromigration performance. Cu-CMP (chemical mechanical polishing) is one of the most important techniques for damascene interconnect process.

The performance of Cu-CMP, however, has not been satisfactory, since many scratches, large dishing [1] and easy-to-corrode characteristics are observed when conventional slurries are used. It is presumed that the scratches on dielectric layer surface as well as on Cu surface is due to hard alumina abrasive commonly contained in conventional slurries. In addition, the alumina abrasive makes the post-CMP cleaning very difficult, and the patterned wafers were not thoroughly cleaned by an ordinary post-CMP cleaning process. Dishing is considered to be strongly dependent on chemical characteristics and abrasive particle size of the slurries.

A silica-abrasive-based slurry using glycine and hydrogen peroxide (H_2O_2) was proposed by Hirabayashi *et al.* to reduce Cu dishing [2]. The glycine was added in order to form Cu-glycine chelate complex that is soluble in water. It was effective to reduce Cu dishing, however Cu-CMP rate was rather small.

Thus, new slurries are required to improve the CMP performance with a large removal rate. We have clarified the mechanism of Cu-CMP process, and propose a surface-reaction-controlled, silica-abrasive-based slurry. The performance of the Cu-CMP process using the new slurry is also shown.

EXPERIMENTAL

CMP conditions

A dead weight type CMP machine was used. A wafer carrier with a backing pad (Rodel R201-80J) to support wafers is rotated with almost the same rotation speed of the platen. A foamed poly-urethane, hard polishing pad (Rodel IC1000) with 15 mm pitch lattice groove is used. The relative speed between the wafer center and the platen and the down force are 34 m/min and 220 g/cm^2, respectively. The slurry is supplied to the platen center at a rate of about 50 cc/min. When CMP is finished, deionized water is supplied at a rate of approximately 3000 cc/min for 15 seconds in order to rinse the polished wafer surface.

Sample preparation

The polishing characteristics of the slurries were examined using 5" size wafers on which blanket Cu and/or TiN films are formed. The blanket Cu film samples were prepared by sputter-depositing TiN film of 50 nm thickness and a Cu film of thickness 800 nm continuously, on 200 nm thick SiO_2 films (PECVD films using tetraethoxy-silane; p-TEOS).

The Cu metal lines were formed by the damascene process using test device wafers with interconnect test patterns. A silicon nitride (SiN) film of 200 nm and a 500 nm thick p-TEOS film were formed on a silicon substrate, and grooves for interconnects were fabricated. After depositing a TiN layer of 50 nm thick as a barrier layer, a Cu film of 800 nm thick was continuously deposited by a long-throw sputtering. Vacuum heat annealing (reflow process) was also performed at 430°C for 3 min in the sputtering machine for Cu filling.

CMP characteristics evaluation

Two types of Cu slurries were evaluated. One is a commercially available slurry using alumina abrasive and H_2O_2 as an oxidizer. The other is a new slurry consisting of silica abrasive, oxidizer, acid media and inhibitor. For abrasive, a neutral silica dispersive solution (Cabot Cab-O-Sperse SCE; 15 wt% silica, pH 5) is adopted. For oxidizer, commercially available 30% aqueous solution of H_2O_2 is used. For acid media, citric acid is used as a representative of chelating agent for Cu ion, and BTA is used for inhibitor.

The CMP rate was evaluated by measuring the sheet resistance change of the Cu film before polishing and after polishing for 2 min. The etching rate was also measured by the sheet resistance change of the Cu film before and after etching for 10 min dipping in the slurry. The sheet resistance was measured using a four probe method (Napson RG-7).

Change of the CMP rate, etching rate and chemical characteristics of the slurries were measured by optimizing the composition of the chemicals. Then, the influences of the ratio of CMP rate to etching rate of the slurry on the CMP performance were evaluated using the patterned wafers. The corrosion and scratches on the polished surface of Cu was evaluated by visual observation under a floodlight or an optical microscope. The detailed observation was performed using scanning electron microscope (SEM, Hitachi S-900). Abrasive size of both alumina and silica was measured using a particle size distribution analyzer (CPS Disc Centrifuge) and SEM.

RESULTS AND DISCUSSION

CMP mechanism of the new slurry

A CMP model for Cu has been proposed by Hirabayashi, *et al.* [2], which is modified from the model for tungsten [3]. It consists of oxidation and passivation of Cu by H_2O_2, polishing of oxide layer by silica abrasive, and change to water-soluble Cu-glycine chelate complex. The pH-potential value of the slurry is not inside of the corrosion domain of Cu but inside of the CuO passivation domain, that is indicated in Pourvaix diagram [4]. The CMP rate was very sensitive to the H_2O_2 concentration, and the nominal CMP rate was around 40 nm/min or less.

The new slurry was prepared according to the following model. An oxide layer is formed on Cu surface by H_2O_2 and is protected by BTA as shown in Fig. 1. Then, the protection layer on protruded regions is removed by the mechanical abrasion and the oxide is dissolved by the citric acid. The exposed Cu surface is oxidized again, while the recessed regions remain protected. Since the silica abrasive is rather soft comparing with alumina abrasive, a reduced surface scratching is expected.

Optimization of the composition

Influence of composition of citric acid, H_2O_2 and BTA on the CMP characteristics was evaluated. Figure 2 shows the dependence of slurry characteristics on the amount of citric acid. In this experiment, 0.1 wt% BTA and 9 vol% H_2O_2 concentration were used. The CMP / etching rate ratio is also shown in the figure. By increasing the citric acid concentration, both the CMP rate and the etching rate increase linearly, while the CMP / etching rate ratio was about 4 and independent of citric acid concentration. Since the slurry containing 0.15 wt% citric acid shows a high CMP rate of about 240 nm/min, it was applied to fabricating the Cu damascene patterns. As a result, large Cu dishing occurred, i.e., most of Cu layer in the grooves disappeared as shown in Fig. 3. It is considered to be due to insufficient CMP rate / etching rate ratio.

Thus, dependence of the large CMP / etching rate ratio on the amount of BTA was evaluated as shown in Fig. 4. In this experiment, 0.15 wt% citric acid and 9 vol% H_2O_2

concentration were used. Both CMP rate and etching rate decrease according to the increase of BTA in the slurry. The CMP / etching rate ratio increases, however, due to the significant decrease of the etching rate. Slurries containing more than 0.3 wt% BTA show saturated CMP / etching rate ratio of about 25. In the case of 0.3 wt% BTA, a large CMP / etching rate ratio of around 25 was obtained, while maintaining a practically applicable CMP rate of around 80 nm/min. The CMP rate can be increased further by increasing the down force during the CMP. The adequate concentration of the BTA is, therefore, considered to be 0.3 wt%, to provide both a large CMP rate and CMP / etching rate ratio.

Figure 5 shows the CMP rate dependence of Cu and TiN on silica abrasive concentration. The TiN was used as a barrier metal. It is seen that practically large CMP rate for TiN as well as for Cu (more than 150 nm/min under the condition of down force 220 g/cm^2) is obtained when the slurry contains 1 wt% or higher concentration of the silica abrasive. Since the pH-potential values of the slurry are 2.9 and 450 mV respectively, the property of the slurry is inside the Cu corrosion domain of the Pourvaix diagram [4].

Interconnect pattern fabrication

Next, we polished a patterned wafer using the slurry containing 1 wt% silica abrasive, and it was compared with the patterns obtained using the conventional slurry. Figure 6 shows SEM images of Cu lines obtained by the new slurry. A very smooth, scratch-free, anticorrosive surface is obtained. It was also found that most of the silica abrasive was removed only by scrubbing with DI water, as seen in the same figure. In addition, no settling of the abrasive in the slurry was observed during the process. This improved settling characteristic also stabilized the CMP process.

Figure 7 shows a typical example of damascene Cu interconnects formed using a commercially available, alumina-abrasive-based slurry and a popular post-CMP cleaning process (brush-cleaning using diluted ammonium hydroxide). Numbers of scratches on both Cu and SiO_2 surface can be seen. In addition, numerous residue of alumina particles are also seen especially on/in Cu lines.

Figure 8 indicates a comparison of abrasive size distribution (agglomerated particle size) between the silica dispersion (SCE) and the conventional slurry. The silica abrasive size is almost a half of the alumina abrasive size. The smaller size of the silica abrasive is advantageous both in scratch reduction and less-settling characteristics.

CONCLUSIONS

A silica-abrasive-based slurry for Cu-CMP was proposed. The slurry consists of 1 wt% silica abrasives, 0.15 wt% citric acid as the etchant, 9 vol% H_2O_2 as the oxidizer and 0.3 wt% BTA as the inhibitor. The CMP process using the new slurry, with the no-settling characteristic, was shown to provide a very clean, scratch-free, anticorrosive polished surface. This new slurry is, thus, very promising for improving both yields and performance of the damascene Cu interconnects fabrication process.

Acknowledgments

We thank Mr. Owada Nobuo, Mrs. Hizuru Yamaguchi, Mr. Tatsuyuki Saito, Mr. Junji Noguchi, Mr. Toshinori Imai, Mr. Nezu Hiroki, Mr. Takeshi Kimura of Hitachi Device Development Center and Dr. Kenji Hinode, Mr. Ken-ichi Takeda and Mr. Masayuki Nagasawa of Hitachi Central Research Laboratory for their help in making samples and valuable discussions.

References

[1] J. M. Steigerwald, R. Zirpoli, S. P. Murarka, D. Price and R. J. Gutmann, J. Electrochem. Soc: 141, 2842 (1994).

[2] H. Hirabayashi, M. Higuchi, M. Kinoshita, H. Kaneko, N. Hayasaka, K. Mase and J. Oshima, *Proceedings of 1996 CMP-MIC Conference*, p. 119.

[3] F. B. Kaufman, D. B. Thompson, R. E. Broadie, M. A. Jaso, W. L. Guthrie, D. J. Pearson, and M. B. Small, J. Electrochem. Soc. 138, 3460 (1991).

[4] Pourbaix: "*Atlas of Electrochemical Equilibria in Aqueous Solution*," NACE, Houston, TX (1975) p. 387.

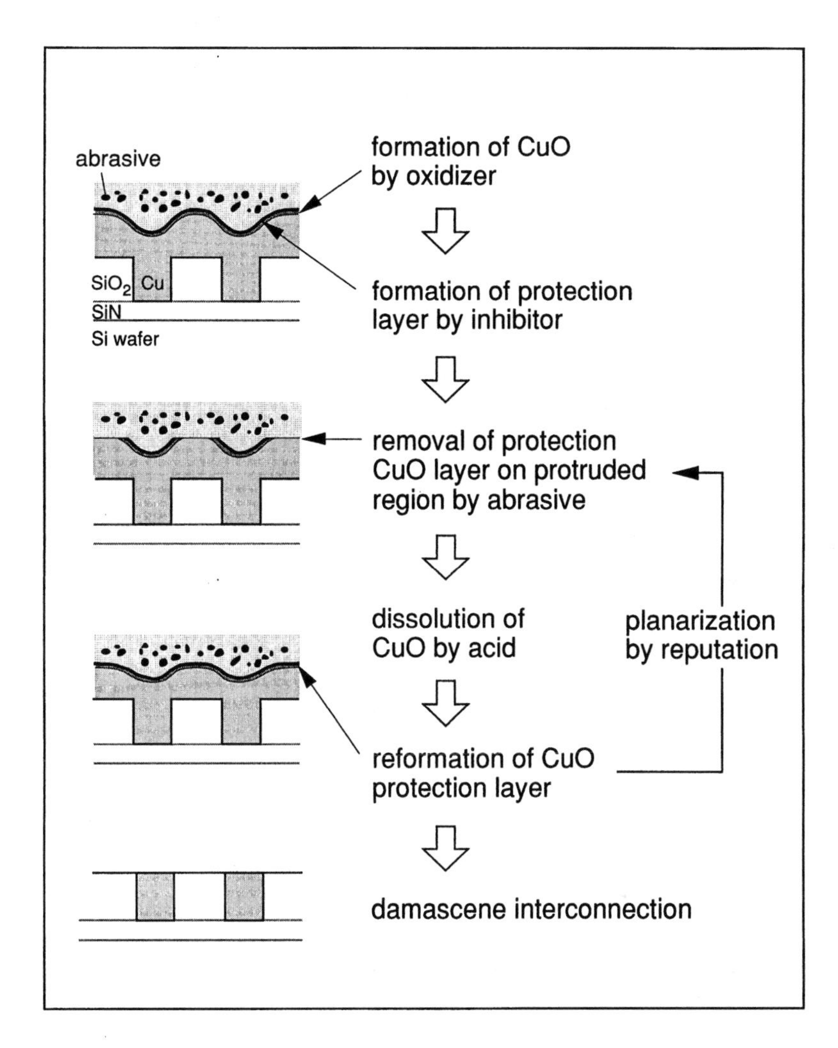

Fig. 1 The mechanism of Cu CMP

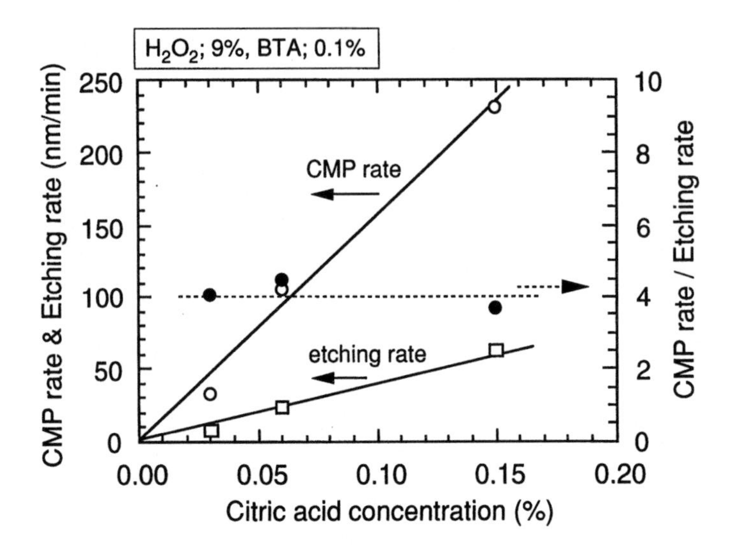

Fig. 2 Dependence of CMP rate and etching rate of Cu on citric acid concentration.

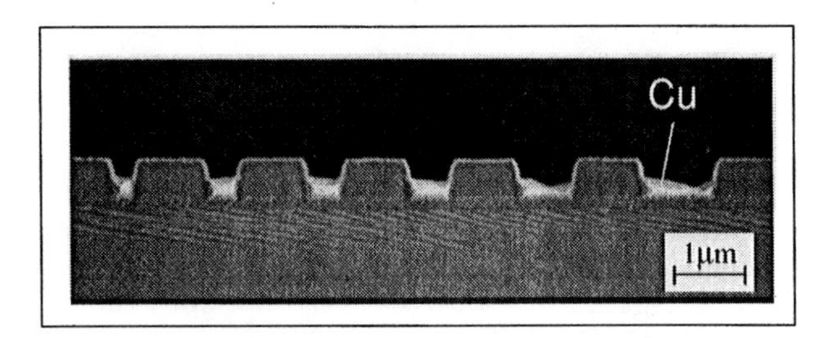

Fig. 3 An SEM image of Cu interconnects using slurry of CMP rate / etching rate ratio of 3.7.

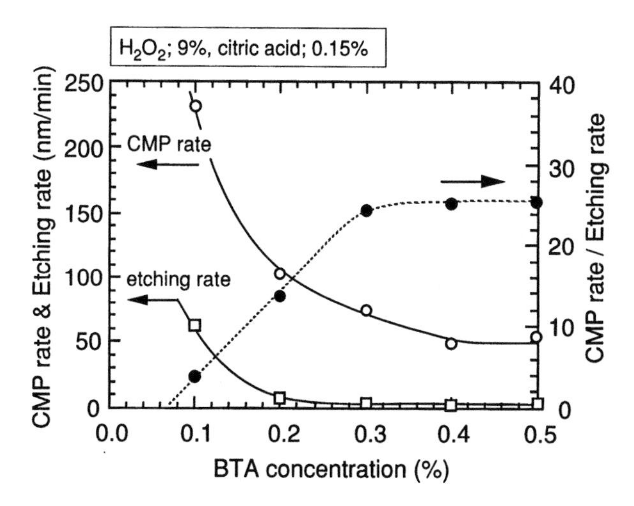

Fig. 4 Dependence of CMP rate and etching rate of Cu on BTA concentration.

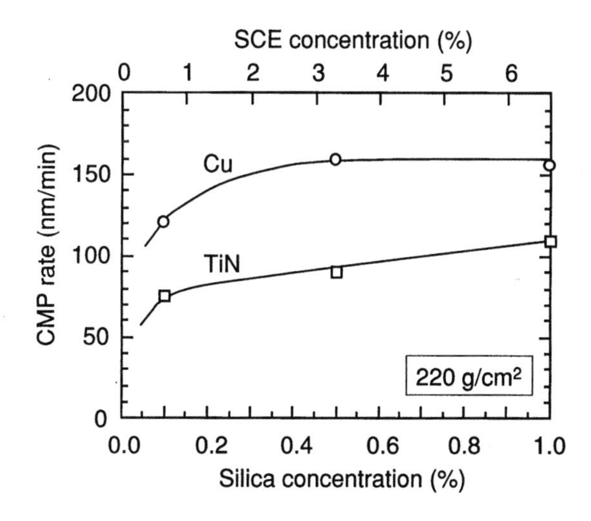

Fig. 5 Dependence of Cu and TiN CMP rate on silica abrasive concentration.

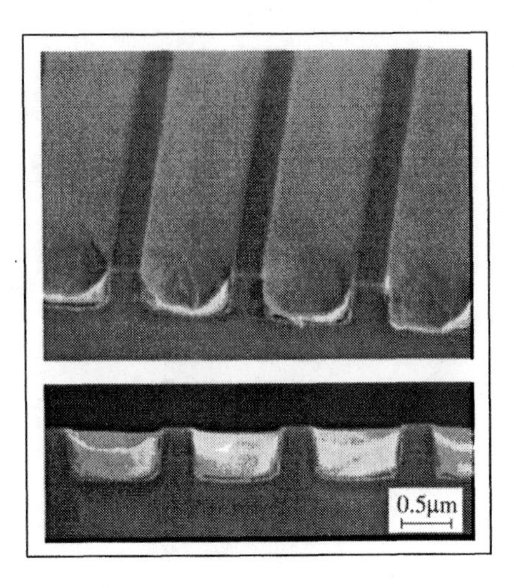

Fig. 6 Damascene Cu interconnects obtained using the silica slurry.

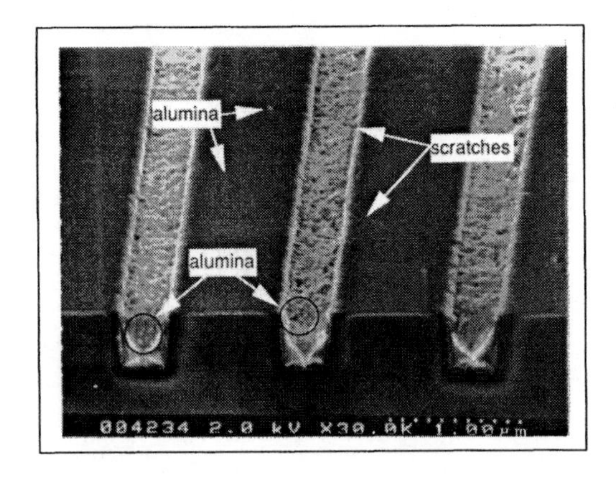

Fig. 7 Damascene Cu interconnects obtained using a conventional alumina slurry.

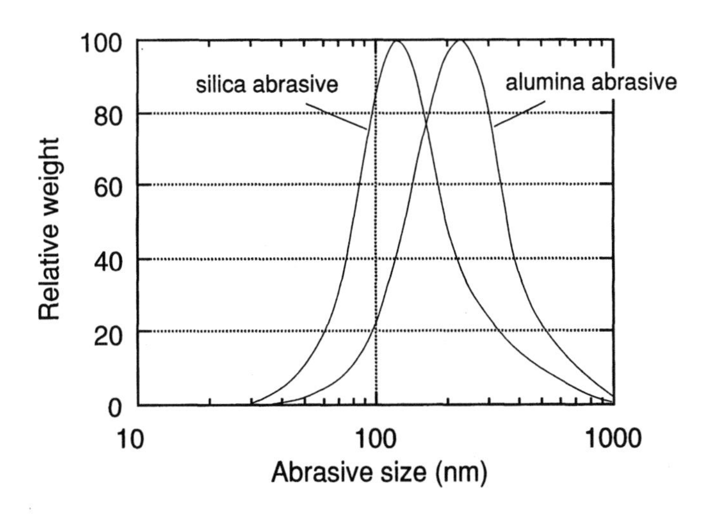

Fig. 8 Abrasive size distribution.

ELECTROCHEMICAL, CONTACT ANGLE, AND SPECTROSCOPIC CHARACTERIZATION OF METAL FILMS DURING COPPER DAMASCENE PROCESSING

Ronald A. Carpio, Steve Hymes, and Robert Mikkola
SEMATECH
2706 Montopolis Dr., Austin, TX 78741

Michael Pavlov
ECI Technology
1 Madison Street, East Rutherford, NJ 07073

ABSTRACT

Localized cathodic chronopotentiometric measurements were utilized for determining the thickness and type of oxide as well as for detecting the presence of reducible organic films on copper surfaces. It was found that Cu_2O is the principal oxide formed by CMP using hydrogen peroxide based slurries. The oxide thickness is generally in the 20 to 30 Å range after CMP, but this thickness is reduced to approximately half by a typical brush cleaning process. Anodic chronopotentiometry was utilized for Cu thickness measurements. Other DC and AC electrochemical measurements were performed to characterize the corrosion properties of processed copper and barrier metal film surfaces. The corrosion properties of copper were dependent upon the electrolyte utilized for the measurements. Sessile contact angle measurements were valuable for studying the spatial variation of the wetting properties of processed copper films. FTIR spectroscopy is shown to have the capability for monitoring the Cu_2O thickness.

INTRODUCTION

One of the key challenges of copper processing is to design and conduct the processing steps so that metal corrosion will not occur during or subsequent to that step. Achieving this goal requires that copper corrosion is understood and the appropriate monitoring tools are available for process control. Oxidation or more generally corrosion is a greater problem with Cu than Al interconnects since Cu does not form a self-passivating oxide layer on its surface. The copper surface oxides do not totally prevent reaction with atmospheric oxygen or other oxygen containing environments. In multi-level interconnect applications with the geometries of interest (CDs of 0.18 microns or less and film thicknesses of the order of 1.0 micron), loss of the metal film surface to corrosion cannot be tolerated. Many studies have been performed on copper corrosion and its prevention, but few with copper thin films [1-5] and none, known to the authors, in the area of copper damascene processing.

In this work the spatial variation of the corrosion properties of copper thin films are studied as a function of oxidant type and concentration in slightly basic and acidic electrolyte solutions. The corrosion properties of Cu are compared with metal films which are candidates for barrier applications. Of interest is the surface copper oxide layer which affords some degree of corrosion protection and whose characteristics, especially thickness, provide insight into the nature of the preceding process step. An important role is played by the trace surface organic films on Cu which originate from additives in plating baths and/or from additives in CMP slurries which function as corrosion inhibitors. Our interests lie in detecting the presence of these organics and measuring the coverage, and not in their identification. Aside from gaining more insight into copper corrosion, these studies provide insight into the uniformity of processing.

There is a growing need for process control techniques in semiconductor manufacturing which are sensitive, accurate, easy and rapid to employ, capable of small area measurements and which are cost effective. The principal approach employed here was to study blanket films after processing using electrochemical, contact angle, and infrared reflection spectroscopy. These techniques meet the requirements for process control tools. Space limitation prevents a detailed presentation of infrared reflection-absorption spectroscopy and its applications, but along with other optical techniques, it possesses the attraction of being noncontact.

The electrochemical techinques which were employed are well known techniques used for studying corrosion processes. In addition to studies of corrosion, the use of chronopotentiometry for Cu metal thickness measurement is covered.

From a processing standpoint, contact angle measurements can be used to assess the efficacy, uniformity, and reproducibility of the microelectronics fabrication process, whether it be electroplating, CMP, or post CMP cleans, for example. Static liquid drop measurements have been used extensively by others for studying copper corrosion inhibitors [1-4]. Newly developed image based instrumentation, which is described in two recent publications [6,7], were utilized for automated sessile and dynamic contact angle measurements on full wafers. This new instrumentation permits maps of the contact angle on multiple sites of the wafer surface to be quickly and accurately measured. It was shown by Thomas et al. [8] that the receding contact angle obtained from dynamic contact angle measurements performed by the Wilhelmy slide technique correlated with the corrosion rate of copper. Thus, the degree of hydrophilicity of the surface can be used as a measure of the tendency of copper to corrode.

Infrared reflection absorption spectroscopy assisted by the use of polarized radiation at angles near the Brewster angle has great sensitivity for the study of thin films on metal surfaces [9].

EXPERIMENTAL

An ECI Technology (East Rutherford, NJ) Surface Scan QC-100 Instrument was utilized for the surface electrochemical measurements. A cross section of the electrochemical cell assembly is shown in Figure 1. The working electrode is the copper or other metal suface of the wafer. The actual working electrode area for all studies was 0.08 cm^2. The auxiliary electrode was stainless steel, and the reference electrode was a Ag/AgCl reference electrode. This instrument was designed specifically for performing Sequential Electrochemical Reduction Analysis of the oxides on plated metal surfaces, which is a patented technology from Rockwell International. SERA is essentially reductive chronopotentiometry in which a constant cathodic current is applied to the metal surface of interest and the potential response is measured as a function of time. Plateaus occur on this curve which are characteristic of the species being reduced. The duration of each reduction plateau can be related to the amount of reducible species present on the surface. The oxide thickness and type was evaluated from the oxide reduction charge [10,11].

A borate buffer having a pH of 8.4 was made from DI water and reagent grade sodium borate ($Na_2B_4O_7 \cdot 10H_2O$) and boric Acid (H_3BO_3). The electrolyte was purged prior to and during use with nitrogen. Some studies were also performed with a dry air purge. A special electrolyte, TMCu, was developed for the copper thickness measurements, and it is discussed in the Anodic Chronopotentiometry Section.

Since the ECI QC-100 only has capability for chronopotentiometric measurements, other types of DC and AC electrochemical measurements were performed by coupling the ECI QC-100 electrochemical cell to an EG&G Princeton Applied Research Model 6310 Impedance Analyzer. Three software packages were utilized to perform the various types of electrochemical measurements; The Model 398 Software was used for impedance measurements; Model 352 Software was used for corrosion measurement; Model 270 Software was employed for general electrochemical mesurements.

The impedance measurements were performed with a 5 mv sine wave over the frequency range of 100 kHz to 10 mHz with 10 measurements per decade. For the potentiodynamic scans, the potential was ramped from a potential cathodic of the open circuit potential to a potential anodic of this open circuit potential. A scan rate of 10 mV/s was employed. The potentiodynamic data was represented by means of semilog plots. The Tafel regions can be found on either side of the corrosion potential where the current increases linearly with potential. From these plots and PARCcal software available in the Model 352 Software package, the corrosion potential and current were obtained.

Contact angle measurements were conducted with an AST Products (Billerica, MA) VCA 3000 System. The VCA-3000 operates as a fully automated contact angle measurement instrument. The instrument performs contact angle measurements on any number of points selected by the user or in a previously programmed pattern. In these measurements the shape of the sessile drop of DI water is observed on the surface. The shape is defined by the contact angle. The drop size was typically 1.0 microliter, and the measurements were performed in a clean room area where the temperature and humidity were controlled. The accuracy of the contact angle measurements is 0.1 degree.

Fourier transform infrared reflection absorption measurements were performed using p-polarized infrared radiation at an approximate 75 degree incident angle using a Harrick Refractor Accessory and a Nicolet Model 800 Infrared Spectrophotometer, equipped with an mercury cadium telluride detector. To insure good signal-to-noise ratios each spectrum was obtained by co-adding 128 scans.

RESULTS AND DISCUSSION

E_{corr} vs. Time Measurements

Figure 2 shows that the E_{corr} increases over time for a Cu surface after CMP in an air purged borate buffer solution, indicating that is becoming more passive. If the borate buffer is purged with nitrogen, the potential drops over time, revealing that the copper is becoming more reactive, due to the solubilization of the surface oxide layer. Notice that three sites tested on the surface of the wafer show different decay curves. This type of across wafer variability was typical. Ta after CMP, on the other hand, even in a nitrogen purged borate buffer, shows an increase in oxidation potential. It is probable that oxide growth on the Ta surface can occur due to the presence of trace oxygen in the electrolyte. The behavior of Cu and Ta indicates that it is difficult to remove O_2 completely by nitrogen purging.

Impedance Measurements

Impedance spectroscopy has been useful for investigating both the mechanism and kinetic aspects related to the corrosion processes of metal surfaces. This a convenient technique for Ta surfaces, since the surface oxide cannot be reduced in aqueous solutions. Figures 3 and 4 are the Bode amplitude and phase plots, respectively , for a Cu CMP surface at open circuit in a nitrogen purged borate buffer. The initial measurement was performed on the surface after the CMP process and a DI water rinse followed by a spin dry. The surface oxide, which was approximately 25 Å of Cu(I) oxide, was then reduced potentiodynamically, and a second impedance scan was performed on the same area of the wafer as the first. The sensitivity of both impedance parameters at frequencies below 1 kHz to the surface oxide is apparent.

Using the same electrolyte, the Bode phase plots in Figure 5 for a Cu and Ta film after CMP with a different pad than used for Figure 3 and 4, reveal a different low frequency behavior. The lowest frequency peak in the phase is thought to be attributable to the presence of a surface corrosion inhibitor while the higher frequency peak to that of copper oxide.

The Nyquist impedance plot in Figure 6 for Cu after CMP in an aerated borate buffer show a single semicircle which may be attributed to a charge transfer control of the corrosion reaction which takes place on the metal surface. This can be contrasted with the Nyquist plot in Figure 7 for deposited Cu at open circuit in an acidic copper sulfate bath used for plating. The real and imaginary components of the impedance have been normalized for the electrode area, and the magnitude of the corresponding impedance parameters are much larger in the case of the borate buffer. The highest frequency data points are on the left, and the frequency decreases as one goes to the right in the Nyquist plots.

Cathodic Chronopotentiometric Measurements

The reductive chronopotentiograms in the Figure 8 compare the reduction processes for an electroplated Cu wafer prior to CMP and after CMP. The nitrogen purged buffer was utilized. The current density was 30 $\mu A/cm^2$ for these measurements. The surface of the electroplated wafer shows a more complicated reduction behavior than than after CMP. This is thought be attributable to the presence of trace organic plating bath additives which are present on the surface after the completion of the plating and rinsing process. The initial portion of the chronopotentiogram for the Cu surface after CMP shows evidence of a short arrest which is which is thought to be due to an organic corrosion inhibitor.

Figure 9 provides overlays of other typical chronopotentiograms of Cu surfaces after CMP. The initial portion of the dashed curve at low negative potentials shows evidence of a corrosion inhibitor. The curve with the solid line has no features due to a corrosion inhibitor, but there is evidence of both Cu_2O and CuO. All of these wafers were processed with the same slurry, but using different CMP process conditions. The wafers were rinsed with DI water and spun dry. It is apparent that the corrosion inhibitor has been removed and the oxide has been reduced in thickness to 14 Å, or approximately half its initial value in the brush cleaning process. This result suppports the previous report that copper oxide build up on PVA brushes will occur during cleaning and special proprietary chemistry is required to dissolve the copper oxide in the PVA brushes. [12]. It should be noted that studies of Cu surfaces after CMP process variations revealed that the surface oxide was principally Cu_2O and ranged in thickness from 19 to 28 Å. In general an across wafer variation of several angstroms in the surface oxide thickness and a variability in the surface coverage of the corrosion inhibitor was found for wafers tested after CMP.

Anodic Chronopotentiometry

This analyzer was used in the anodic chronopotentiometric stripping mode for copper thickness measurements as well as to compare the stripping characteristics of copper films. Faraday's Principle enables the copper film thickness, T, to be derived from the following equation:

$$T = (MIt10^8)/(nFSD)$$

I= current (amperes)
M= molecular weight (grams)
t = time (seconds)
n= number of electrons
F= Faraday's constant (96,498 amp-sec)
D= density (g/cm^3)
S= surface area (cm^2)
T= thickness of film (Å)

The main factors which governed the selection of the electrolyte was that it could not etch the copper surface during the time scale of the measurement. Passivation of the copper surface could also not occur in the electrolyte during the anodic process, even when conducted over a range of current densities. A special electrolyte, which is named TMCu, was developed for this application. To guarantee the optimium accuracy the current density was selected so that the anodization was completed in 2 to 3 minutes.

Examples of anodic chronopotentiograms of Cu in the TMCu electrolyte are shown in Figure 10. The endpoint for the stripping process can be accurately measured.

The thicknesses derived from measurements at 4 different sites on the wafer are given in Table 1. The measurements were obtained without delay and after the copper surface was in contact with the electrolyte for 10 minutes. The results show consistent thickness values (Table 1), and they confirm that the electrolyte does not react chemically with the copper surface.

The samples with known thickness of electrodeposited copper have been also tested. These samples were prepared by the standard weighting technique. According to the supplier, the coating is uniform and the thickness variations should be within ±5%. The results of five consecutive tests are given in Table 2 and reveal good repeatability and accuracy.

Table 1

Sample	Thickness, Å			
	1	2	3	4*
Vacuum deposited	1020	1005	1034	1010
Electroplated	16020	16100	16050	16040

* with 10 min. delay

Table 2

Expected Thickness, Å	Averaged Measured Thickness, Å	Standard Deviation, Å
2500	2492	45
5300	5314	63

The next test was performed to verify accuracy of the thickness measurements at different current densities. The current density has to be optimized for each specific type of copper coating, and it represent the rate of stripping. If the current density is too low, the test will take a long time and the inflection point (end of the stripping process) might not be properly defined. At high current density, the inflection is defined better. However, if the current density is too high, the reproducibility of the measurements will suffer. For chronopotentiometric measurement, the optimum duration time is about 2-3 minutes. Table 3 shows the results obtained from the standard sample with known thickness (2500 Å) using different current densities.

Table 3

Normalized Current Density		
1	0.5	2

2518 Å	2488 Å	2450 Å

Potentiodynamic Measurements

Figure 11 shows that there is a prononunced dependence of the corrosion properties of a CMP Cu surface in the borate buffer upon the type of purge used (ie., nitrogen or clean, dry air) and the time delay prior to measurement in the air saturated electrolyte. This finding is consistent with the E_{corr} vs. time plots of Figure 2 in which the copper surface which has been subjected to CMP becomes progressively passivated in the air purged borate buffer solution over time. This is manifested in an increase in the corrosion potential over time. Corrosion parameters measured after a 15 minute equilibration time reveal that the corrosion potential and the corrosion current are higher and lower, respectively, than the corresponding values obtained by potentiodynamic measurements without any delay. This indicates that the corrosion rate diminishes over time in the air saturated electrolyte. In the case of the nitrogen purged borate buffer, the E_{corr} vs. time curves reveal that the E_{corr} becomes more negative over time. However, the measured corrosion current after 15 minutes is lower than in the case of the air purged buffer. This can be attributed to the deficiency of oxygen which is required for the cathodic reaction.

The potentiodynamic curves in Figure 12 illustrate the dependence of the corrosion characteristics of copper (both corrosion potential and corrosion current) upon the electrolyte medium. The data are listed in Table 4.

Table 4 Comparison of the Corrosion Parameters of Cu in Different Electrolytes

Electrolyte	E_{corr} vs. Ag/AgCl (mv)	I_{corr} ($\mu A/cm^2$)
A- Borate buffer +2 vol% H_2O_2	290	30
B- $H_2SO_4/CuSO_4$ plating bath	79	1847
C – Air-saturated DI H_2O	14	8
D – N_2 purged borate buffer	-149	0.3

The highest corrosion rate occurs in the acidic plating bath (B) in which the copper surface is oxide free. However, in such plating baths organic films are known to be present to serve as suppressors and brighteners. These films block the surface of copper to some degree and should, thus, impact the corrosion rate. The passivation behavior in the anodic segment of Solution B is thought to be due to the formation of CuCl or Cu_2O. The high corrosion potential is Solution A can be attributed to the presence of hydrogen peroxide. Hydrogen peroxide can participate in both anodic and cathodic half-reactions. This explains the large Tafel slopes. The low corrosion potential of Cu in Solution D can be attributed to the low concentration of oxygen. In Solution A the reduction of hydrogen peroxide is expected to be the major cathodic reaction, while in C and D the reduction of oxygen is the dominant cathodic reaction. Cu^{+2} reduction can occur in B. These corrosion parameters obtained in DI water (C) can be compared to those reported previously by Brusic et al. [5].

The variation of the corrosion properties is dependent upon the concentration of hydrogen peroxide in the borate buffer as illustrated Figures 13 and 14. This is of particular interest since many commerical CMP slurries use hydrogen peroxide as the oxidizing agent. Thus, the removal rate of the slurry as well as post CMP corrosion may be affected by the hydrogen peroxide concentration in contact with the metal surface. The slope change in both the Figures 13 and 14 which occurs at about 12 vol% is thought to be due to a mechanism change for the electrochemical processes.

Barrier metals were also studied, since galvanic coupling can be a possible cause for increased material loss. The requirement, it appears, for galvanic corrosion to occur is the coupling of Cu with a metal with a higher corrosion potential which can raise the potential of the copper. Potentiodynamic polarization measurements were performed on Cu, TiN, Ta, and TaN to compare the corrosion rates and corrosion potentials of the individual metal films. The results for the borate buffer electrolyte are shown in Figure 15. TiN is the only barrier metal which would raise the corrosion potential of Cu. It is clear that Cu has a higher corrosion rate than barrier metals being considered in this study. It also can be inferred that there will be a large difference in the polishing rate between Cu and these metals, which has indeed been found to be the case in the slurries tested to date.

Contact Angle Measurements:

Table 5 shows the wide range of contact angles for DI water which can be obtained on processed copper surfaces. Also, the standard deviation provides an indication of the nonuniformity across the surface. The high contact angle after deposition is thought to be related to the organic additives in the plating bath. The high contact angle seen on some CMP processes is thought to be due to the presence of a corrosion inhibitor in the slurry. Post CMP cleaning significantly reduce the contact angle.

Table 5 Summary of DI Water Contact Angle Study of Cu Surfaces after Processing
(9- Point Maps)

PROCESS	AVE. (DEG)	STD. DEV
CU AFTER DEPOSITION (NEW BATH)	76.1	2.5
CU AFTER NORMAL. CMP	63.5	3.0
CU AFTER CMP PROCESS WITH DIFFERENT PAD	30.3	4.3
CU AFTER BRUSH DI CLEAN	22.1	2.4
CU AFTER BRUSH CLEAN USING SPECIAL CHEMICALS	16.3	1.8

Figure 16 is an illustrative contact angle wafer map of copper surfaces after CMP processing and brush cleaning with proprietary chemicals. The brush cleaning process has caused a significant reduction in contact angle. The range in the contact angle of 15.5 to 34.8 degrees reveals considerable nonuniformity. It was shown previously by chronopotentiometry that the corrosion inhibitor was removed by this process and the copper (I) oxide thickness was reduced.

Figure 17 illustrates the change in the contact angle of freshly plated Cu surfaces (after a DI water rinse and spin dry) during bath usage. The contact angle decreased in a linear fashion. It is known that at least one of the organic additives is consumed and the by-products build up in the plating bath over time. This may be a reflection of this accumulation of organic by-products over time. Also, shown is the contact angle measured for a new prepared plating bath. It should be noted that this value is significantly lower than the average value of a wafer when plated in a second new plating bath of 76° as listed in Table 5. Further studies will be required to correlate the measured contact angles with the plating bath compositional changes.

In addition, the wetting properties of the surface can be measured in a direct dynamic mode in which the drop volume is either increased or decreased and the image viewed over time. The dynamic wetting of the surface can be related to the corrosion properties of copper surfaces, as shown earlier by Thomas et al [8], using the Wilhelmy plate technique which does not lend itself to whole wafer analysis. Advancing and receding drop angles were studied by using the motorized syringe accessory with the dynamic capture software. By programming the dynamic capture software to capture images while the motorized syringe dispenses or withdraws the test liquid, which was DI water, the drop can be studied as it advances or recedes across the surface. Additional details can be found in Reference 7.

P-Polarized, grazing angle FTIR Study of Copper Surfaces

FTIR spectra can be readily determined at grazing incidence using p-polarized radiation. The band at 668 cm^{-1} shown in Figure 18 is proportional to the copper (I) oxide thicknesss measured by the cathodic chronopotentiometric technique. The thickness range covered by this graph covers the range of oxide thickness encountered in this study. The lowest thickness was obtained for Cu after CMP and post CMP clean in a brush scrub system. The midrange thickness were obtained for Cu films after CMP and a spin/rinse dry process using DI water. Thicknesses at the high end of this calibration curve near 40 Å were obtained for Cu films after electrodeposition. Our band assignment agrees with that of Persson and Leygraf [13].

CONCLUSION

In this paper localized electrochemical, contact angle, and FTIR techniques were outlined and illustrative examples were provided of results obtained on processed Cu and barrier metal films. These techniques are conveniently applied to full wafer surfaces. The results show good consistency across the various techniques.

Several processing steps where corrosion can be an issue in Damascene processing are highlighted in this study. The oxidation can occur in the plating process in the Cu surface if allowed to remain in contact with the plating bath without an externally applied potential or after plating if the plating solution is not removed quickly and completely by rinsing. Chemical mechanical polishing is based upon continuous oxidation and mechanical removal of the oxide layer. It follows that it is essential to achieve complete and rapid removal of the slurry chemicals after chemical mechanical polishing is complete. Post CMP cleaning processes are of special interest, since these processes generally use surface treatment with brushes and chemicals to remove slurry particles and trace levels of chemicals. Storage and shipment of Cu coated wafers must also be closely controlled to insure that atmospheric corrosion is minimized or totally prevented. No attempt was made to discuss copper plating bath, CMP slurry, or wafer cleaning formulations. Although it must be recognized that the specifics of such formulations will have a large influence upon the amount of corrosion which will be realized during and subsequent to processing. The value of localized electrochemical and sessile drop contact angle measurements using DI water at multiple sites on the wafer surface for process monitoring and control of the corrosion properties of copper surfaces during copper Damascene processing has been illustrated.

REFERENCES

[1] Y. Yamamoto, H. Nishihara, and K. Aramaki, J. Electrochem. Soc., vol. 140, 436 (1993)
[2] H. Itoh, H. Nishara, and K. Aramaki, J. Electrochem. Soc., vol. 141, 2018 (1994)
[3] M. Itoh, H. Nishihara, K. Aramaki, J. Electrochem. Soc., vol. 142, 1839 (1995);
[4] M. Itoh, H. Nishihara, and K. Aramaki, J. Electrochem. Soc., vol. 142, 3696 (1995)
[5] V. Brusic, G. S. Frankel, C.-K. Hu, M. M. Plechaty, and G. C. Schwartz, IBM J. Res. Develop. **37**, 173(1993).

[6] R. Carpio, J.S. Petersen and D. Hudson in Proceedings. Second International Symposium on Process Control, Diagnostics, and Modeling in Semiconductor Manufacturing, eds., M.

Meyyappan, D. Economou, and S. W. Butler, Electrochemical Soc.Proc., **PV 97-9**, 222 (1997).
[7] R. Carpio and D. Hudson in 9th Annual IEEE/SEMI Advanced Semiconductor Manufacturing Conference and Workshop Proceedings, pp. 272-277 (1998).

[8] R. R. Thomas, V. Brusic, and B. M. Rush, J. Electrochem. Soc. **139**, 678(1992)]
[9] N. J. Harrick and M. Milosevic, Applied Spectroscopy **44** , 519 (1990
[10] H. D. Speckmann, M. M. Lohrengel, J. W. Schultze and H. H. Strehblow, Ber. Bunsenges. Phys. Chem., **89**, 392(1985).
[11] Y-Y. Su and M. Marek, J. Electrochem. Soc. **141**, 940 (1994).
[12] D. Hymes, H. Li, E. Zhao, J. de Larios, Semiconductor International, **21**, 117(1998).
[13] D. Persson and C. Leygraf, J. Electrochem. Soc. **140**, 1256 (1993).

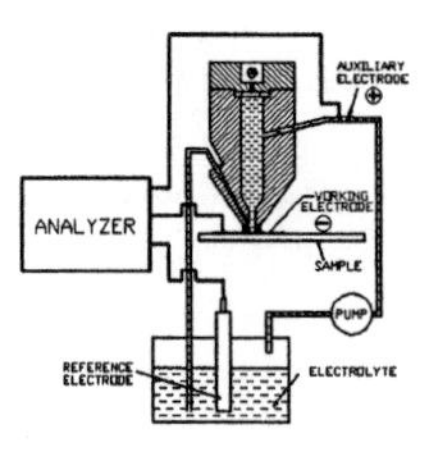

Fig. 1. Schematic of ECI Cell

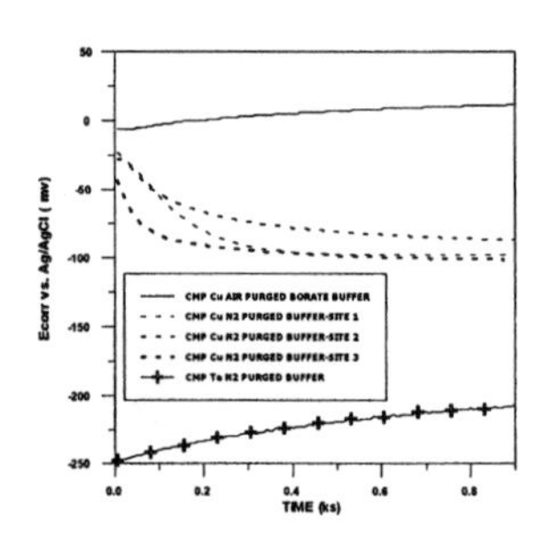

Fig. 2. Ecorr vs. Time Plots for Cu and Ta after CMP in borate buffer

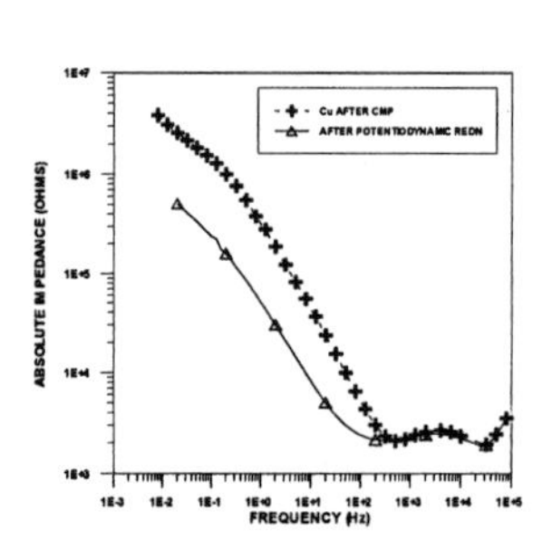

Fig. 3. Bode amplitude plots of Cu surface after CMP and then after reduction of surface oxide.

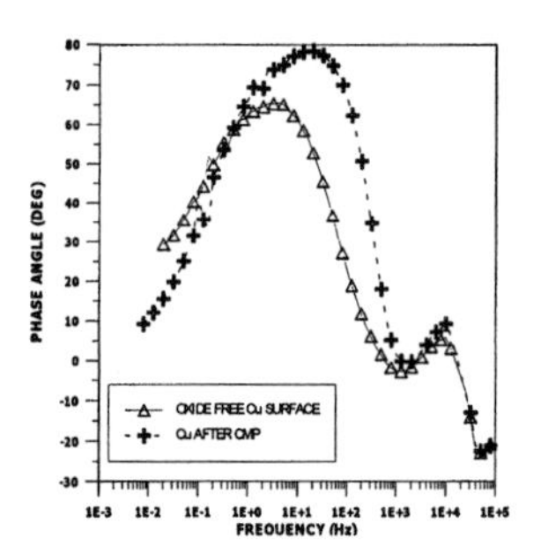

Fig.4. Bode phase plots of Cu surface ater CMP and then after reduction of surface oxide layer.

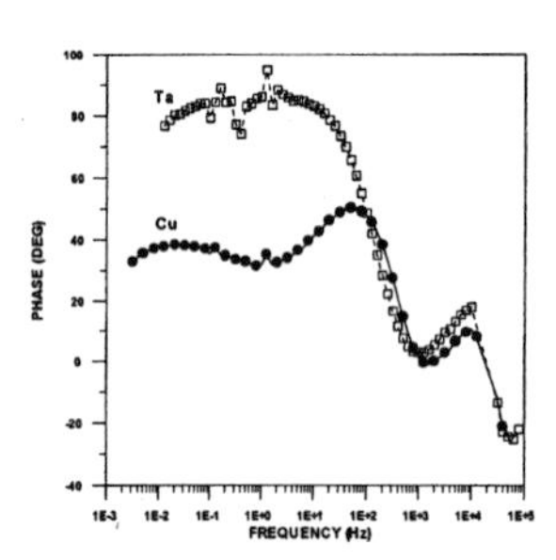

Fig. 5. Bode phase comparisons of Ta and Cu surfaces after CMP. Different polishing pad than used for Fig. 3.

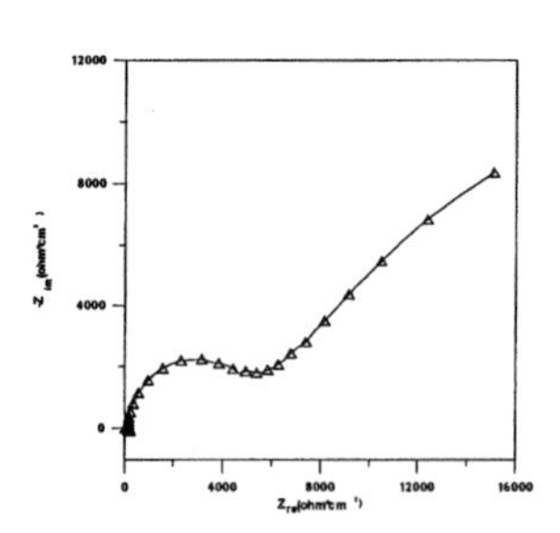

Fig. 6. Nyquist plot of Cu surface after CMP in borate buffer.

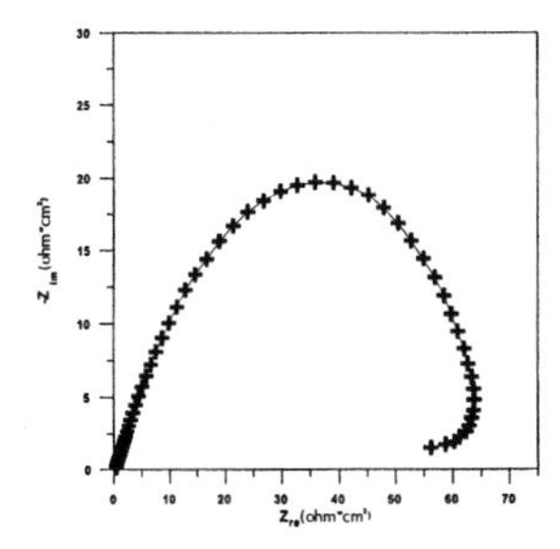

Fig. 7. Nyquist plot for Cu in plating bath solution.

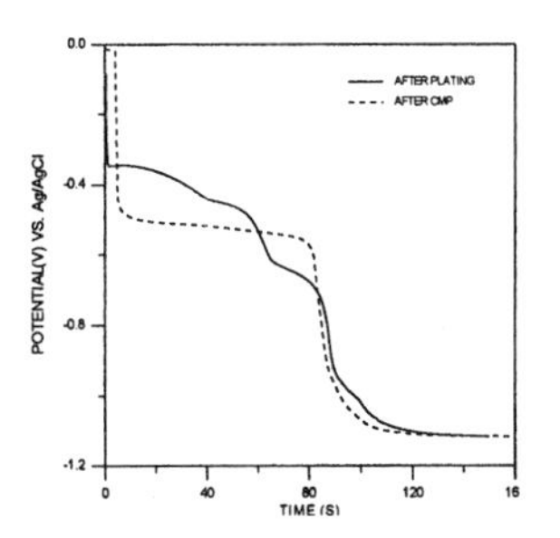

Fig. 8. Chronopotentiogram of Cu surface after electroplating and CMP.

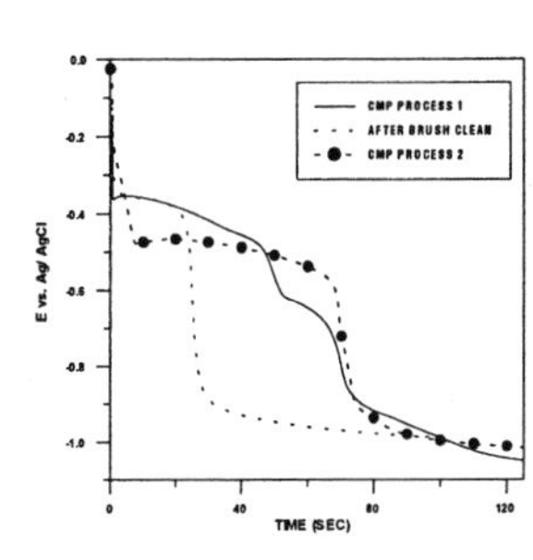

Fig 9. Reductions of CMP Cu films in borate buffer at 30 μA/cm2.

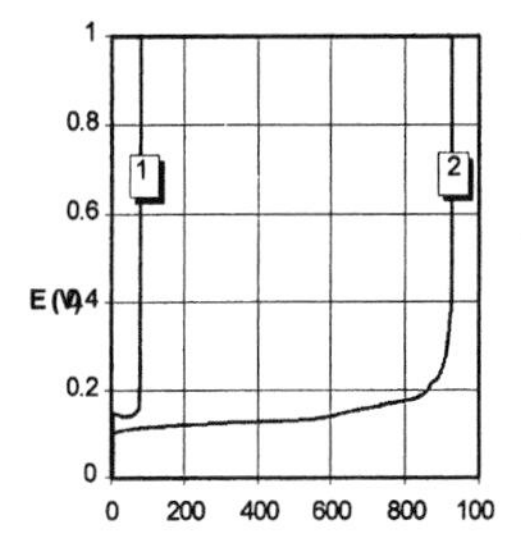

Fig. 10 Anodic chronopotentiograms of Cu seed layer (1) and electrodeposited Cu (2) in TMCu electrolyte.

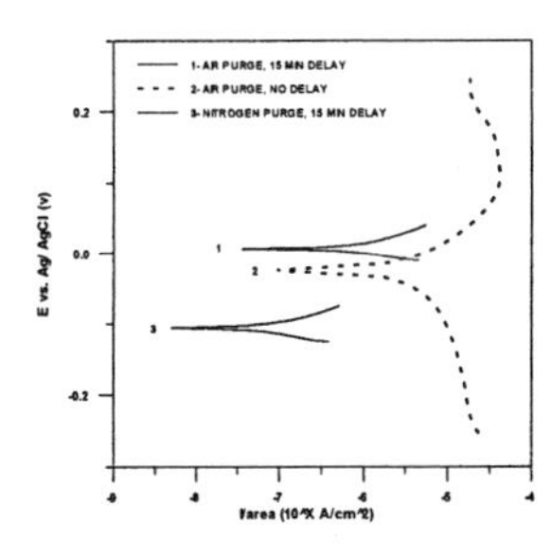

Fig. 11. Potentiodynamic scans of Cu surface in borate electrolyte using different conditions for purge and delay time.

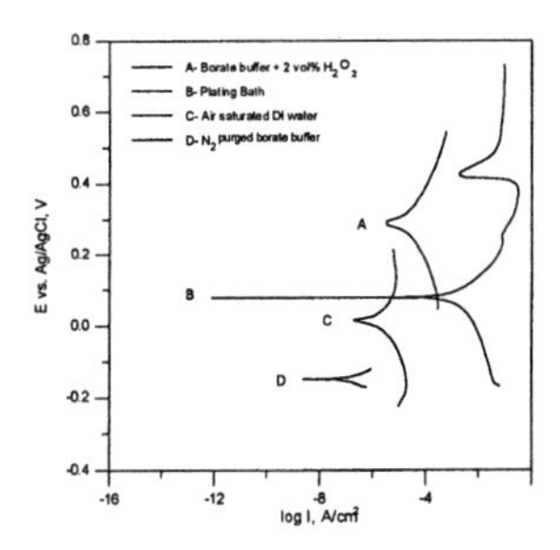

Fig. 12. Potentiodynamic scans of Cu films in different electrolytes.

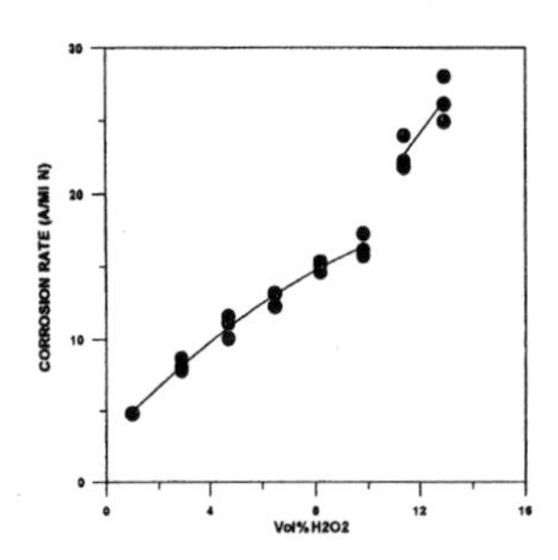

Fig. 13. Dependence of corrosion current of Cu upon H_2O_2 concentration in borate buffer

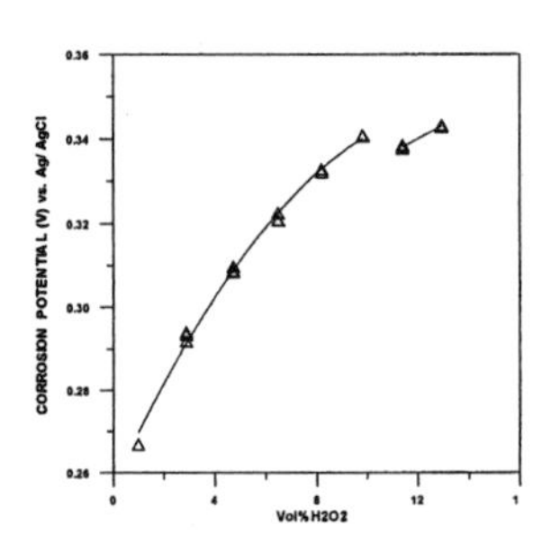

Fig. 14. Dependence of corrosion potential of Cu upon H_2O_2 concentration in borate buffer.

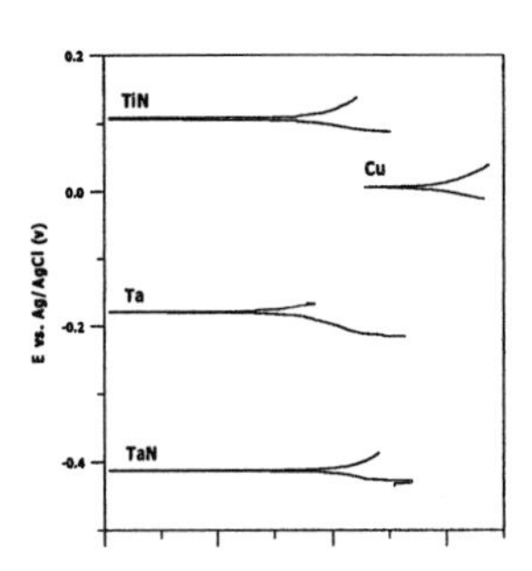

Fig. 15. Potentiodynamic plots for different metal films in borate buffer.

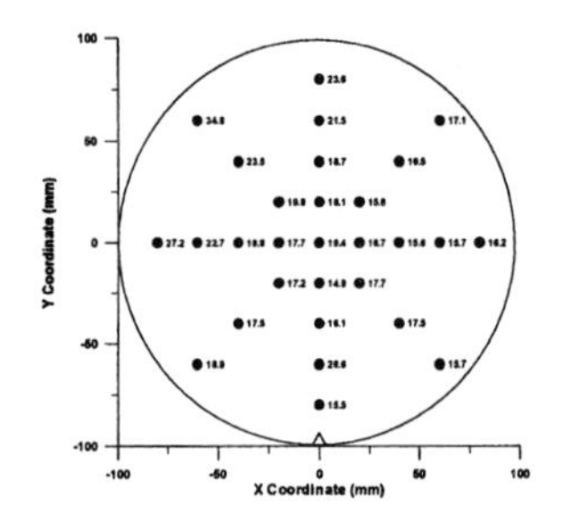

Fig. 16. Contact angle map for Cu surface after post CMP brush clean.

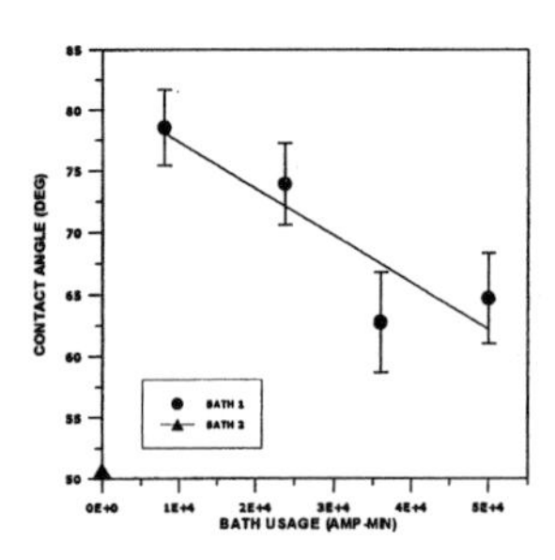

Fig. 17. Dependence of DI water contact angle of electroplated Cu surface upon plating bath usage.

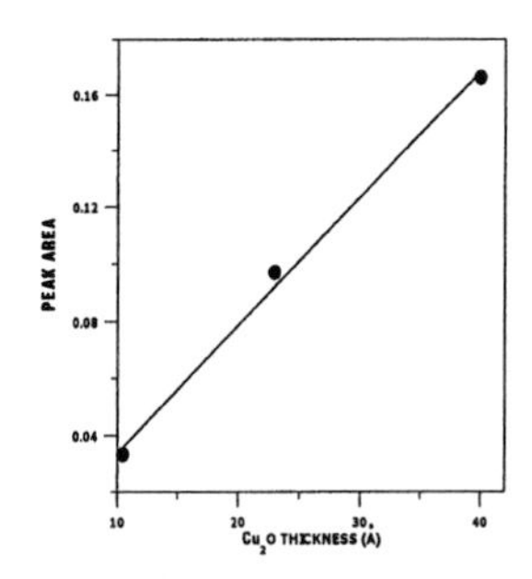

Fig. 18. Dependence of 668 cm^{-1} p-polarized FTIR peak area of Cu surface upon Cu_2O thickness.

PROGRAMMED RATE CVD PROTOCOLS

D. Yang[1], J. J. Kristof[2], R. Jonnalagadda[3], B. R. Rogers[4], J. T. Hillman[5], R. F. Foster[5], and T. S. Cale[1]
[1]Department of Chemical Engineering, Rensselaer Polytechnic Institute, Troy, NY 12180
[2]Motorola, 3501 Ed Bluestein Blvd., Austin, TX 78721
[3]Allied Signal Inc. 3520 Westmoor St. South Bend, IN 46628-1373
[4]Department of Chemical Engineering, Vanderbilt University, Nashville, TN 37325
[5]Tokyo Electron Arizona, 2120 W. Guadalupe Rd., Gilbert, AZ 85233-2805

We review two programmed rate chemical vapor deposition (PRCVD) studies; PRCVD of tungsten, and PRCVD of aluminum. During PRCVD, operating conditions are systematically changed during deposition to enhance some property of the film or some aspect of the process. In the tungsten study, PRCVD provided significantly greater throughput than conventional, constant rate CVD (CRCVD). PRCVD of aluminum from tri-isobutyl aluminum (TIBA) can produce films with better film properties that are considered important to electromigration performance than CRCVD.

INTRODUCTION

In this paper, we review two studies of what we call programmed rate chemical vapor deposition (PRCVD); PRCVD of tungsten, and PRCVD of aluminum. During PRCVD, reactor set-points are systematically changed during the process in order to improve one or more specific process parameters or film properties. Though these studies use PRCVD for very different reasons, they both focus on using our understanding of the relative rates of processes that occur during deposition. They help set the stage for using similar processes throughout IC fabrication, with its ever-increasing number of processing steps and layers (interfaces). In general, as the wafer state changes during individual process steps (deposition, etch, etc.), we cannot expect constant process conditions to be optimal in terms of time or film properties.

PRCVD of Tungsten

In the tungsten study, we showed that PRCVD can provide significantly greater throughput than conventional, constant rate CVD (CRCVD). CVD tungsten has been used for several years to fill vias [1]. The concept of using programmed temperature trajectories during CVD processes was first proposed by Cale and co-workers [2] as a means to increase wafer throughput in single wafer reactors. The basis for the application is that as features fill, their aspect ratios naturally increase. PRCVD can produce films which are conformal, in less total time than is required for traditional processing methods; *i.e.*, CRCVD. For PRCVD of tungsten, we present our experimental results of tungsten deposition and discuss some of the difficulties associated with bringing this protocol to realization.

PRCVD of Aluminum

We have shown that PRCVD of Al from tri-isobutyl aluminum (TIBA) can produce films with better film properties that are considered important to electromigration performance, than CRCVD processes. CVD Al has been investigated as one way to deposit metal into tight contact holes and vias, and processes that use CVD Al have been established. One of the drawbacks to CVD Al is its poor nucleation performance on different types of substrates that results in rough surface morphologies. In our Al PRCVD work, temperature ramping is used in Al CVD to enhance nucleation rather than to improve throughput. Essentially all studies on Al CVD have focused on deposition at constant conditions; *i.e.*, CRCVD protocols. Not much is known about the growth behavior and film properties when process conditions are varied, as during PRCVD. However, Masu *et al.* [3] showed that CVD Al nucleation is very sensitive to process conditions, and the same depositions that yield optimal film growth (post-nucleation) may not yield optimal film nucleation. Our goal from Al PRCVD is to control the nucleation stage of the CVD process because the initial stages of deposition set the stage for growth. We want to grow smooth highly (111) textured CVD films, because they are apparently resistant to electromigration [4].

EXPERIMENTAL

PRCVD of Tungsten

A modified single wafer, cold wall, lamp heated SPECTRUM 202 LPCVD reactor shown schematically in Figure 1 was used to deposit tungsten onto five inch wafers. This reactor is equipped with a load-lock, which enables loading/unloading of the wafer without exposing the reaction chamber to the ambient atmosphere. This reactor can be used for multi-step processes; *e.g.*, a nucleation step, then one or more deposition steps. The

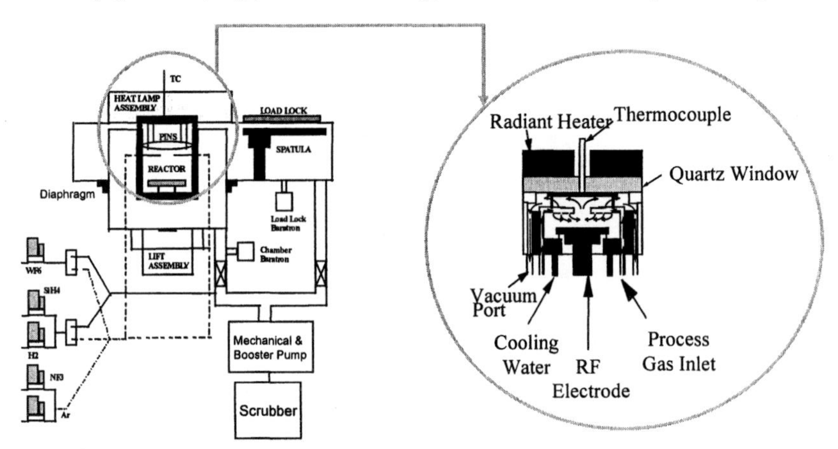

Figure 1. Schematic diagram of the SPECTRUM 202 LPCCVD reactor.

substrate temperature was measured using two thermocouples at the center of the backside of the wafer. An Eurotherm (model 821) temperature controller is used to program a series of temperature set points and ramp rates. Patterned SiO_2 wafers with sputtered TiN were used for step coverage experiments. TiN substrates require a two-step blanket tungsten deposition process to achieve both good throughput and good feature fill [5]. Silane reduction of WF_6 was used for the 'nucleation layer' on TiN followed by tungsten film deposition by hydrogen reduction of WF_6.

The deposition process conditions for both CRCVD and PRCVD during the nucleation layer deposition were 100 sccm WF_6, 500 sccm H_2, 10 sccm SiH_4 flow rate, 698 K substrate temperature, 0.5 Torr pressure, and 30 s deposition time. The desired nucleation layer thickness was 100 nm. The conditions for CRCVD (after the nucleation layer) were 500 sccm H_2, 38 sccm WF_6, 10 sccm SiH_4 flow rate, 723 K substrate temperature, and 180 seconds deposition time. The process conditions for PRCVD after the nucleation layer were 570 sccm H_2, 38 sccm WF_6, 10 sccm SiH_4 flow rate, 823-723 K starting temperature, 723-633 K ending temperature ramp. PRCVD pressures and times varied.

PRCVD of Aluminum

In our PRCVD Al study, the same type of reactor as shown in Figure 1 was used to deposit Al from tri-isobutyl aluminum (TIBA) onto four-inch silicon wafers coated with sputtered TiN_x. Semiconductor grade TIBA, a liquid at room temperature, was delivered to the reactor using a "bubbler" evaporator system. The following process conditions were common to all experiments; 1 Torr total pressure, 60 sccm carrier gas (Ar), 318 K bubbler temperature, 353 K line temperature, and -100 or -200 K/min temperature ramp rate. The partial pressure of TIBA in the inlet stream was approximately 0.019 Torr.

The process sequence for temperature ramp down experiments began by stabilizing the substrate temperature at the desired initial value. Then, the precursor flow and temperature ramp down were started and stopped after the desired time. Table I summarizes some conditions for each experiment. For pulse deposition experiments (PU1 and PU3), the precursor flow was stopped at the end of a 5 s or 10 s pulse time, while the

Table I. Conditions used for continuous deposition during temperature ramp down (CN), pulse deposition (PU), and constant temperature (CO) depositions.

ID	Pulse time (s)	Total dep. time (s)	Initial temp. (K)	Final pulse temp. (K)	Final dep. temp. (K)
CN1	0	5	673	N/A	656
CN2	0	5	623	N/A	606
CN3	0	30	673	N/A	573
CN5	0	1200	673	N/A	573
PU1	5	30	673	656	573
PU3	10	610	673	640	573
CO1	0	5	573	N/A	573
CO2	0	30	573	N/A	573
CO3	0	900	573	N/A	573

temperature continued to ramp down to 573 K. Once the temperature stabilized at 573 K, the precursor flow was turned on for a specified growth time. Figure 2 shows the substrate temperature and precursor flow trajectories for sample PU3.

Weight gain measurements were used to estimate the average deposited film thicknesses. Film resistivities were determined using a 4-point probe. Crystal orientations were determined using x-ray diffraction (XRD). Film morphologies were analyzed by field emission scanning electron microscopy (FESEM). Contact mode atomic force microscopy (AFM) was used to observe the nucleation densities and measure the surface roughnesses. Film reflectivities were determined using a Nanospec/AFT at a wavelength of 520 nm using a polished Si (100) wafer as the reference. Auger electron spectroscopy (AES) was used for chemical analysis.

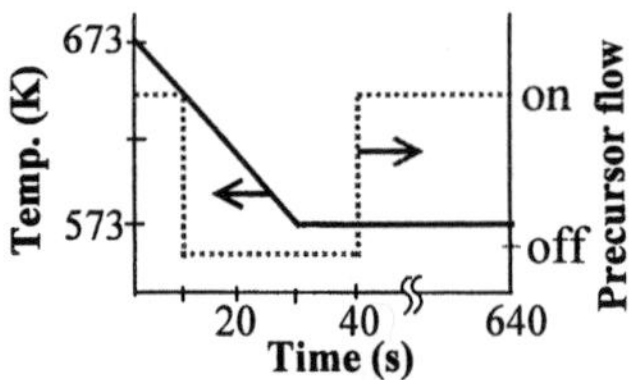

Figure 2. PRCVD protocol: precursor flow was pulsed for10 s while ramping from 673 K to 573 K, then flow was restarted for 10 min at 573 K [sample PU3].

RESULTS AND DISCUSSION

PRCVD of Tungsten

The desire to maintain acceptable throughput as single wafer reactors were being adopted in place of multiple wafer reactors led Cale and coworkers to introduce PRCVD of tungsten [2]. To date, temperature has been used to control the instantaneous deposition rate as a function of time through the process, because the reaction is thermally activated. We take advantage of the inherent increase in the aspect ratios of vias as they fill, and the knowledge that the conditions (rates) used in CRCVD processes are set to maintain high conformality as the vias close. We start the deposition process at a much higher rate (temperature) than that for the comparable CRCVD processes, then decrease the temperature during deposition. The deposition temperature at the end of the process is a little lower than the CRCVD process temperature. The net effect is a higher average deposition rate than realized in a CRCVD process.

Step coverage was analyzed using patterned TiN wafers. Some of the deposition conditions resulted in poor coverage, while some showed good, conformal deposition. The main result from these experiments is the successful feature fill for some PRCVD samples compared to similarly processed CRCVD samples. The SEM micrograph shown in Figure 3 is selected from among the best films deposited by CRCVD. This film was deposited at 2 Torr. As noted above, a two-step deposition process was applied: silane reduction on TiN for a nucleation layer, followed by 180 seconds of tungsten deposition by hydrogen reduction of WF_6 at 723 K. The film completely filled the feature with

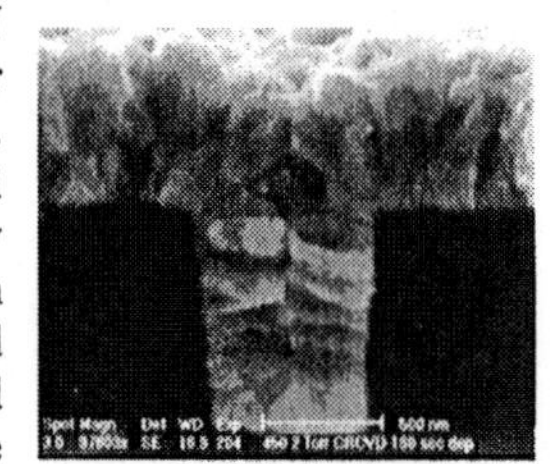
Figure 3. SEM micrograph of a film from a CRCVD process (see text).

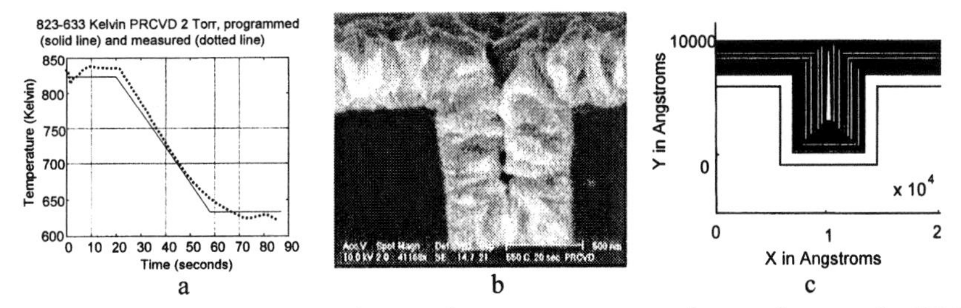

Figure 4. RPCVD process results: a. substrate temperature trajectory, b. example SEM micrograph, and c. feature fill predicted by a simplified model [6], using the measured temperature during deposition (as in Figure 4a).

no voiding.

Figure 4 shows some PRCVD results for a deposition pressure of 2 Torr, and Ar flow during initial thermal stabilization. The temperature trajectory is shown in Figure 4a. The measured substrate temperature trajectory (dotted line) is reasonably close to the programmed temperature trajectory (solid line). The programmed trajectory consists of 20 seconds at 823 K, a temperature ramp down of 5 K/s until the temperature reaches 623 K, and finally the temperature is held constant at 623 K for 33 seconds. Figure 4b shows the film deposited by this PRCVD protocol (87 s). This is about a 50% time savings over the 'reference' 723 K, 180 s CRCVD process (see Fig. 3), with very little voiding. However, this same PRCVD process protocol resulted in larger voids in films deposited in features having less than 0.8 μm width. The control model's prediction in Figure 4c, using the temperature measured during deposition (in Figure 4a), shows slight voiding and closure in 23 seconds. After closure, film growth will continue. The control model used is the simplified model used by Song [6], which is based on a kinetic model and a one dimensional, continuum transport and reaction model [2]. It was used to predict the deposition rates and step coverages.

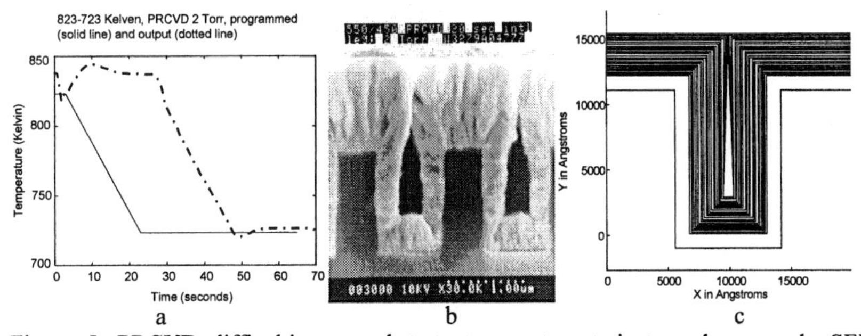

Figure 5. PRCVD difficulties: a. substrate temperature trajectory, b. example SEM micrograph, and (c) model prediction using the measured temperature trajectory (as in Figure 5a).

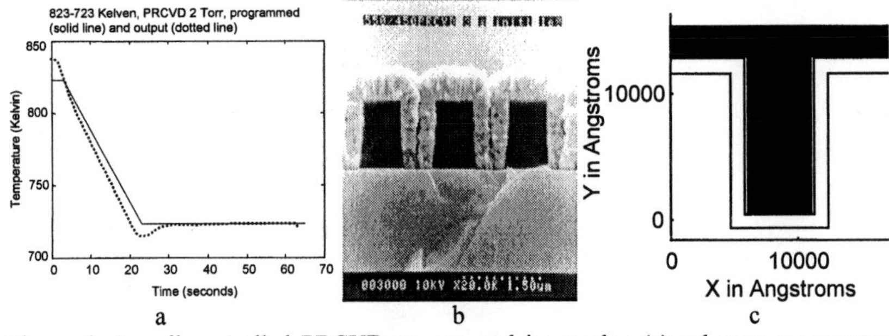

Figure 6. A well-controlled PRCVD process and its results. (a) substrate temperature trajectories, (b) SEM micrograph of well-controlled film, and (c) model prediction

Figure 5 shows one difficulty we encountered with the PRCVD process. In Figure 5a, the programmed temperature trajectory consists of a 5 second constant temperature leg, followed by ramp down of 5 K/s for 20 seconds, then the temperature is kept constant at 723 K for 45 seconds. Essentially, we tried to speed up the ramp down process in order to fill narrower features without voiding. However, the reactor process controller held the temperature above the desired initial temperature for 20 seconds longer than desired. Figure 5b shows an example SEM from the film that results from this badly controlled temperature. Figure 5c shows the simplified model's prediction of feature fill. It shows a significant void using the measured temperature data. The simplified model also indicates poor conformality; however, the model is expected to be less accurate as conformality degrades [2,6].

In order to improve upon the initial thermal stabilization, the process was modified to use hydrogen during the stabilization step, instead of argon. Figure 6 shows an example of a well-controlled PRCVD process. The measured temperature trajectory followed the programmed trajectory fairly well, as shown in Figure 6a. Figure 6b shows an example SEM from the resulting film. Fill was accomplished in 67 seconds; *i.e.*, a 63% time savings over the reference CRCVD process at 723 K, 180 s (see Fig. 3). The model prediction shown in Figure 6c matches the data fairly well. The experimental results of selected PRCVD tungsten experiments are summarized in Table II.

Modeling results indicate that throughput enhancements of about a factor of ten

Table II. Summary of PRCVD tungsten study.

Program	823 to 633 K 90 seconds	823 to 723 K 67 seconds	823 to 723 K 67 seconds
Actual time of run	87 seconds	69 seconds	64 seconds
Initial lag, prg	20 seconds	5 seconds	5 seconds
Measured initial	23 seconds	33 seconds	3 seconds
Comments	Void for less than 0.8 μm feature size	Modified program to use H_2 instead of argon after thin run	Excellent fill, as predicted by the control model

can be achieved in principle. We have achieved throughput increases of about a factor of three, with more improvement easily obtainable [5,7]. In addition to the increase in throughput, the properties of the PRCVD films were equal to, or superior to, CRCVD films. In short, PRCVD shows promise for increasing throughput in similar applications.

PRCVD of Aluminum

In an effort to achieve smooth films with small (111) grained structure, we evaluated protocols in which the deposition temperature was varied at the start of deposition in order to change the relative rates of nucleation and growth. Our Al PRCVD process is based upon reports that the activation energy for steady growth is lower (by a factor of three) than that of nucleation [8]. We take advantage of the higher activation energy for nucleation by ramping temperature from a higher initial temperature to a lower temperature for growth, and produce films that possess some superior properties relative to CRCVD films.

Figure 7 presents plan view FESEM micrographs of nuclei after 5 second TIBA exposures for three temperature trajectories (sample numbers CN1, CN2 and CO2 as described in Table I) and nucleation size distributions determined from the micrographs. The nuclei size distributions captured in these micrographs were determined using the public domain image analysis software, NIH Image v. 1.60, available from the National Institute of Health [9]. The resulting histograms are presented below the FESEM

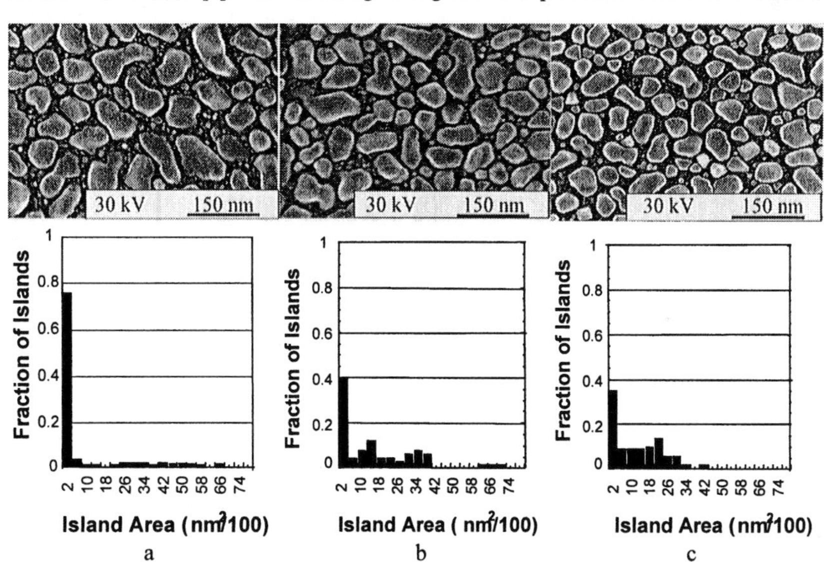

Figure 7. Field emission SEMs of nuclei and size distributions resulting from 5 second depositions using: a. temperature ramping from 673 K [CN 1], b. temperature ramping from 623 K [CN 2] and c. a constant temperature of 573 K [CO 1].

micrographs in Figure 7. These histograms plot the fraction of the total number of nuclei measured, binned according to their top-down cross-sectional area. Note that the three temperature trajectories result in significantly different nuclei size distributions. Ramping the temperature down from either 673 or 623 K (CN1 and CN2 PRCVD protocols), resulted in a larger fraction of small nuclei compared to deposition at 573 K (the CO1 CRCVD protocol).

Plan view FESEM micrographs and AFM images of nuclei resulting from different 30 second depositions (sample numbers PU1, CN3 and CO2 as described in Table I) are presented in Figure 8. Again, the island sizes vary significantly for various deposition conditions. For these samples the deposition using a 5 second pulse initiating at 673 K (PU1) resulted in a higher nucleation density (50 nuclei/μm^2) compared to deposition during ramping from 673 to 573 K (CN3) (7 nuclei/μm^2) and the deposition at constant 573 K (CO2) (4 nuclei/μm^2). Film thickness and roughness data for these same 30 second depositions are also presented in Figure 8. Notice that PU1 film's surface roughness is lower, both as an absolute number and as a percentage of average film thickness, compared to all the other films. This figure also presents the ratio of Al (111) grains to Al (200) grains, determined by XRD, for the three films discussed above. Note that the two films (PU1 and CN3) deposited using 673 K as the initial substrate

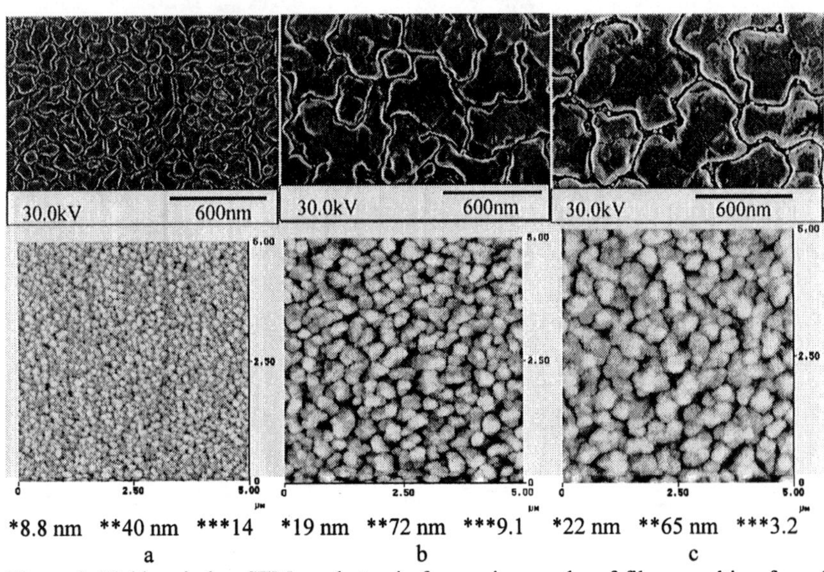

Figure 8. Field emission SEMs and atomic force micrographs of films resulting from 30 second depositions using: (a) 5 second ramping from 673 K/ 25 seconds at 573 K [PU 1], (b) 30 second ramping from 673 K [CN 3] and (c) 30 seconds at 573 K [CO 2] temperature trajectories. (*Surface roughness; **Film thickness; ***$I_{Al(111)}/I_{Al(200)}$)

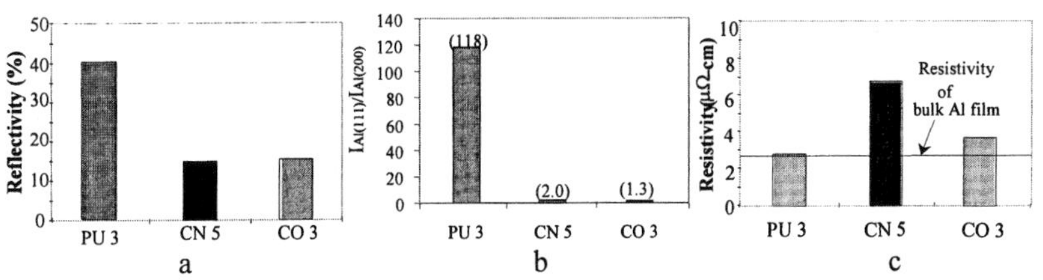

Figure 9. Film properties deposited about 450 nm thickness: (a) reflectivity, (b) Al (111) to Al (200) ratios and (c) resistivity.

temperature have higher ratios of Al (111) grains to Al (200) grains, compared to that of the film deposited at a constant substrate temperature of 573 K (CO2). However, the deposition in which the precursor flow was pulsed (PU1), resulted in films with an approximately 53% higher ratio of Al (111) to Al (200) grains, compared to that of the films from depositions in which the temperature was continuously ramped (*e.g.*, CN3).

Figure 9a shows the reflectivities of films with average thicknesses of about 450 nm. The deposition using a pulse initiating at 673 K (PU3) resulted in a film with the highest reflectivity. Figure 9b presents grain orientation ratios of Al (111) to Al (200) for the same films shown in Figure 9a. The deposition using a pulse initiating at 673 K (PU3) resulted in a much higher fraction of Al (111) grains in the film compared to films deposited using the other protocols. The resistivities of the same films are presented in Figure 9c. Films deposited with the precursor flowing continuously during temperature ramping (*e.g.*, CN5), which are therefore the ones deposited using the highest average temperature, had the highest resistivities, due to carbon incorporation. However, notice that films using a precursor pulse for a short time at elevated temperature (PU3) have resistivities very close to that of bulk aluminum. The result of the AES analysis of this film (PU3), shown in Figure 10, indicates the absence of carbon and oxygen in the bulk aluminum film.

CONCLUSIONS

PRCVD of Tungsten

Improving wafer throughput for single wafer reactors is a considerable challenge, since process changes needed to achieve high throughput may compromise film property constraints. However, conventional LPCVD processes do not have adequate degrees of freedom to achieve high throughput and maintain good film properties. We have studied the film properties resulting from selected PRCVD protocols. Results from this study of blanket tungsten PRCVD show consistent, excellent adhesion to silane based nucleation layers (on TiN). Experiments to

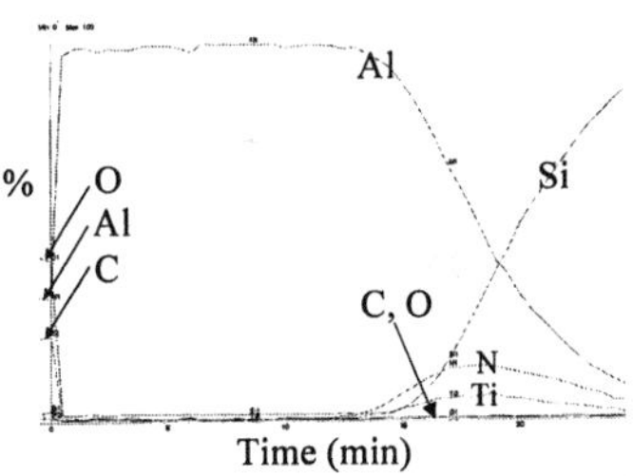

Figure 10. AES depth profile of sample PU3. The broad AES transitions between layers (Al/TiN/Si) are probably due to drift of sampling position within the sputter crater.

determine the step coverage and the time savings for PRCVD blanket tungsten produced some very good results. The time saved for the PRCVD processes used was between 50-63%, compared to the reference CRCVD control samples. Good step coverage is important, as poor step coverage results in keyhole formation at feature closure. Voids and keyholes can act as traps for corrosive materials that can negatively impact device reliability and performance. The results from this study indicate that PRCVD may be a viable protocol, and that the potential time savings for this process is significant.

PRCVD of Aluminum

In an effort to achieve small (111) grained structure, we studied programmed rate chemical vapor deposition (PRCVD) protocols for CVD Al from TIBA. In particular, we evaluated protocols in which the deposition temperature was varied at the start of deposition in order to change the relative rates of nucleation and growth. Control over the ratio of rates has apparently been achieved. We studied both relatively short deposition times (5 and 30 s) to look into the early stages of deposition, and longer deposition times (10 min and longer) to look at films that were thick enough to be of practical interest. For the short depositions, we studied films deposited while varying substrate temperature for 5 and 30 s. Nuclei size distributions of films deposited using PRCVD protocols were significantly different from those of films deposited using constant process protocols. In particular, there was a higher fraction of small nuclei for depositions performed during temperature ramping from 673 to 656 K, compared to constant temperature protocols. Pulsing (starting and stopping) the precursor flow at the start of temperature ramp down (–200 K/min) from 673 K (for 5 s), followed by deposition at 573 K (for 25 s) yielded films with larger nuclei densities, higher fractions of (111) orientated Al, and lower surface roughnesses. The most interesting 'thick film' results were obtained by pulsing the precursor flow while ramping the substrate temperature from 673 K to 573 K over 10 s, then depositing at 573 K for a total of 10 min. This protocol produces films with larger fractions of Al (111) orientation, higher reflectivities and higher deposition rates. Also, the resistivities of these films were close to that of bulk aluminum, and there was no carbon detected by AES analysis of the films. The results of this PRCVD work show that we can apparently control film microstructure and properties.

Overall Conclusions

More generally, fundamental understanding of the separate processes that occur during deposition will help in efforts to develop optimal trajectories of deposition conditions. As our understanding of the processes improve, we should be able to implement PRCVD (programmed rate protocols in general) that optimize the desired property of the final wafer state. We use temperature as it provides the best available control variable for the reactor design used in our experiments; other variables such as pressure and precursor flow could be used in other equipment.

ACKNOWLEDGMENTS

The authors gratefully acknowledge support for this project from the Semiconductor Research Corporation and National Science Foundation. We thank Dr. A. M. Yates, Dr. B. L. Ramakrishna, and Mr. T. Karcher for their help with XRD, AFM, and AES analyses, respectively.

REFERENCES

1. S. Sivaram, *Chemical Vapor Deposition*, VNR, New York, 1995, p. 168 - 169.

2. T. S. Cale, M. K. Jain and G. B. Raupp, J. Electrochem. Soc. **137**, 1526 (1990)

3. K Masu and K. Tsubouchi, in Advanced Metallization for ULSI Applications in 1994, R. Blumenthal and G. Janssen, Editors, p. 477, Mater. Res. Soc., Pittsburgh, PA (1995)

4. S. Vaidya, D. B. Fraser and A. K. Sinha, Proc. of the 18th Annual Reliability Physics Symposium, IEEE, New York, (1980) p.165-170.

5. K. Tracy, M.S. Thesis, Arizona State University, 1996.

6. L. Song, Ph.D. Dissertation, Arizona State University, 1997.

7. J. Kristof, M.S. Thesis, Arizona State University, 1998 (expected).

8. K.-I. Lee, Y.-S. Kim and S.-K. Joo, J. Electrochem. Soc. **139**, 3578 (1992).

9. Image Analysis Software version 1.60, free application software available on the website of the Division of Computer Research and Technology, National Institute of Health (at http://www.nih.gov).

MODELING OF ELECTROMIGRATION-INDUCED FAILURE OF METALLIC THIN-FILM INTERCONNECTS

Dimitrios Maroudas, M. Rauf Gungor, and H. S. Ho
Department of Chemical Engineering, University of California, Santa Barbara
Santa Barbara, CA 93106-5080, USA
and
Leonard J. Gray
Computer Sciences and Mathematics Division, Oak Ridge National Laboratory
Oak Ridge, TN 37831-6367, USA

ABSTRACT

Failure of metallic thin-film interconnects driven by electromigration is among the most challenging materials reliability problems in microelectronics. One of the most serious failure mechanisms in these films is the current-driven propagation of transgranular voids. In this paper, we present a comprehensive theoretical analysis based on self-consistent simulations of void dynamics under electromigration conditions and the simultaneous action of mechanical stress. For unpassivated films, our simulations predict void faceting, wedge-shaped void formation, propagation of slit-like features from void surfaces leading to failure, and propagation of surface waves on the voids prior to failure. For passivated films, void morphological instabilities can lead to film failure by propagation from the void surface of either faceted slits or finer-scale crack-like features depending on the strength of the electric and mechanical stress fields. More importantly, we demonstrate that in textured films, there exists a narrow range of conditions over which failure due to slit propagation can be inhibited completely.

INTRODUCTION

Failure of polycrystalline aluminum and copper thin films, which are used for device interconnections in integrated circuits, is one of the most serious materials reliability problems in microelectronics [1]. Ultra-large-scale integration (ULSI) and miniaturization of devices has pushed the widths of these films to submicron scales. Failure mechanisms in these systems are mediated mainly by the mass transport phenomenon of electromigration [1]. In passivated metallic films that are mechanically confined due to their encapsulation by a dielectric material, thermomechanical stresses are induced during film cooling after passivation. Voids usually nucleate at the film edges as a mechanism of thermal stress relaxation. After their cooling and aging, the metallic films are in a state of hydrostatic tension [1,2]. This residual stress is another important driving force of void morphological evolution that may cause film failure.

It has been established experimentally that transgranular voids are common sources of failure in bamboo films, where grain boundaries are oriented almost perpendicularly to the length-direction of the film [3,4]. Such voids are not intersected by grain boundaries and may form after detachment from grain boundaries and further migration into the grain under the action of an applied electric field. In spite of recent theoretical studies of electromigration-induced transgranular void dynamics [5-10] and of our understanding of surface instabilities in mechanically stressed solids [11], the complex nonlinear phenomena associated with transgranular void evolution are not fully understood. More importantly, the combined effects of electromigration and mechanical stress on void morphological evolution and interconnect failure have not been studied systematically and are much less understood.

In this paper, we present a comprehensive account of the nonlinear phenomena associated with transgranular void evolution under electromigration conditions and their relation to failure of metallic thin films. Numerical simulations in unpassivated films, taking into account the diffusivity anisotropy on the void surface, reveal void faceting, wedge-shaped void formation, propagation of slit-like features from void surfaces leading to failure, and propagation of surface waves on the voids prior to failure. Our predictions are in good agreement with experimental observations from accelerated electromigration tests [3]. In addition, we examine theoretically the effects of the simultaneous action of applied mechanical stress and electric field on transgranular void dynamics. Based on self-consistent numerical simulations of void morphological evolution, failure is predicted to occur by the coupling of two modes of surface morphological instability: one that is current-driven and another that is stress-driven. The void surface morphology associated with these instabilities consists of faceted slits and crack-like features, respectively, and is consistent with recent experimental observations [4]. Most importantly, it is predicted that appropriate tailoring of the current and stress conditions in textured films can stabilize the void surface, thus inhibiting failure.

THEORETICAL FORMULATION AND COMPUTATIONAL METHODS

Our analysis is based on a continuum formalism of surface mass transport under the action of external fields [7,10,12]. The total mass flux, $\boldsymbol{J}_s$, on the void surface is given by

$$\boldsymbol{J}_s = -\frac{D_s \delta_s}{\Omega k_B T} \left(q_s^* \boldsymbol{E}_s + \nabla_s \mu\right), \tag{1}$$

where D_s is the surface atomic diffusivity, Ω is the atomic volume, δ_s / Ω is the number of surface atoms per unit area, k_B is Boltzmann's constant, T is temperature, q_s^* is a surface effective charge [1], $\boldsymbol{E}_s$ is the local electric field component tangent to the void surface, μ is the chemical potential of an atom on the void surface, and ∇_s is the surface gradient operator. In Eq. (1), $-q_s^* \boldsymbol{E}_s$ expresses the local electromigration force that acts on the void surface. Both surface free energy and elastic strain energy contribute to μ, which is expressed by

$$\mu = \mu_0 + \Omega \left(\frac{1}{2} tr(\sigma \cdot \varepsilon) - \gamma \kappa \right). \tag{2}$$

In Eq. (2), μ_0 is a reference chemical potential for a flat stress-free surface, σ and ε are the local stress and strain tensors, respectively, γ is the surface free energy per unit area, and κ is the local surface curvature. Mass conservation gives the evolution of the local displacement normal to the void surface, u_n, through the continuity equation, i.e.,

$$\frac{\partial u_n}{\partial t} = -\Omega \, \nabla_s \cdot \boldsymbol{J}_s. \tag{3}$$

The electric field, $\boldsymbol{E}$, in the metallic conductor can be written as $\boldsymbol{E} \equiv -\nabla\Phi$, where the electrostatic potential, Φ, obeys Laplace's equation, i.e., $\nabla^2\Phi = 0$. Cauchy's mechanical equilibrium equation, $\nabla \cdot \boldsymbol{\sigma} = \boldsymbol{0}$, also is satisfied. The mechanical deformation of the solid is addressed within the framework of isotropic linear elasticity and in the limit of infinitesimal displacements: $\varepsilon = (1/2)\left[\nabla \boldsymbol{u} + (\nabla \boldsymbol{u})^T\right]$, where $\boldsymbol{u}$ is the displacement field and ^{T} denotes the transpose of a tensor. Our analysis is limited in two dimensions, x (length) and y (width), which implies that the void extends throughout the film thickness (in z) [10,12]. In this 2-D model, plane strain conditions are assumed. The void surface is considered to be traction free, while the boundaries of the 2-D computational domain are subjected to a constant applied stress, σ_0, imposed as hydrostatic tension. This loading resembles the strain state of interconnect films after thermal processing and aging. The void surface and the film's edges are modeled as electrically insulating boundaries, while a constant electric field, $\boldsymbol{E}_\infty = E_\infty \hat{\mathbf{x}}$, is imposed far away from the void. Surface adatom diffusivity is expressed by

$$D_s = D_{s,\min} f(\theta), \tag{4}$$

where $D_{s,\min}$ is the minimum surface diffusivity corresponding to a specific surface orientation, and $f(\theta) \geq 1$ is an anisotropy function of the angle θ formed by the local tangent to the surface and $\boldsymbol{E}_\infty$. In the present treatment, we have adopted the simple functional form

$$f(\theta) = 1 + A \cos^2[m(\theta + \phi)], \tag{5}$$

where A, m, and ϕ are dimensionless parameters that determine the strength of the anisotropy, the grain symmetry, and the misorientation of a symmetry direction of fast surface diffusion with respect to the applied electric field, respectively. On the other hand, γ is assumed to be isotropic in Eq. (2); this assumption is justified by recent atomistic simulations, according to which the dependence of γ on θ is weaker by orders of magnitude than that of D_s [13]. Specifically, for the materials and temperature range of interest the strength of the anisostropy, A, can reach values as high as 10^4; the anisotropy strength increases with decreasing temperature [13]. Finally, q_s^* also is assumed to be independent of surface orientation.

Dimensional analysis of Eqs. (1) - (3) yields three important dimensionless parameters: the surface electrotransport number [7], $\Gamma \equiv E_\infty q_s^* w^2 / (\gamma\Omega)$, the scaled strain energy, $\Sigma \equiv \sigma_0^2 w / (E\gamma)$, and the dimensionless void size, $\Lambda \equiv w_t / w$. Γ scales electric forces with capillary forces and Σ scales elastic strain energy with surface energy; E is the Young's modulus of the material, w is the width of the film, and w_t is the initial extent of the void across the film. The resulting time scale is $\tau \equiv k_B T w^4 / (D_{s,\min} \delta_s \gamma\Omega)$. In an Al film of $w \approx 1$ μm, a value of Γ=50 corresponds to a current density of about 2 MA/cm^2, which is typical of accelerated electromigration experiments [3,4]. For the same film, a value of $\Sigma = 1$ corresponds to an applied stress of about 140 MPa; this is typical of residual stresses in interconnect lines after cooling and aging [1,2]. Finally, for the same film τ is estimated on the order of 10^4 hours.

The electrostatic potential and the displacement vector on the domain boundary are computed using a Galerkin Boundary Element (BEM) formulation. The corresponding electric and stress fields on the boundary are computed using the formulation of Ref. 14. The surface flux divergence is calculated based on a centered finite-difference scheme and Eq. (3) is integrated using an Adams-Bashforth algorithm to compute the evolution of the normal surface displacement, $u_n(t)$. Our BEM discretization is adaptive and employs several hundred nodes along the void surface. In most of our computations the void surface is assumed to intersect the film's edge at 90°, as suggested by experimental observations [6]. This choice of boundary condition does not alter qualitatively any of the results that we present below; we have established this by comparison with results of more involved simulations that we performed, where the angle of intersection is not set and the diffusional propagation is extended over the entire length of the film's edge [12]. In the simulations, the initial void shape is taken to be semi-circular, i.e., a configuration that includes all possible surface orientations with respect to $\boldsymbol{E}_\infty$.

RESULTS AND DISCUSSION

We have analyzed transgranular void dynamics based on a systematic exploration of a six-dimensional (6-D) hyperspace defined by the six dimensionless parameters Γ, Σ, Λ, A, m, and ϕ; note that there is only one dimensionless parameter expressing strength of applied stress, because the loading mode is taken to be purely hydrostatic. This parametric search is the key to unraveling the most interesting nonlinear dynamical behavior. We have focused on three grain symmetries: six-fold ($m=3$), four-fold ($m=2$), and two-fold ($m=1$), which are representative of $\langle 111\rangle$-, $\langle 100\rangle$-, and $\langle 110\rangle$-oriented grains, respectively [6]. For these three symmetry cases, the parameter range that we have investigated is given by $0 \leq \Gamma \leq 500$, $0 \leq \Sigma \leq 2$, $0.1 < \Lambda \leq 0.8$, $1 < A \leq 10^4$, and $|\phi| \leq \pi/(2m)$. In our discussion below, we focus first on void dynamical behavior in unpassivated metallic thin films, where the stress level can be approximated by Σ=0. Next, we discuss the effects of mechanical stress in passivated films by exploring the dynamics as the level of mechanical stress in the film increases, i.e., along the Σ-axis.

Void Faceting and Migration

A very common response of transgranular voids under the action of electric fields is faceting of the void surface because of the anisotropy of surface diffusion; this faceting effect is more pronounced at high grain symmetry, $m > 1$, and under strong electric fields for high anisotropy strength. Results representative of this type of response from our numerical simulations are shown in Fig. 1 for voids that are located both in the middle and at edge of the film. For voids in the interior of the film such as those shown in Figs. 1(a) and 1(b), the projection of the void morphology can be described accurately as a closed polygon with $N = 2m$ sides under the conditions described above; the number of facets on the void surface also is affected strongly by the phenomenon of facet selection described below. For voids at the edge of the film, the corresponding shape is semi-polygonal, as shown in Figs. 1(c) and 1(d). The orientation of the facets on the void surface depends on the strength of the applied field and the anisotropy. In general, the facets do not develop along symmetry directions, which is a well understood current-induced drift effect [10].

We have derived analytical expressions for the mobility of stable faceted voids that migrate at constant speed, v_m, where $\boldsymbol{v}_m = v_m \hat{\mathbf{x}}$ is the void migration velocity [15]. Specifically, we have derived a useful scaling result for characterization of stable faceted void migration: $v_m \propto 1/l$, where l is a measure of the average facet length. This expresses for the case of faceted voids the well known result that small voids migrate faster than large voids. This scaling result can be rewritten as $v_m \propto 1/(A_v)^{1/2}$, where A_v is the area of the void in the xy-plane that is proportional to the void volume for the cylindrical voids considered in our 2-D modeling. We have used the results of our numerical simulations to examine the validity of this scaling relation. Some representative results are shown in Fig. 2, which demonstrates that indeed the constant migrating speed of a stable faceted void scales with $(A_v)^{-1/2}$. The above results extend and generalize the classical result derived by Ho, according to which a circular void with isotropic surface properties migrates under electromigration conditions at constant speed inversely proportional to its radius, R [16]. A linear analysis for the morphological stability of faceted voids migrating at constant speed also has been presented in Ref. 15.

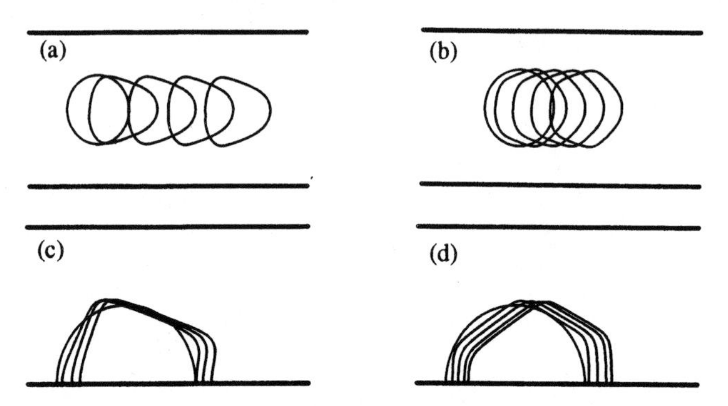

FIG. 1. Evolution of transgranular voids in a metallic film under the action of an electric field directed from left to right. The corresponding parameters and times for the different morphologies from left to right are: (a) Γ=50, Λ=0.45, A=10, m=2, ϕ=45°; t=0, 0.702, 1.946, 3.136, 4.233$\times 10^{-4}$ τ, (b) Γ=50, Λ=0.5, A=10, m=3, ϕ=0°; t=0, 0.412, 1.131, 1.847, 2.545$\times 10^{-4}$ τ, (c) Γ=50, Λ=0.5, A=10, m=2, ϕ=45°; t=0, 0.235, 0.706, 1.259$\times 10^{-4}$ τ, and (d) Γ=50, Λ=0.5, A=10, m=3, ϕ=-15°; t=0, 0.462, 1.075, 1.687, 2.094$\times 10^{-4}$ τ. In (a) and (b), the void is initially circular and placed in the middle of the film. In (c) and (d), the void is initially semi-circular and its surface intersects the edge of the film.

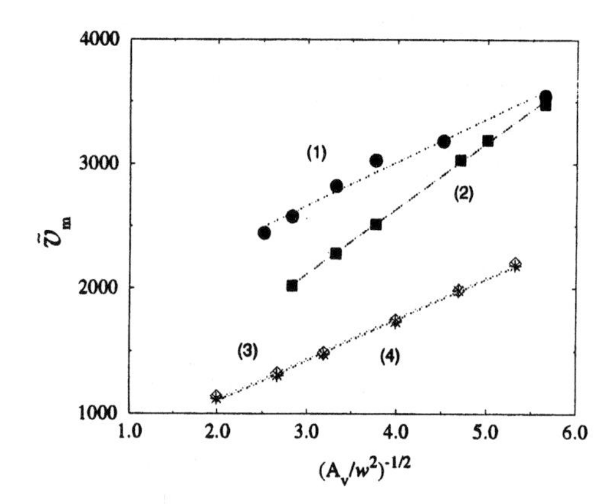

FIG. 2. Dependence of void migration speed on void size for stable faceted voids under electromigration conditions. The scaling relation $v_m \propto 1/(A_v)^{1/2}$ discussed in the text is verified for the cases: (1) Γ=50, A=10, m=2, ϕ=45°; (2) Γ=50, A=10, m=3, ϕ=0°; (3) Γ=50, A=10, m=2, ϕ=45°; (4) Γ=50, A=10, m=3, ϕ=-15°. In (1) and (2), the void is initially circular and is placed in the middle of the film. In (3) and (4), the void is initially semi-circular and is placed at the edge of the film. A dimensionless migration speed [15] is plotted.

Facet Selection and Faceted Slit Formation

A commonly observed dynamical phenomenon in our simulations for high grain symmetry, $m = 2$ and $m = 3$, is facet selection, i.e., a process during which one or more of the void surface facets get eliminated accompanied by the simultaneous growth of their neighboring facets. This is demonstrated in Fig. 3(a) for parameters $m = 3$, $\Gamma=50$, $\Lambda=0.65$, $A=10$, and $\phi=30^o$. In this case, the current crowding effect is very pronounced at the tip of the void close to the upper edge of the film. The combined action of this effect and capillarity drive a failure mechanism, which is characterized by facet selection that eliminates ultimately the originally formed right tilted facet and proceeds simultaneously with extension of the void tip toward a wedge-like void morphology. This facet selection mechanism also is demonstrated in Fig. 3(b) for parameters $m = 3$, $\Gamma=50$, $\Lambda=0.5$, $A=10$, and $\phi=-15^o$. In this case, the initially developed semi-hexagonal faceted void evolves toward a wedge-shaped void by elimination of the right tilted facet; the wedge-shaped void is morphologically stable and migrates along the film at constant speed.

An interesting case of failure initiated by a facet selection process in demonstrated in Fig. 3(c), for a parameter set identical to that of the previous case, Fig. 3(b), except for the applied electric field that is three times stronger, $\Gamma=150$. Here, elongation of the left tilted facet and complete destabilization of the right vertical facet are observed in addition to the facet selection dynamics. As a result, a slit forms that consists of two parallel planar surfaces. The faceted slit propagates fast and causes the failure of the film. This type of failure, predicted for grains of high symmetry ($m = 3$), suggests that slit-like failure is possible even in $\langle 111\rangle$-oriented grains in textured interconnect lines; however, such a failure mechanism would require the action of very strong electric fields.

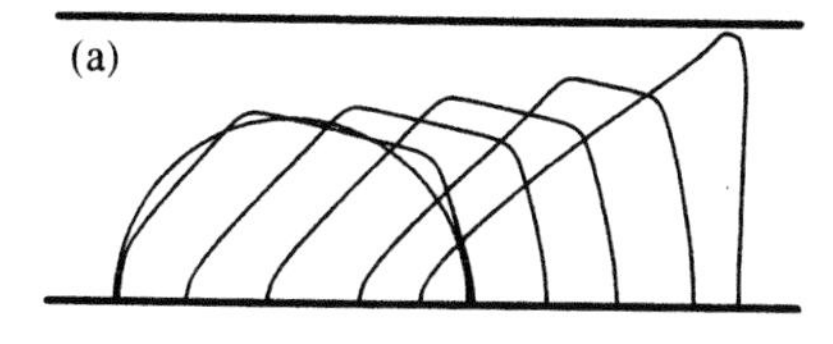

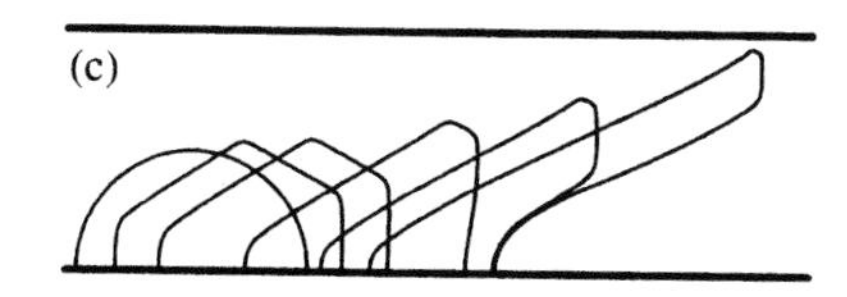

FIG. 3. Evolution of initially semicircular transgranular voids in a metallic film with six-fold symmetry, $m = 3$, under the action of an electric field directed from left to right. The corresponding parameters and times for the different morphologies from left to right are: (a) $\Gamma=50$, $\Lambda=0.65$, $A=10$, $\phi=30^o$; $t=0$, 0.034, 0.336, 0.610, 0.864, 1.058×10^{-3} τ, (b) $\Gamma=50$, $\Lambda=0.5$, $A=10$, $\phi=-15^o$; $t=0$, 0.0348, 0.431, 1.041, 1.934×10^{-3} τ, and (c) $\Gamma=150$, $\Lambda=0.5$, $A=10$, $\phi=-15^o$; $t=0$, 0.574, 1.272, 2.486, 3.395, 3.929×10^{-4} τ.

Failure due to Propagation of Slit-like Features

Failure due to slit formation and propagation can occur easily in the lowest-symmetry case that we have examined, $m = 1$, which is characteristic of $\langle 110\rangle$-oriented grains. A simulation result that is very representative of this dynamical behavior is shown in Fig. 4 for A=1000 and ϕ=-45° at Γ=150 and Λ=0.3. In general, the orientation of the slit depends strongly on the anisotropy parameters, A and ϕ, as well as the experimental conditions. Figure 4 demonstrates the propagation of a slit across the film, at an almost vertical orientation, which is promoted for high anisotropy strength. In addition, Fig. 4 demonstrates that large void sizes are not necessary for slit formation. Under the appropriate range of electromigration conditions, expressed by Γ, smaller voids can evolve into narrower slits. Furthermore, slit formation is observed to occur not only across the film but also along the film. The latter response is observed by changing the misorientation angle, ϕ, to positive values and for anisotropy strengths on the order of 10; under these conditions the void collapses into an elongated non-fatal morphology of high aspect ratio.

FIG. 4. Evolution of an initially semi-circular transgranular void in a metallic thin film with two-fold symmetry, $m=1$, under an electric field directed from left to right. The corresponding parameters and times for the different shapes from left to right are: Γ=150, Λ=0.3, A=1000, ϕ=-45°; t=0, 0.672, 3.014, 7.030, and $9.288\times10^{-7}\ \tau$.

In all cases of slit formation that we observed in our numerical simulations, for grain symmetries $m=1$, $m=2$, and $m=3$, the translation of the right (cathode) end of the void practically stops and the evolution is characterized solely by severe morphological change. Clearly, slit formation is the outcome of an instability in the competition for mass transport between electromigration and capillarity, as expressed by the high value of Γ for given values of the other parameters.

Propagation of Surface Waves on the Void Surface

Wedge-shaped voids also can evolve for low grain symmetry, $m=2$ and $m=1$, in certain regions of parameter space. Figures 5(a) and 5(c) show the evolution of two voids toward stable wedge-like shapes that translate along the film at constant speed for Γ=50, Λ=0.5, A=10, and $\phi=\pi/(2m)$ for both grain symmetries. In addition, we have predicted that surface waves can bifurcate from such stable wedge-like void morphologies under conditions that render the above faceted shapes unstable. Propagation of such surface waves on void surfaces is demonstrated in Figs. 5(b) and 5(d) for the same set of parameters, as given above, only with the exception of a larger void size, Λ=0.8. In both cases, we observe appearance of solitary-like waves that travel along the void surfaces driven by the electric field. As this soliton-like feature propagates toward the cathode end of the void, the tip of the void extends and propagates in the same direction along the void surface. The soliton-like feature disappears gradually, while the extension of the void tip continues to be driven by the local current crowding and leads to failure. The main qualitative difference between the evolving morphologies of Figs. 5(b) and 5(d), is that at the higher symmetry case, $m=2$, the void surface is faceted prior to the appearance of the surface wave. Note also that in the case where $m=1$, the void assumes an almost wedge-like shape upon failure. The dynamical behavior of Figs. 5(b) and 5(d) has been obtained starting from the asymptotic states of Figs. 5(a) and 5(c), respectively, marching along the Λ-axis in parameter space and crossing over a critical void size.

The portion of parameter space that we have explored so far corresponds to a 5-D hypersurface of the full 6-D parameter space; this hypersurface represents void dynamical behavior in the absence of simultaneous hydrostatic loading of the metallic films. Our simulation results are in very good agreement with available recent experimental observations for parameters compatible with the reported experimental conditions [3].

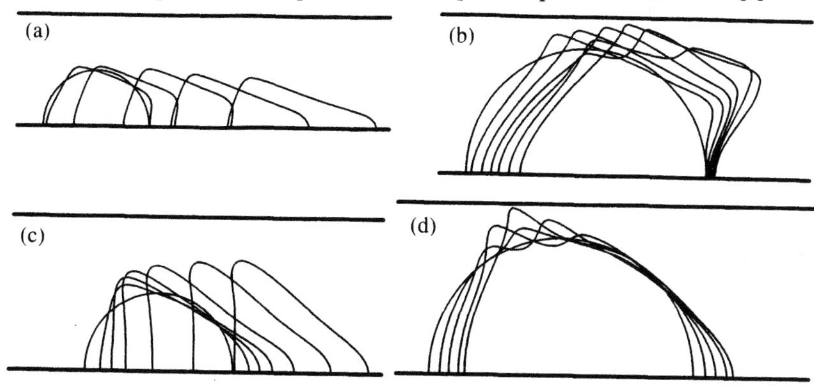

FIG. 5. Evolution of initially semicircular transgranular voids in metallic films with four-fold symmetry, m=2, (a) and (b), and two-fold symmetry, m=1, (c) and (d), under the action of an electric field directed from left to right. The corresponding parameters and times for the different morphologies from left to right are: (a) $\Gamma=50$, $\Lambda=0.5$, $A=10$, $\phi=45^{o}$; t=0, 0.021, 0.243, 0.699, 1.135, 1.645×10^{-4} τ, (b) $\Gamma=50$, $\Lambda=0.8$, $A=10$, $\phi=45^{o}$; t=0, 0.401, 0.813, 1.144, 1.401, 1.743, 2.147×10^{-4} τ, (c) $\Gamma=50$, $\Lambda=0.5$, $A=10$, $\phi=90^{o}$; t=0, 0.500, 0.997, 1.785, 3.575, 6.443, 9.160×10^{-4} τ, and (d) $\Gamma=50$, $\Lambda=0.8$, $A=10$, $\phi=90^{o}$; t=0, 0.381, 0.710, 1.093, 1.346×10^{-4} τ.

Effects of Mechanical Stress on Current-Driven Void Dynamics

The effects of mechanical stress on transgranular void dynamics are examined, first, without the simultaneous action of an electric field. It is demonstrated in Fig. 6 that failure can occur under the action of hydrostatic tension alone and for high-symmetry grains, $m=3$, if the level of the applied stress is higher than a critical level; this is the outcome of a morphological instability that is well understood for surfaces of stressed solids [11]. For a stress level Σ=1.2, Fig. 6 shows that the initially semi-circular void morphology becomes faceted and, later, fine-scale crack-like features start emanating from corners of the faceted shape; propagation of these crack-like features provides the mode of failure in this case. The location of crack formation is determined by the distribution of elastic strain energy on the faceted void surface. The directions of crack propagation are symmetric with respect to both principal stress directions.

Next, the electromigration-induced void dynamical response is examined under the simultaneous action of mechanical stress. Figures 7(a)-(f) demonstrate how the void dynamics changes as the applied hydrostatic tension in the film increases. Specifically, void dynamics is investigated along the Σ-axis of the 6-D parameter space starting from the point that corresponds to the dynamical behavior of Fig. 3(c), i.e., a case of slit-like failure for $m=3$ at Σ=0. In Fig. 7(a), $\Sigma=0.3$, and the dynamical sequence resembles closely that of Fig. 3(c): faceting is followed by facet selection, wedge formation, and faceted slit formation and propagation. In this case, however, after the slit propagates a certain distance, a finer-scale crack-like feature forms and its tip propagates at a speed greater by about two orders of magnitude than the average void migration speed. Thus, open-circuit failure is caused by the fast propagation of the crack-like feature's tip; this feature is narrower than the original slit by at least an order of magnitude. Fig. 7(b) shows the corresponding dynamical sequence at $\Sigma = 0.7$. The void shape evolves again into a wedge but interestingly, subsequent faceted slit formation is not observed in this case. Instead, a fine-scale crack-like feature forms at the tip of the wedge and propagates fast to cause failure. As the applied stress keeps increasing, not even the facet selection process can be completed toward wedge formation. Instead, crack-like features appear at corners of the faceted void shape, where the strain energy density increases abruptly. Such a dynamical behavior is shown in Fig. 7(c) at $\Sigma = 1.05$; again, failure occurs by tip propagation of the crack-like feature.

The dynamical behavior of the transgranular void changes drastically if the applied stress is increased slightly to $\Sigma = 1.075$, as demonstrated in Fig. 7(d). Specifically, there is a narrow range of stress within the interval $1.05 < \Sigma < 1.1$, where neither electromigration-induced facet selection nor stress-induced crack formation are initiated. Instead, the faceted void is stable and it migrates along the film at constant speed maintaining its semi-hexagonal shape. This response implies that film failure due to electromigration-induced slit-like morphological instabilities of the void surface can be inhibited by the simultaneous action of a mechanical stress. The applied stress level must be just enough to balance the destabilizing effect of the electric field without causing crack-like morphological instabilities that lead to a faster mode of failure.

Increasing the applied stress above the narrow range of stability leads to formation of crack-like features that emanate from the faceted void surface. This is shown in Fig. 7(e) for $\Sigma = 1.1$. Facet selection can be initiated again as in the case of Fig. 7(c); however, the strain energy distribution on the void surface is different here and leads to formation of a crack-like feature at the top of the void. At higher applied stress, crack-like failure can occur even faster than facet selection. This is demonstrated in Fig. 7(f) for $\Sigma = 1.5$; crack-like features form at the two corners of the faceted shape where the strain energy density along the void surface increases most abruptly. At such high stresses, the crack propagation directions are almost symmetric with respect to both principal stress directions.

The predicted general mixed mode of failure, where the void surface exhibits electromigration-induced features, such as faceted slits, coexisting with stress-induced crack-like features is consistent with recent experimental observations. For example, such void morphologies have been observed by Joo and Thompson through SEM measurements in passivated single-crystalline Al films [4]. Generally, the location of crack formation is correlated strongly with the distribution of elastic strain energy on the faceted void surface.

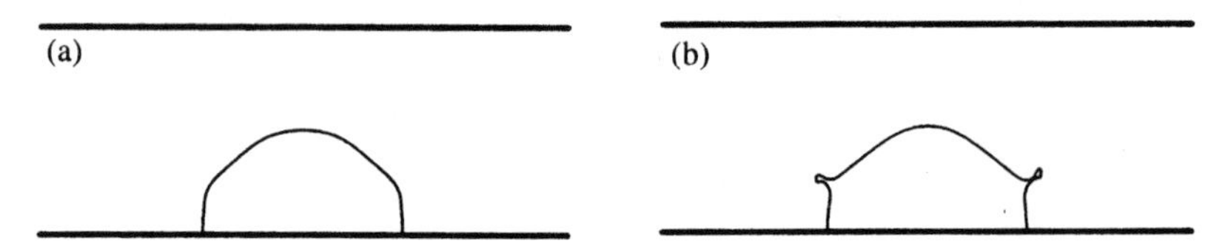

FIG. 6. Evolution of an initially semi-circular void in a metallic film under the action of hydrostatic tension. The corresponding parameters are Γ=0, Σ=1.2, Λ=0.5, A=1000, $m = 3$, and ϕ=0° and the times are (a) t=1.411×10^{-6} τ and (b) t=2.983×10^{-6} τ.

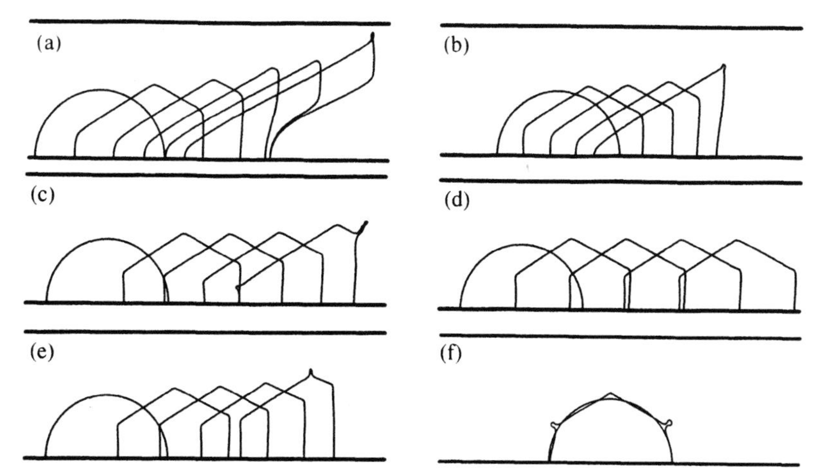

FIG. 7. Evolution of transgranular voids in a $\langle 111 \rangle$-oriented grain of a metallic film under the action of both hydrostatic tension and an electric field directed from left to right. The corresponding parameters are Γ=150, Λ=0.5, A=10, $m=3$, ϕ=-15° and (a) $\Sigma=0.3$, (b) $\Sigma=0.7$, (c) $\Sigma=1.05$, (d) $\Sigma=1.075$, (e) $\Sigma=1.1$, and (f) $\Sigma=1.5$. The times corresponding to the different morphologies from left to right are: (a) t=0, 1.047, 2.061, 2.776, 3.228, 3.608×10^{-4} τ, (b) t=0, 0.672, 1.453, 2.176, 2.773×10^{-4} τ, (c) t=0, 1.901, 2.974, 4.012, 4.959×10^{-4} τ, (d) t=0, 1.314, 2.702, 4.089, 5.477×10^{-4} τ, (e) t=0, 1.778, 2.882, 3.976, 5.125×10^{-4} τ, and (f) t=0, $2.041\times10^{-6}\tau$.

For low-symmetry grains, $m=1$, the propagation of slits is generally enhanced under the action of mechanical stress. This is demonstrated in Fig. 8, for the same parameters as those in Fig. 4, but for nonzero stress. Again, the coupling is shown of the two modes of instability driven by electromigration and stress, respectively. Formation of crack-like features is shown to occur at the tip of the void; these crack-like features form earlier as the level of the applied stress increases. The synergistic effects of stress and electromigration toward failure in this case of grain symmetry also explain the important role of film texture in interconnect reliability, as a means of controlling grain orientation. Finally, the simultaneous action of mechanical stress inhibits the propagation along the void surface of the solitary-like waves shown in Fig. 5. Specifically, faster failure occurs aided by stress-induced void tip extension before completion of the wave propagation toward the cathode end of the void.

CONCLUSIONS

In conclusion, we have presented a systematic theoretical study of transgranular void dynamics in metallic thin films and modeled failure mechanisms mediated by such void dynamical phenomena driven by electromigration and applied mechanical stress. Our modeling study has been based on self-consistent numerical simulations of void morphological evolution, where the anisotropy of surface diffusivity has been taken into account explicitly. In unpassivated metallic films, our simulations are captured a wide variety of nonlinear dynamical phenomena that are important for understanding metallic thin-film failure. These phenomena include void faceting, migration at constant speed of stable faceted voids, facet selection and wedge-shape void formation, formation of faceted slits, propagation of narrow slits across the film's width to cause failure, as well as propagation of surface waves on void surfaces prior to failure.

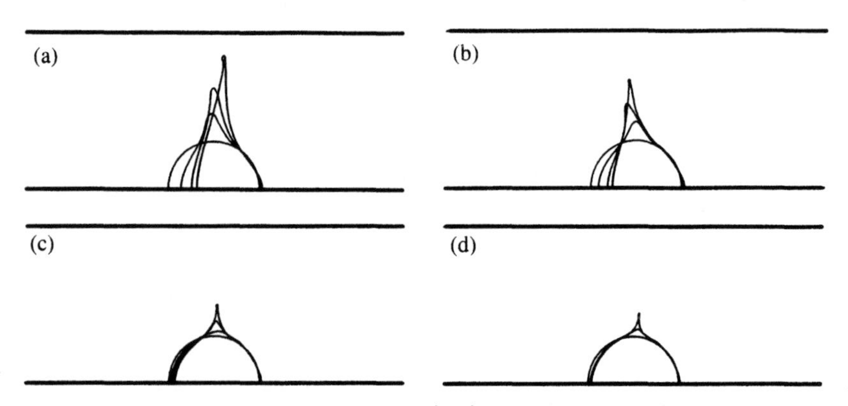

FIG. 8. Evolution of transgranular voids in a $\langle 110 \rangle$-oriented grain of a metallic film under the action of both hydrostatic tension and an electric field directed from left to right. The corresponding parameters are $\Gamma=150$, $\Lambda=0.3$, $A=1000$, $m=1$, $\phi=-45^{\circ}$ and (a) $\Sigma=0.3$, (b) $\Sigma=0.7$, (c) $\Sigma=1.2$, and (d) $\Sigma=1.5$. The times corresponding to the different morphologies from left to right are: (a) $t=0$, 0.977, 2.216, 3.178×10^{-7} τ, (b) $t=0$, 0.563, 1.327, 1.929×10^{-7} τ, (c) $t=0$, 1.889, 3.408, 4.691×10^{-8} τ, and (d) $t=0$, 2.112, 2.781×10^{-8} τ.

The simultaneous effects of mechanical stress on the electromigration-induced void dynamics also have been explored in passivated metallic films that are mechanically confined. Under such conditions, failure occurs through the coupling of two modes of instability: one is driven by electromigration and leads to slit formation, while the other is driven by stress and leads to formation of finer-scale crack-like features. Interestingly, for textured films of high grain symmetry, there exists a narrow range of electromechanical conditions over which the effects of the two driving forces can be balanced by the stabilizing effect of capillary forces; therefore, the faceted void surface becomes stable and failure is prevented. This possibility has very important implications for the reliability of metallic thin-film interconnects: thermal processing strategies of these films can be developed in order to tailor the resulting state and level of stress in the films to inhibit electromigration-induced failure.

Atomistic simulations can contribute significantly toward predictive modeling in this area. First, an accurate materials property database can be developed for the dependence of thermodynamic and transport properties on void surface orientation, which has been demonstrated to play important roles in determining void dynamics. In addition, the length scale characteristic of the width of the stress-induced crack-like features that we have predicted is on the order of nanometers, i.e., very fine for purely phenomenological analysis. Molecular-dynamics simulations of stress-driven void tip phenomena can elucidate the possible role of dislocation-mediated mechanisms and upgrade the present mesoscopic-level modeling accordingly. This atomic-scale simulation work is currently underway and the results will be reported in forthcoming publications.

ACKNOWLEDGMENTS

The authors acknowledge fruitful discussions with C. V. Thompson, S. J. Zhou, and D. R. Clarke. This work was supported by the Frontiers of Materials Science Program of the UCSB Materials Research Laboratory and Los Alamos National Laboratory (Award No. STB-UC:97-63), by the National Science Foundation through a CAREER Award to DM (ECS-95-01111) and grant PHY94-07194 with the Institute for Theoretical Physics, UCSB, by ORISE through a Faculty Research Fellowship to DM, and by the US Department of Energy under contract DE-AC05-96OR22464 with Lockheed Martin Energy Research Corporation.

REFERENCES

[1] P. S. Ho and T. Kwok, Rep. Progr. Phys. **52**, 301 (1989); C. V. Thompson and J. R. Lloyd, MRS Bulletin **18**, No. 12, 19 (1993), and references therein.
[2] see, e.g., A. I. Sauter and W. D. Nix, J. Mater. Res. **7**, 1133 (1992).
[3] J. E. Sanchez, Jr., L. T. McKnelly, and J. W. Morris, Jr., J. Electron. Mater. **19**, 1213 (1990); J. Appl. Phys. **72**, 3201 (1992); J. H. Rose, Appl. Phys. Lett. **61**, 2170 (1992); O. Kraft, S. Bader, J. E. Sanchez, Jr., and E. Arzt, Mater. Res. Soc. Symp. Proc. **309**, 199 (1993); E. Arzt, O. Kraft, W. D. Nix, and J.E. Sanchez, Jr., J. Appl. Phys. **76**, 1563 (1994).
[4] Y.-C. Joo and C. V. Thompson, J. Appl. Phys. **81**, 6062 (1997).
[5] Z. Suo, W. Wang, and M. Yang, Appl. Phys. Lett. **64**, 1944 (1994); W.Q. Wang, Z. Suo, and T.-H. Hao, J. Appl. Phys. **79**, 2394 (1996).
[6] O. Kraft and E. Arzt, Appl. Phys. Lett. **66**, 2063 (1995); O. Kraft and E. Arzt, Acta Mater. **45**, 1599 (1997).
[7] D. Maroudas, Appl. Phys. Lett. **67**, 798 (1995); D. Maroudas, M. N. Enmark, C. M. Leibig, and S. T. Pantelides, J. Comp.-Aided Mater. Des. **2**, 231 (1995).
[8] L. Xia, A. F. Bower, Z. Suo, and C. F. Shih, J. Mech. Phys. Solids **45**, 1473 (1997).
[9] M. Schimschak and J. Krug, Phys. Rev. Lett. **80**, 1674 (1998).
[10] M. R. Gungor and D. Maroudas, Appl. Phys. Lett. **72**, 3452 (1998).
[11] see, e.g., D. J. Srolovitz, Acta Metall. **37**, 621 (1989); W. H. Yang and D. J. Srolovitz, Phys. Rev. Lett. **71**, 1593 (1993).
[12] M. R. Gungor and D. Maroudas, Surf. Sci. **418**, L1055 (1998).
[13] C.-L. Liu, J. M. Cohen, J. B. Adams, and A. F. Voter, Surf. Sci. **253**, 334 (1991).
[14] L. J. Gray, D. Maroudas, and M. N. Enmark, Comp. Mech. **22**, 187 (1998).
[15] H. S. Ho, M. R. Gungor, and D. Maroudas, Mater. Res. Soc. Symp. Proc. **529**, in press (1998).
[16] P. S. Ho, J. Appl. Phys. **41**, 64 (1970).

QUANTITATIVE OPTIMIZATION OF THE DOPANT CONCENTRATION RANGE IN BPSG FILMS FOR ULSI CIRCUIT TECHNOLOGY

V.Y. Vassiliev, J.Z.Zheng, and C. Lin
R&D Department, Chartered Semiconductor Manufacturing Ltd.
60 Woodlands Industrial Park D, Street 2, Singapore 738406
Tel: (65)-360-4733 Fax: (65)-362-2945

Films of borophosphosilicate glass (BPSG) are widely used in modern ULSI devices as a planarized interlayer dielectric material. BPSG films are prone to absorb moisture from ambient. This leads to defect formation and to the worsening of device metallization reliability. Approaches for quantitative characterization of the optimized dopant concentration range are developed based on summarized investigations of glass flowing properties on the device steps, BPSG moisture absorption and defects formation phenomena. Empirical equations for the calculation of the optimized dopant concentration range in BPSG films are presented.

INTRODUCTION

BPSG films are widely used in multilevel VLSI and ULSI circuit technology as a planarized interlayer dielectric (ILD) between the polysilicon or silicide gate and first level of device metallization. There are a few published papers that summarized the general information about BPSG films [1-4]. Based on our own work and published data, we proposed in previous paper [5] that the basic properties of BPSG films prepared by different deposition methods are similar and they can be generalized for full characterization of BPSG films. Quantitative characterization of BPSG planarization for isothermal anneal conditions and characterization of defect formation in BPSG films was discussed in the previous paper. We also developed an approach for characterization of an optimized dopant concentration range in BPSG films for their usage in modern multilevel ULSI circuit technology.

The investigation of BPSG films prepared by different deposition techniques has shown that BPSG films are unstable especially with high boron and phosphorus concentrations. Moisture absorption phenomenon and, as a result of this, defect formation phenomenon was found to be stronger in as-deposited films especially with high boron concentration [3,6]. Recently, important results of moisture absorption studies using Pressure Cooker Test (PCT) and/or Thermal Desorption (TD) method have been published [7,8]. Yoshimaru and co-workers found different behavior of boron and phosphorus interaction with the ambient water vapor [7]. The increase of boron concentration in BPSG films was found to enhance the moisture penetration depth in the film, but phosphorus was found to limit the absorbed water on the film surface. BPSG films were revealed to be a water getter due to moisture uptake from under and/or cap layer silicon dioxide films [8]. An analysis of published data about borosilicate glass (BSG) films shows that BSG films easily absorb moisture from ambient [3,9]. Thus, the impact of boron in moisture absorption phenomenon in BPSG film seems to be much more than that of phosphorus.

There are two approaches to reduce these undesirable effects. The first one is to decrease the total dopant concentrations in BPSG films. The second one is to introduce a thin undoped silicon dioxide film on top of the BPSG film to decrease the speed of moisture absorption and, as a result of this, to increase the time of defect appearance in films [6]. Unfortunately both of these approaches are in contrast with the necessity to improve BPSG planarization for smaller circuit geometry using either well known isothermal furnace anneal or relatively recently employed Rapid Thermal Processing (RTP) [10].

Summary of published data shows that in addition to glass flowing and defect formation, there is an issue of optimization of boron concentration in BPSG film to improve film stability itself without decreasing the planarization capability of glass. This issue becomes more serious as device spacing becomes smaller and temperature of film anneal becomes closer or lower than glass transition temperature. In this paper we present an approach for quantitative optimization of the dopant concentration range based on the summary of investigation on glass flowing properties, defects formation and moisture absorption phenomena in BPSG films.

RESULTS AND DISCUSSION

Quantitative Characterization of BPSG Planarization

Simplified schemes of device step with as-deposited and flowed BPSG film are presented in Fig.1a and Fig.1b. Based on the cross-section scanning electron microscopy analysis of single model device steps (height of polysilicon step-0.6 μm, width-2.5 μm, thickness of BPSG-1 μm), it was shown [5,11-13] that flowing properties of BPSG films at isothermal furnace conditions were described by the following empirical equation:

$$A = \arctan\left\{\tan A_0 \times 10^{-562(T_f - 973)/973/T_f \times 10^{0.045(3.2[B]+2.3[P])}}\right\}, \quad (1)$$

where A_0 and A are the initial and post-flowed angles (see Fig.1), T_f is the temperature of flowing in Kelvin degrees, $[P]$ and $[B]$ are the concentrations of dopants in films in terms

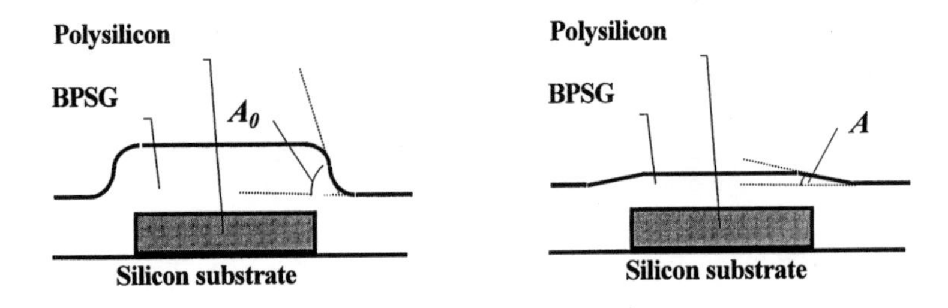

Fig.1. Scheme of device step with as-deposited (a) and flowed (b) BPSG film.

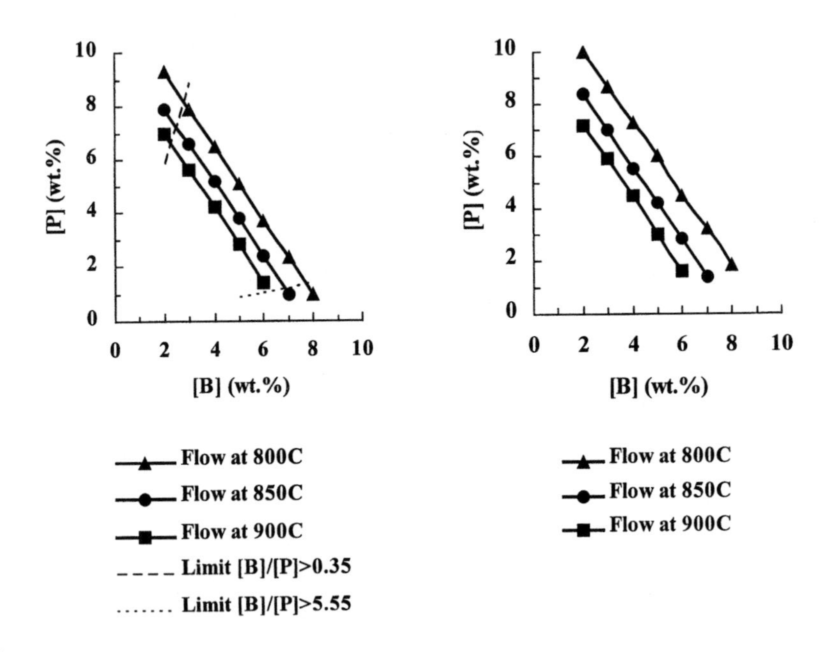

Fig.2. Calculated concentration dependencies for anneal in furnace (a) and RTP (b) conditions. (See text for details).

of weight percent (wt.%) of elements. The equation (1) is valid for the following initial range of parameters: the limits of dopant concentration range: $5.2<1.4[B]+[P]<13$ and $0.35<[B]/[P]<5.55$; the initial step coverage angle $A_0 < 85°$; thickness ratio of BPSG films to polysilicon step is more than 1.6; the time of flowing is 30 minutes in furnace; ambient of flowing are argon, or nitrogen, or oxygen. Fig.2a shows the calculated curves in accordance with equation (1) as a function of the flow temperature and dopant concentration for planarization of a single model device step to have a final planarization angle of 15°. Note that second concentration limit seems to be flexible and can be extended because no experimental points with lower or higher values have been checked.

For time dependence estimation, the modified equation from [14] can be used:

$$A_x = \arctan\left\{\tan A_t \left(t/t_x\right)^{0.42}\right\}, \qquad (2)$$

where A_t is the experimentally observed angle corresponding to time of flowing t; A_x is the unknown angle corresponding to time of flowing t_x.

Equations (1) and (2) allow us to understand a quantitative impact of parameters to the flowing properties of BPSG films. This data are presented in Fig.3 for the following initial conditions: A_0=75°; T_f=1123K (850°C); $[B]$=4.5 wt.%; $[P]$=4.5 wt.%; t=30 minutes. Data in Fig.3 shows that anneal temperature has the highest impact on flowing properties of BPSG films and the duration of anneal has the lowest impact on flowing properties. The impact of initial step coverage shows opposite behavior, namely the higher initial step coverage the worse flow properties of film on the step of device. Finally, for chosen example the impact of parameters T_f, A_0, $[B]$, $[P]$, t can be drawn quantitatively using tangents of line angles in Fig.3 as follows: -27; +16; -7; -5; -1.

It is believed that similar approach can be used for evaluation of BPSG film planarization with RTP anneal. Based on experimental data presented in [10] the empirical equation (3) was drawn. The correlation of calculated curves and experiment is shown in Fig.4. Fig.2b shows the calculated flowing curves using 40 second RTP conditions for different dopant concentration in film for planarization of a single device step with A_0=50° to have a final planarization angle A=30°.

$$A = \arctan\left\{\tan A_0 \times 10^{-143(T_f - 998)/998/T_f \times 10^{0.0506(3.2[B]+2.3[P])}}\right\}, \qquad (3)$$

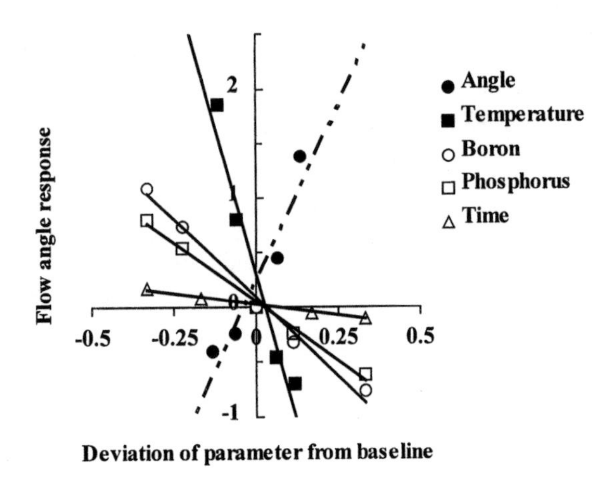

Fig.3. Calculated impact of parameters in BPSG flowing properties (See text for details).

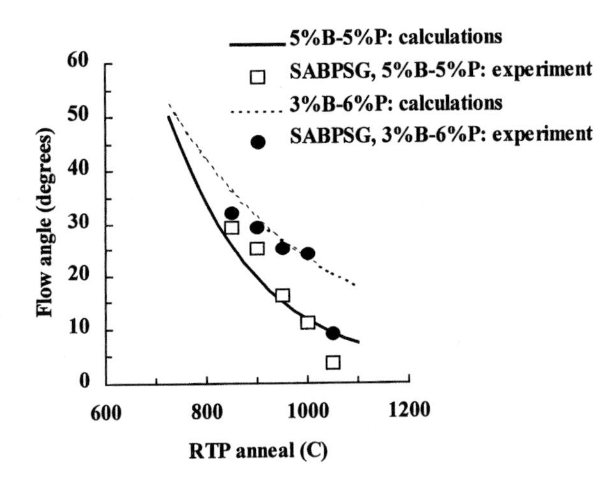

Fig.4. Correlation of experimental data [10] for 40 second RTP anneal of BPSG film and calculated curves in accordance with equation (3).

Quantitative Characterization of Defect Formation in BPSG Films

The problem of defect formation in BPSG films is well known [1-4]. Usually all types of defects in BPSG films are generalized as solid state BPO_4 and/or boric acid crystals. First assumption seems to be true for high-doped films annealed at high temperature [15]. To understand the defect formation phenomenon we have investigated the process of defect appearance on the BPSG film surface in detail [16]. It was found that defects formation is due to film moisture absorption followed by appearance of small liquid drops on the surface of BPSG, which grow in size due to merging of small defects. After that, transformation of liquid defects into different types of solid state defects have been found, which depends on the dopant concentrations in film and thermal anneal conditions [17-19]. It was observed during stability study of BPSG films [17-20] that the increase of boron concentration enhanced the trend of surface crystallization but no surface interaction with ambient moisture has been seen. At the same time, clear interaction of moisture with film surface and defects at high phosphorus concentration in BPSG film has been seen after long storage time of BPSG film [17-19].

A review of defect investigation in BPSG films prepared by several methods has been done in our previous papers [2,4,5] both for as-deposited and annealed films. A summary of published experimental data about defect formation in BPSG films [1,6,16,20-35] and the calculated concentration boundary of BPO_4 compound (using data [15] for ternary system SiO_2-B_2O_3-P_2O_5) are presented in Fig.5. Using optical microscopy for defect observation, two boundaries either for 1-hour or for 24-hours defect-free areas in as-deposited BPSG films were drawn [23,25]. Fig.6 shows a combination of 24-hours defect-free area boundary with the calculated curves in Fab.2a for BPSG flow at temperature 850°C and 900°C and isothermal condition. The optimized range of BPSG dopant concentrations is placed in between the above mentioned curves and it can be quantitatively characterized using empirical equations (4) and (5) for anneal time of 30 minutes in dry gas ambient:

$$0.35 < [B]/[P] < 5.55 \tag{4}$$

$$30.4 - 1.39[B] - 0.023T_a < [P] < -0.17[B]^2 - 0.19[B] + 9.6, \tag{5}$$

where dopant concentrations are presented in elemental weight % and T_a is the temperature of film anneal in Celsius degrees. Equations (4) and (5) allow us to calculate the dopant concentration in BPSG films in a wide dopant range that can meet the device planarization requirements and reduce the problems of defect appearance in film. Empirically optimized experimental data points in Fig.6 [4], which represent the actual BPSG film dopant concentrations in device production, were found to be in good agreement with the above calculations. According to these data, most of practically used BPSG films consist of approximately equal boron and phosphorus concentration.

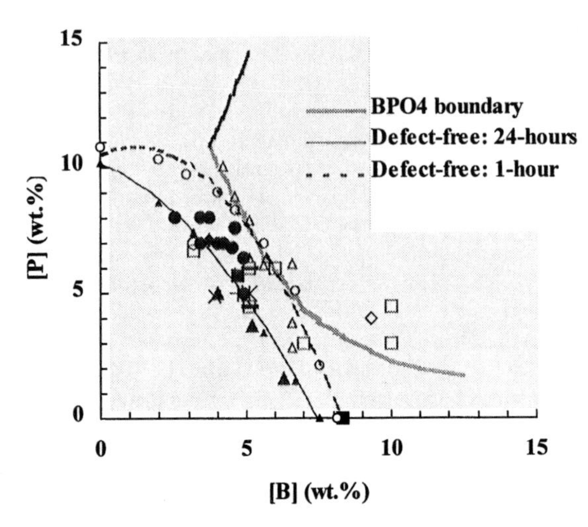

Fig.5. Summarized BPSG defect data (details see in [5]) and calculated boundary of BPO_4 Compound after high temperature anneal.

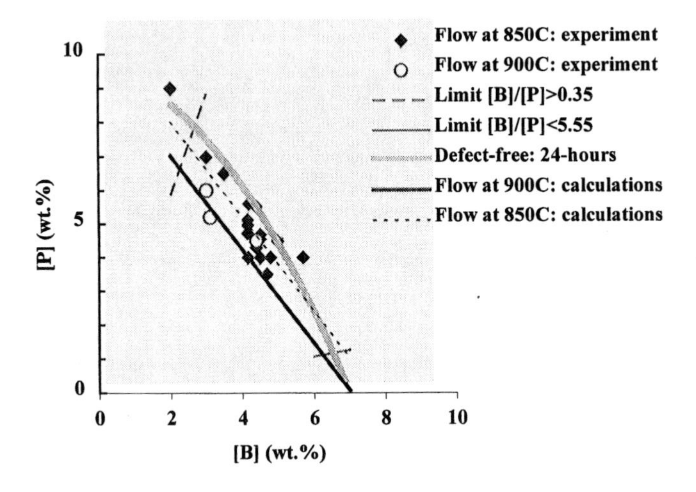

Fig.6. A combination of 24-hours defect-free area with the calculated curves for BPSG flow at isothermal condition (see text).

Quantitative Characterization of Moisture Absorption Phenomenon

As it was mentioned above, BPSG films are prone to absorb moisture from ambient that eventually causes the formation of defects in film. Interaction of water vapor with phosphorus in film is usually described as: $(\text{-Si-O-})_3P=O+H_2O \rightarrow (\text{-Si-O-})_3P(\text{-O-H})_2$, but until recently there was no water interaction scheme found for boron-contained silicon dioxide films. The ability of BPSG films to absorb moisture from ambient has been studied using Fourier Transform Infra Red spectroscopy (FTIR). This phenomenon, as well as defect formation phenomenon, was found to be more pronounced for highly boron-doped [3,6], and as-deposited films [36,37]. We found a significant increase of moisture-related peaks with the increase of boron content in BPSG films using FTIR monitoring of as-deposited sub-atmospheric pressure tetraethylorthosilicate-ozone BPSG films (SABPSG) during film storage at clean-room condition. At the same time those films were found to be much more stable after anneal at 850-900°C [36]. To prevent these effects, undoped silicon dioxide film with thickness of ~50Å are recommended to be used on the top of BPSG film [6,38,39]. Nevertheless, we found almost linear correlation between needed silicon dioxide cap-layer thickness and the total dopant concentration in SABPSG films [40]. More than 200Å oxide thickness was found to be needed to suppress defects in films with ~ 5 wt.% of each dopant.

The moisture absorption by film is a relatively slow process and it takes a few days to see some change in FTIR spectra. To accelerate the investigation of moisture-film interaction, a PCT is usually used [7]. This test together with the quantitative TD method allowed the observation of a principal difference in behavior of boron and phosphorus in film with respect to the moisture. The increase of boron concentration in BPSG films was found to enhance the moisture penetration depth in the film. At the same time the increase of phosphorus concentration provides a limit of water on the film surface. These data give us a possibility to compare defect formation and moisture absorption phenomena quantitatively in terms of boron share [B]/([B]+[P]) in the total dopant concentration in BPSG films. Thus, a correlation of 24-hours defect-free area boundary from Fig.5 [25] and the moisture penetration depth into the film during PCT (calculated using experimental data [7]) is presented in Fig.7. Data shows that the increase of boron share above ~ 30 % causes a sharp increase in moisture penetration depth during PCT (see open symbols). The decrease in total concentration of 24-hours defect-free area boundary can be seen in Fig.8 at about the same value of boron share in film (see triangles and solid line).

Therefore, in order to make BPSG films highly reliable by decreasing their ability to absorb moisture from ambient either during exposure at normal condition, or under the PCT, an addition limit of boron concentration in BPSG films can be drawn:

$$[B] \leq\sim 0.4[P] \tag{6}$$

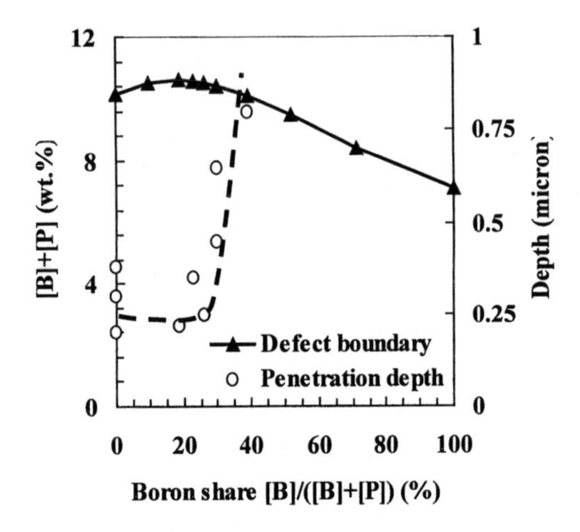

Fig.7. Calculated dopant concentration boundary of 24-hours defect-free area (Fig.5) and water penetration depth [7] vs. boron share in BPSG films.

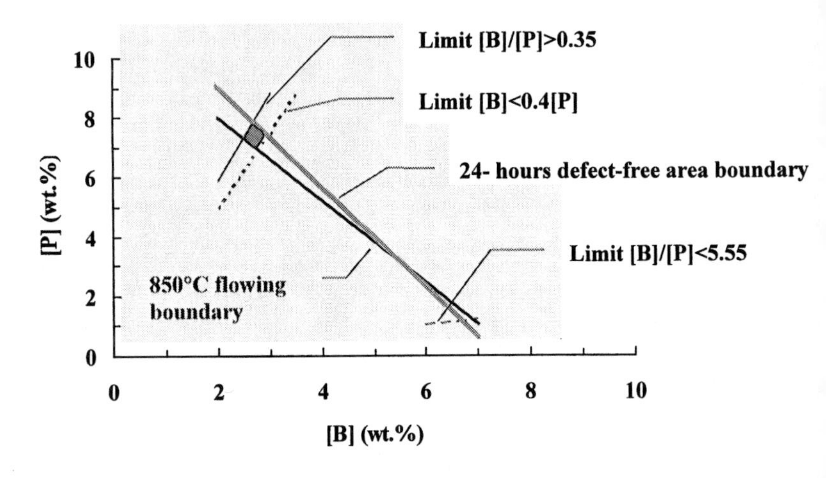

Fig.8. Example of optimized range of dopant concentration in BPSG films (shaded area). Conditions are the same as those of Fig.6.

Optimized Range of Dopant Concentration for BPSG Films

Summarized BPSG film defect data show that total dopant content in BPSG films can not exceed a certain limit, which can be chosen based on device and technology needs. Approaches for characterization of BPSG films presented above allow us to define the optimized range of dopant concentration which predicts good ILD planarization without moisture absorption and defect problems during the selected time of storage. For other initial conditions of step coverage and anneal, a similar optimized dopant concentration ranges can be easily obtained.

To achieve highly reliable BPSG films for usage in ULSI circuit technology the calculation of the optimized dopant concentration range can be easily done using empirical equations (4)-(6). Optimized range of dopant concentrations in BPSG films is shade in Fig.8. It is small, but as we mentioned above, the concentration limit $[B]/[P]>0.35$ can be slightly extended. The better film stability is expected to be at the boron share in total dopant concentration less than ~30%. Compare to films with approximately equal boron and phosphorus content, BPSG films with reduced boron share can be effectively employed in ULSI technology even with anneal at temperatures near or below the glass transition temperature.

CONCLUSIONS

The above data show that the flow properties, moisture absorption and defect formation phenomena in BPSG films can be summarized independently from the method of glass film deposition to provide a full material characterization. The approach for quantitative definition of optimized dopant concentration range of BPSG films for their usage in modern ULSI technology is originated. Based on analysis of glass flow properties on the device steps in isothermal or RTP conditions, as well as on a correlation of moisture absorption and defects formation phenomena, the optimized dopant concentration field attributed to the highly reliable BPSG films is outlined quantitatively. The empirical equations for quantitative characterization of BPSG films can be used to define the optimized range of dopant concentration, which predicts good planarization without moisture absorption and defect problems during the selected time of storage. Similar optimized dopant concentration ranges can be easily calculated for other conditions.

There are two unclear phenomena at the moment that makes our understanding of processes in BPSG films non-completed. First phenomenon is the enhancement of moisture absorption and defect formation with the increase of boron share in BPSG film higher than 30% that approximately corresponds to the boron share in BPO_4 compound. Second one is the mechanism of water penetration in highly boron doped BPSG films. These issues would be a matter of our future work.

ACKNOWLEDGMENTS

The authors would like to thanks Mr. S.K. Tang for his assistance.

REFERENCES

[1] W. Kern and G. L. Schnable, RCA Review, **43**, 423 (1982).
[2] T. G. Duhanova, V.Y. Vassiliev, and Y. I. Veretenin, in Reviews on Electronic Technology / Series 3, **1370,** Institute "Electronics", Moscow, Russia (1988).
[3] C.H.Ting, in *Handbook of Multilevel Metallization for Integrated Circuits*, S.R. Wilson, C.J.Tracy and J.L.Freeman, eds., Noyes Publications, 202 (1993).
[4] V.Y. Vassiliev,"Modern BPSG Film Technology", *in Visual booklet of tutorial course at the Fourth Dielectrics for ULSI Multilevel Int. Conf. (DUMIC),* 112 (1998).
[5] V.Y.Vassiliev and J.Z.Zheng, Electrochem. Soc. Proc., **PV 97-25,** 1199 (1997).
[6] K. Nauka and C. Liu, J. Electrochem. Soc., **138**, 2367 (1991).
[7] M. Yoshimaru and H. Matsuhashi, J. Electrochem. Soc., **143**, 3032 (1996).
[8] K. Ikeda, Y. Okazaki, S. Nakayama, and T. Tsuchiya, Proc. 1996 Symp. on VLSI Techn., Honolulu, 116 (1996).
[9] K. Fujino, Y. Nishimoto, N. Tokumatsu, and K. Maeda, J. Electrochem. Soc., **138,** 3019 (1996).
[10] R. Iyer, R.P.S. Thakur, H. Rhodes, R. Liao, R. Rosler, and E.Yieh, J. Electrochem. Soc., **143,** 3366 (1996).
[11] V.Y. Vassiliev, T.G. Duhanova, V. I. Kosyakov, and A. I. Shestakov, Electronic Industry (Russian), **5**, 40 (1988).
[12] V.Y. Vassiliev and T.G. Duhanova, Electron. Technics (Russian), Series 3, **143**, 38 (1991).
[13] V.Y. Vassiliev and T. G. Duhanova, J.Chem. Phys. (Russian), **11**, 1699 (1992).
[14] J.S. Mercier, Solid State Technol., **30**, 85 (1987).
[15] W.J. Englert and F.A. Hummel, J.Soc.Glass.Technol., **39,** 126T (1955).
[16] V.Y. Vassiliev, L. Wei, Z. J .Zheng, H. R. Wang, and L. Chan, Proc. of 3rd Int. Dielectrics for ULSI Multilevel Interconnection Conf.(DUMIC), 171 (1997).
[17] S.K. Tang, S. Mridha, V.Y.Vassiliev, J.Z. Zheng, and L. Chan, Proc. 7th Int. Symp. IC Technol., Singapore, 601 (1997).
[18] S.K. Tang, V.Y. Vassiliev, S. Mridha, and L. Chan. Abst. 44th Int. Symp. Amer. Vac. Soc., San Jose, 189 (1997).
[19] S.K. Tang, V.Y. Vassiliev, S.Mridha, and L. Chan, Proc. of 4th Int. Dielectric for ULSI Multilevel Interconnection Conf.(DUMIC), 287 (1998).
[20] A. Learn and B. Baerg, Thin Solid Films, **130**, 103 (1985).
[21] F. S. Becker, D. Pawlik, H. Schafer, and G. Staudigl, J. Vac. Sci. Technol. B, **4**, 732 (1986).
[22] M. Susa, Y. Hiroshima, K. Senda, and T. Takamura, J. Electrochem. Soc., **133**, 1517 (1986).

[23] Watkins-Johnson Company information material, Solid State Technol., **29**, 100 (1986).
[24] D. S. Williams and E. A. Dein, J. Electrochem. Soc., **134**, 657 (1987).
[25] V.Y. Vassiliev and T.G. Duhanova, Electronic Industry (Russian), **3**, 31 (1988).
[26] G. L. Schnable, A. W. Fisher, and J. M. Shaw, J. Electrochem. Soc., **137**, 3973 (1990).
[27] K. Ahmed and C. Geisert, J.Vac.Sci. Techn. A., **10**, 313 (1992).
[28] S. Imai, Y. Yabuuchi, Y. Terai et al., Appl.Phys.Lett., **60**, 2761 (1992).
[29] J. Coniff, M. Shenasa, L.Krott, and S. Woessner, Electrochem. Soc. Proc., **PV 93-25,** 84 (1993).
[30] S. M. Fisher, H. Chino, K. Maeda, and Y. Nishimoto, Solid State Technol., **36,** 55 (1993).
[31] J. Wilson and T. Stephens, Mat. Res. Soc. Symp. Proc., **284**, 211 (1993).
[32] J. Li, B. Fardi, M.D. Thach, M. Moinpour and F. Mogdaham, Proc. of 1st Int. Dielectrics for ULSI Multilevel Interconnection Conf. (DUMIC), 124 (1995).
[33] M. Yoshimaru and H. Wakamatsu, J. Electrochem. Soc., **143**, 666 (1996).
[34] S. Matsumoto, T. Hattori, Y. Hata, and H. Ogawa, Proc. of 13th Int. VLSI Multilevel Interconnection Conf. (VMIC), 122 (1996).
[35] J.C.S. Chu, T. Tu , K.C. Chen, W. Su and T. Chang, Proc. of 14th Int. VLSI Multilevel Interconnection Conf. (VMIC), 499 (1997).
[36] W.Lu, V.Y. Vassiliev, Z. J. Zheng, and L. Chan, Proc. of 3rd Int. Dielectrics for ULSI Multilevel Interconnection Conf. (DUMIC), 219 (1997).
[37] V.Y. Vassiliev, S.K. Tang, J.Z. Zheng, and M. Liao. Proc. 4th Int. Dielectric for ULSI Multilevel Interconnection Conf. (DUMIC), 253 (1998).
[38] D.S. Pyun and S.B. Kim, Proc.of 12th Int. VLSI Multilevel Interconnection Conf. (VMIC), 151 (1995).
[39] H. Bar-Ilan and I. Rabinovich, Proc. of 12th Int. VLSI Multilevel Interconnection Conf. (VMIC), 76 (1995).
[40] V.Y. Vassiliev, W. Lu, J.Z. Zheng, and Y.S. Lin, Proc.14th Int. VLSI Multilevel Interconnection Conf. (VMIC), 538 (1997).

PROCESS CONSIDERATIONS IN THE INTEGRATION OF HDPCVD FILMS

Rao V. Annapragada and Subhas Bothra
VLSI Technology, Inc.
1109, McKay Dr., MS 02, San Jose, CA 95131

HDPCVD films can be used in a number of different applications in the IC manufacturing. Void free gap fill is the main advantage of the process. The main applications of these films are shallow trench isolation, premetal dielectric, intermetal dielectric and passivation. The processing requirements for these applications, however, are significantly different. In this paper the process considerations in the integration of HDPCVD films for these different applications are presented.

SHALLOW TRENCH ISOLATION

For shallow trench isolation (STI) application, it is important to have wet etch rate closer to thermal oxide, little shrinkage, seam free bottom fill films, avoidance of the clipping of nitride and no trench side wall damage. To avoid the corner clipping problem it is important to deposit a thin layer with no bias power. However, as the wafer temperature is low due to the lack of sputtering component, the film deposited can be high in hydrogen content and may lead to hazing of the film. Therefore, it is important to optimize this step so that the film thickness is small and yet provide enough protection against the sputtering of the SiN layer. A thickness of about 200 to 300 Å would be reasonable at this step. The bias power then can be increased to an intermediate value of about 1000 W to enable initial gap fill and avoid possible sputtering of the initial thin layer. The bias power can then be increased to the final value to get good film quality. The aspect ratio being not very aggressive, relatively high deposition to sputter ratio can be used for gap fill. However, to maintain good quality of the oxide, mainly the wet etch rate closer to thermal oxide, there should be sufficient densification of the film which can be achieved by increasing the sputter component via the bias power or by running the process at high temperature without helium backside cooling. By running the process at a higher temperature, typically about 600 ^{0}C the quality of the deposited oxide is improved. STI being a front end process, has high thermal budget and it is possible to run it at these high temperatures. The deposition rate, however, is inversely proportional to the temperature. At higher temperatures as the desorbtion of the surface species results in lower surface concentration and hence lower deposition rate, there is a net decrease of about 5 Å/min deposition rate for each ^{0}C increase in temperature, decreasing throughput [1]. In addition, at higher deposition temperatures there is an increased wafer-oxide thermal expansion mismatch resulting in increased stress. Considering these factors an optimum recipe would be a two step process with depositing about one third of the thickness at higher temperature with helium backside cooling turned off and the rest of the thickness deposited at lower temperature with helium backside cooling turned on. To achieve good film quality bias power can be increased in the second step so that the bombardment of the film by the ions densifies the film making the wet etch rate of the film closer to thermal oxide. Fig. 1 shows the variation of the topography of STI wafer.

In the case of STI, contamination resulting from the sputtering of the chamber walls is also an important consideration. To avoid the contamination, prior to depositing the wafers the chamber walls need to be coated with oxide so that the chamber walls are not sputtered during the deposition step and the sputtered material from the chamber walls is prevented from contaminating the film. Based on our experiments with different seasoning times ranging from 10 sec to 30 sec, it was found that a seasoning time of about 10 sec (about 1000Å) is adequate to prevent any metallic contamination (Fig. 2). The level of metallic contamination was determined by performing carrier life time measurements on the films deposited with varying amounts of film deposited on the chamber walls.

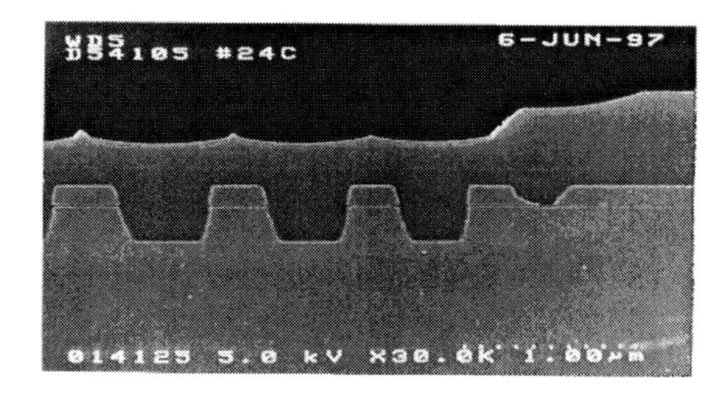

Fig. 1. SEM picture showing STI gap fill and variation of topography.

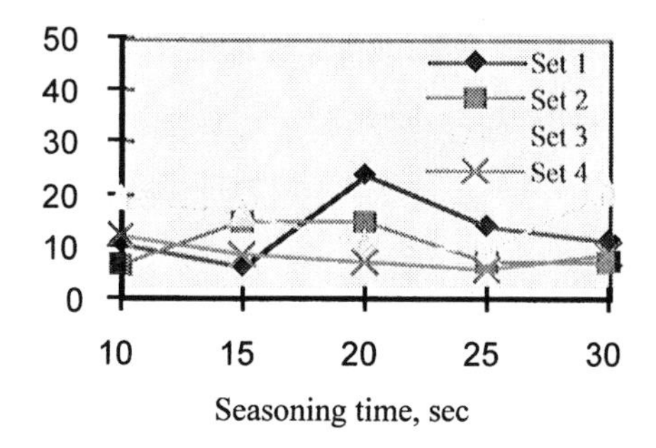

Fig. 2. Variation of Fe concentration (x E10/cu.cm) in the film with seasoning time.

PREMETAL DIELECTRIC

The current industry standard for premetal dielectrics includes an LPCVD $TEOS/O_3$ oxide liner deposited at about 750 °C, followed by a CVD $TEOS/O_3$ BPSG layer deposited at about 550 °C. The CVD deposited $TEOS/O_3$ BPSG layer absorbs moisture, and thus requires a high temperature anneal at 900 °C.

Both LPCVD $TEOS/O_3$ layer and CVD $TEOS/O_3$ BPSG provide an excellent gap fill for high aspect ratio gaps. However, the use of $TiSi_2$ silicide requires a thermal budget no more than 750 °C in order to maintain the stability of silicides and avoid agglomeration of the silicide. When the device geometry further shrinks to sub-0.25 μm geometry, non-traditional silicides such as $CoSi_2$ and NiSi is desirable over $TiSi_2$ in order to further minimize the sheet resistance variation on the narrow poly lines. These silicides require even a lower thermal budget.

The superior gap fill capability and low deposition temperature makes HDPCVD technology an excellent choice as the premetal gap fill dielectric. To avoid any gate oxide damage effects by high density plasma on the thin gate oxide, a PECVD nitride liner is deposited prior to the deposition of HDPCVD oxide [2]. The PECVD liner serves as a protection layer for the gate oxide during the HDPCVD deposition, also as an impurity barrier layer to protect active devices.

In the PMD layer, HDPCVD oxide thickness is typically about 3000-4000 Å. The aspect ratio is low, and hence a higher deposition to sputter ratio (DSR) can be used. It is also important to avoid corner clipping of SiN barrier layer. The lower bias component to achieve higher deposition to sputter ratio also helps in avoiding corner clipping problem. As the DSR is higher, the beneficial affect of ion bombardment on the film quality is reduced. To compensate for this, the helium back side cooling can be turned off so that the deposition takes place at a higher temperature improving the film quality.

As the gate is protected from any metallic contamination from the HDPCVD oxide deposition, seasoning time of about 10 sec, as in STI application, to deposit the oxide on the chamber walls is adequate. After deposition of the HDPCVD film a PECVD oxide was deposited as a CMP sacrificial layer to take advantage of the existing PECVD equipment to achieve better throughput.

INTERMETAL DIELECTRIC AND PASSIVATION

Damage to MOS gate oxide during wafer fabrication can lead to high gate leakage or premature oxide breakdown when the devices are used. Some studies have identified the damage mechanism to be charging of gate oxide during plasma processing. A non-uniformity in the plasma causes an imbalance in the ion and electron currents, locally charging the oxide surfaces until either the surface potential increases, bringing the local sheath conduction currents into balance, or an oxide current via Fowler-Nordheim (F-N) tunneling balances the difference locally. The F-N tunneling current causes oxide damage. The denser plasma in the HDPCVD process compared to a conventional PECVD process implies larger ionic and electronic currents to the surface of the wafer. Therefore, any non-uniformity in the HDPCVD processes would result in increased charging damage to the gate oxide when compared to conventional PECVD processes.

If the deposition takes place at high bias power the exposed metal lines would result in charging of the metal layers and the resulting current flow may lead to device damage. In our experiments we found that an initial protection layer of about 1000 Å deposited at low power will be adequate to prevent the charging of the metal layer (Fig. 3). The aspect ratios are also most severe in the IMD layers and HDPCVD is originally developed for the gap fill problem of the IMD layers. Therefore, the throughput is limited by the necessity to have a large sputter component. A two step process can be implemented wherein after the deposition of the initial 1000 Å thick protection layer at low bias power to prevent charging of the exposed metal lines, high bias power can be used to fill the gap. Once the gap fill is achieved the process can be changed to a high deposition to sputter ratio process to deposit the rest of the thickness at higher deposition rate. It is also important to form a protection layer to avoid corner clipping. The corner clipping (Fig. 4) of the metal layers resulting from the bombardment of the metal interconnect corners can result in incorporation of the metal atoms into the growing oxide film. This would result in higher metal to metal leakage.

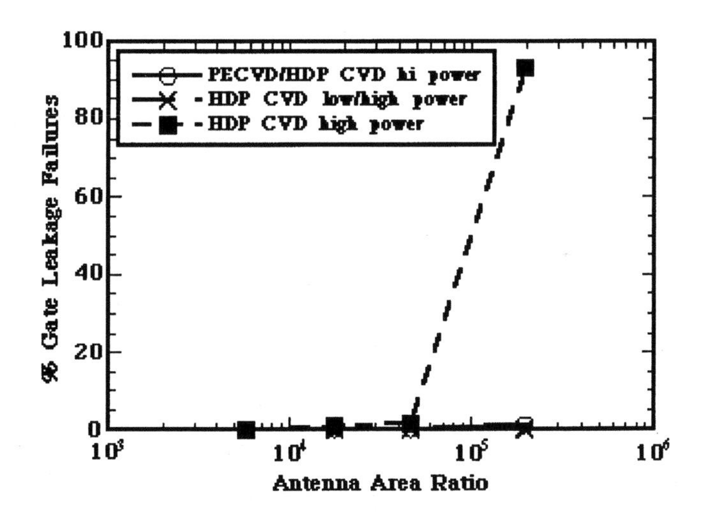

Fig. 3. Gate leakage failures Vs. antenna area ratio for three different deposition conditions.

GENERAL CONSIDERATIONS

Hazing and blistering of the film

Film hazing (Fig. 5) in HDPCVD films may result from the insufficient heating of the wafer, large SiH_4 bursts at the start of deposition or Argon trapping in the film. Large SiH4 bursts and low wafer temperature result in a highly silicon rich film being deposited initially. This film can release large quantities of hydrogen during subsequent deposition and may form hydrogen bubbles. Thus it is important to sufficiently heat up the wafer before starting the deposition. The SiH_4 bursting problem can be avoided by emptying the SiH_4 that may have accumulated after the MFC at the end of the plasma clean step and/or by replacing if the MFC is found to be leaky. The hazing may also result from the non reactive Argon trapping in the film, if only argon is used during the heat up step. This may be alleviated by using Argon and Oxygen together in the heat up step.

Refractive Index and dielectric constant

Oxygen to silane flow ratio of above 1.5 would result in completely oxidized film with refractive index (RI) below 1.47. A ratio within the range of 1.5 to 1.8 would be ideal to achieve a dielectric constant close to that of thermal oxide. Higher ratios result in excessive silanol incorporation and lower ratios would result in excessive hydrogen incorporation both resulting in an increase in dielectric constant. Below a ratio of 1.4 the RI increases sharply and operation in the region should be avoided as any slight changes in the flows of oxygen and silane would result in unstable RI.

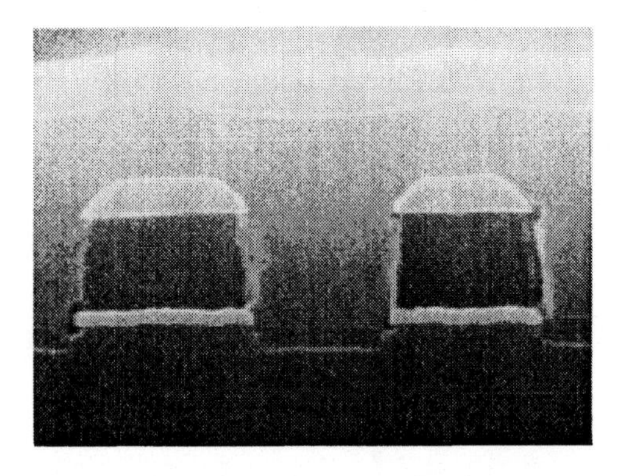

Fig. 4. SEM cross section showing corner clipping of metal for unoptimized process.

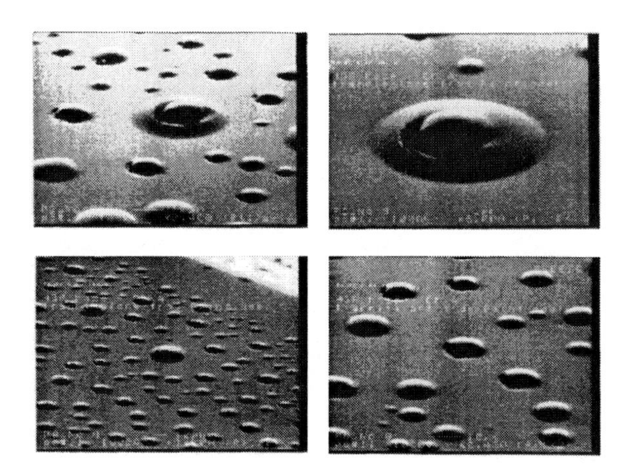

Fig. 5. SEM picture showing film hazing.

CONCLUSIONS

For STI application the films should have seam free gap fill, wet etch rate closer to thermal oxide and avoidance of the corner clipping of the nitride. To achieve these goals a two step process is preferred. For premetal dielectric the aspect ratio is low and hence a high throughput process compatible with salicide process can be implemented. For IMD and passivation gap fill, plasma damage and leakage currents are important and the process needs to be optimized taking into account these considerations.

REFERENCES

1. Ed Korczynski, Solid State Technology, April 1996, p. 63
2. L. Q. Qian, Subhas Bothra et al., 14th International VMIC, June 10-12, 1997.

ZINC INFLUENCE ON COPPER SOLDERING AND PASSIVATION

P.L. Cavallotti*, L. Magagnin*,
V. Sirtori** and F. Zambon**

* Dip. Chimica Fisica Applicata - Politecnico Milano
Via Mancinelli, 7 - 20131 Milano - Italy
** IBM Semea
Via Lecco, 61 - 20059 Vimercate (MI) - Italy

A treatment for the copper substrate of an electronic board to improve the adhesion with solder alloys deposits was developed. The treatment is based on a flash deposition of zinc onto copper substrate. Protection of zinc flash on copper by organic compound was studied and formation of Zn(II)-Benzotriazole steady complex was highlighted. Inhibition of zinc corrosion was obtained as a consequence of the interaction between zinc and BTA, which forms an adsorbed organic film on zinc surface.

Keywords: Adhesion, Benzotriazole, Copper, Corrosion, Soldering, Wettability, Zinc.

1 - Surface treatment for copper substrate

Zinc is known to be very harmful if present in the tin alloy, being preferentially oxidized and giving large quantities of drosses when the solder is stirred. In these studies we examined the importance of the zinc presence at the interface between copper and solder alloy, searching at limiting the influence of zinc to this region.
The use of tin base solder alloys is common in electronic applications, particularly in the manufacture of printed circuit boards (PCB), for the assembly of components onto the boards, providing mechanical and electrical connection. The tin solder alloy are also used to join integrated circuit chips to chip carriers to substrates, and to join circuitization lands and pads in multilayer circuits boards.
Tin-Lead (Sn-Pb) are most used alloys for electronic soldering operations; they were selected for their mechanical strength, low relative cost, electrical conductivity and excellent wetting characteristic; they have also a low melting temperature, that is important in electronic applications, because many components and PCBs use materials easily damaged by exposure to high temperature during manufacture or assembly.
Lead replacement is an actual problem, because of lead health hazard, being toxic for workers and for environment; recently governments have begun to urge the electronic industry to find viable alternatives, in order to reduce the lead exposure of the electronic industry worker and to reduce the amount of lead waste going back into the environment. Lead presence in the soldering alloys is particularly critical in the case of application for manufacturing the last generation of C-MOS; in fact the details are so fine in this case, that the emission of α particles from the emitting radioisotope present in lead can provoke serious problems to the device.
Tin-Bismuth (Sn-Bi) solder alloys were investigated as alternative to Sn-Pb solder alloys. The Sn-Bi electrodeposition is known in the art; electrodeposition with alkyl-sulphonate was patented for electrodeposition onto PCB.
The lead-free alloys known in the art exhibit poor soldering and mechanical properties, that is small peel strength and creep resistance. Also the shear strength of Sn-Bi alloys is normally reported to be smaller than that of Sn-Pb alloys, although the difference in this case is not so high.

Experimentals

Chemicals of reagent grade and double distilled water were used. Solder alloys were electrodeposited from commercial baths and thick deposits, in the range 50-100 μm, obtained. The deposit were of uniform and near eutectic composition, what is important for the reflowing operation.
The binary Sn-Bi solder alloys properties of electrodeposits were improved by adding a zinc salt in the electrolyte (UK Patent). The addition to the bath of zinc methane-sulphonate salt, with zinc content in the range 5-30 g/l, maintaining all other parameters constant, influenced to a great extent the bath behaviour, although zinc was not discharged onto the cathodic surface.
Electrodeposition of a flash zinc layer was done from basic Zn electrolytes on the copper substrate after normal cleaning (5 min in 90 g/l Sodium Persulfate solution); composition was:

Zinc	10 g/l
Iron	0.08 g/l
Caustic Soda	180 g/l

with operation conditions: T = 35 °C; current density 1 A/dm^2; with deposition time from 30 s to 5 min.

Surface deposit morphology and cross section structures were examined after etching with SEM and analysed with EDS ZAF corrected on a Cambridge Stereoscan 250MKII microscope. Deposit microhardness was taken with a Knoop microindenter on a Shimadzu HMV2000 instrument. ESCA-XPS measurements were carried out on PHI5600 with a monochromatic Al Kα X-ray source (1486.67 eV). The pass energy was 29.35 eV, sputtering rate 2 nm/min, the detector/sample angle was 45°.

Mechanical properties were determined by shear and peel tests. The soldering alloys was deposited on a 20x20 mm square area of a rectangular Cu foil, thickness > 50 μm, and overimposed on Sn coated rectangular Cu foils, of 10 mm width, with opposite ends. Reflowing of the joints was made in an industrial oven at 250 °C with N_2 atmosphere.

The loading was made with an Instron machine in controlled conditions: crosshead speed 5 mm/min, humidity 55 %, T = 21 °C, sampling rate 10 points/s. The peeling length was established according to the run, in the range corresponding to a constant peeling load.

Zinc is known to be very harmful if present in the tin alloy; but it is also known that diffusion of Sn in brass, with intermetallic formation, is limited with respect to tin diffusion in pure copper. The zinc presence, at the interface solder/copper, increases the wettability of the tin alloys, improves the peel strength and decreases the growth of the Cu-Sn intermetallic layer with time.

The contact angle of a reflown Sn-Bi electrodeposit on the Cu foil of a PCB, compared with that of a reflown Sn-Pb electrodeposit obtained in similar conditions and with that of Sn-Bi after Zn pretreatment, shows for all electrodeposits good wettability, although Sn-Pb behaves better than Sn-Bi and with Zn pretreatment a slight increase of this property is seen.

Hardness measurements have been carried out on the cross section of thick electrodeposits (load time 10 s) after reflowing. The results are reported in Table 1 and show typical very low hardness values, with an hardness increase for Sn-Bi with respect to Sn-Pb and with an enhancement with the Zn pretreatment.

Table 1: Knoop microhardness of reflown Sn alloys thick electrodeposits on Cu.

Load	Sn-Pb	Sn-Bi	Sn-Bi/Zn pretreat.
5 g	16.5	21.3	25.3
50 g	10.9	13.2	16.8

To evaluate the Sn-Bi soldering alloy we also examined the fracture surfaces of thick electrodeposits on Cu foil without and with Zn pretreatment. Presence of two zones is observed: a Sn rich zone with ductile behaviour and a Bi rich zone with brittle behaviour.

The coherence of the two zones is better after the Zn pretreatment.

Sample of soldered joints were shear tested. The soldering alloys were electrodeposited for 50 min. Results are reported in Table 2, showing that Sn-Bi/Sn joints have a shear strength higher than Sn-Pb/Sn joints and that Zn pretreatment increases this property.

Table 2: Shear strength of Sn solder alloys with Sn on Cu sheets.

Solder alloy	Shear strength
Sn-Pb on Cu sheet	27.5 MPa
Sn-Pb on Zn pretreated Cu sheet	28.3 MPa
Sn-Bi on Cu sheet	28.6 MPa
Sn-Bi(Zn) on Zn pretreated Cu sheet	32.3 MPa

We compared the peeling strength, measuring on a free wheeling rotary test fixture several joints with different soldering. Table 3 reports the results, showing a great increase of the peel strength for the Sn alloys/Cu joints after the Zn pretreatment: the strength is almost doubled for Sn-Bi alloys, rendering them a viable alternative to Sn-Pb alloys. Also the peeling strength of Sn-Pb alloys is significantly increased with the Zn pretreatment, although to a lesser extent.

Table 3: Peel strength of Sn solder alloy (on Cu)/Sn (on Cu) joints.

Sample	Peel strength	Standard deviation
Sn-Pb	2.21 N/mm	± 0.70
Zn (30 s) + Sn-Pb	2.51 N/mm	± 0.22
Sn-Bi	0.55 N/mm	±0.063
Zn (30 s) + Sn-Bi	1.06 N/mm	±0.32
Zn (30 s) + Sn-Bi(Zn)	1.11 N/mm	±0.18
Zn (150 s) + Sn-Bi(Zn)	0.88 N/mm	±0.046

The peeling strength value shows a maximun when the zinc deposit thickness is in the range of 100 nm. Micrographic features of the Sn-Bi/Sn surfaces show void presence at the fracture surface of Sn-Bi/Sn joints, a uniform distribution of small Bi crystals for the Zn pretreated Cu samples and small surface features for Sn-Bi deposits from baths containing zinc, as reported in figure 1 a-b.

Sn-Bi alloys show increased interdiffusion coefficients with respect to Sn-Pb alloys (data are for liquid alloys, but behaviour of solid alloys is similar); thus, the intermetallic layer formation is an increased problem for these alloys.

A very important effect of the zinc pretreatment of the copper surface is observed with regard to the aging behaviour of the coating.

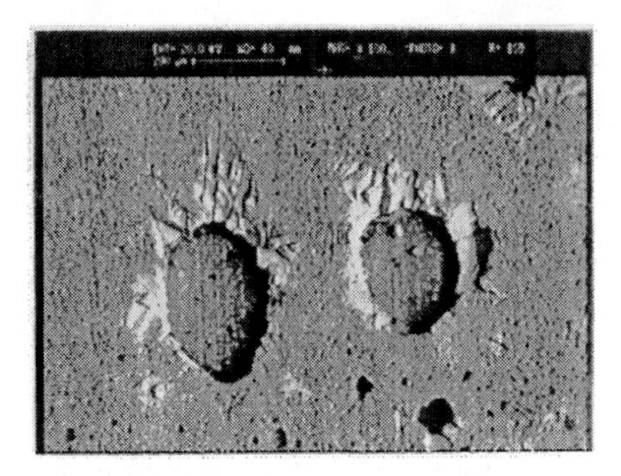

Fig. 1a: Sn-Bi alloy electrodeposit on Cu sheet after peeling test.

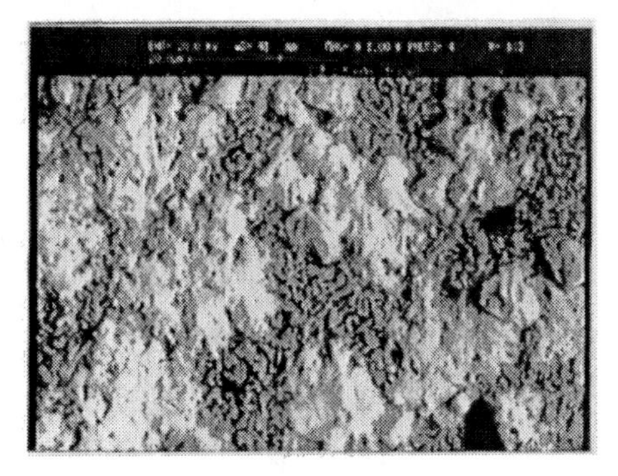

Fig. 1b: Sn-Bi alloy electrodeposit on Cu sheet pretreated with Zn after peeling test.

After soldering the tin alloy with copper at 250 °C, in an industrial oven with N_2 atmosphere, an intermetallic Cu-Sn layer is observed at the interface between the copper substrate and the soldering Sn alloy. This layer is greater when copper is pretreated with the zinc deposit. Problems at the interface can occur with maintenance time if the intermetallic layer increases too much, because the layer is brittle and adhesion can decrease, if the joint is fatigue stressed. We have examined the results obtained after artificial aging of samples, maintaining them at 150 °C, 55 % humidity, for several days. The results are reported in figure 2 and show that, while the growth of the layer was almost linear with the aging time without the pretreatment, it became parabolic with the zinc pretreatment.
After one month of artificial aging the SnBi/Zn/Cu joint had a CuSn intermetallic layer thickness lower than that of the SnBi/Cu joint, comparable with that of the SnPb/Cu joint. The minimum tendency to grow was shown by the SnBi/Zn/Cu joint.

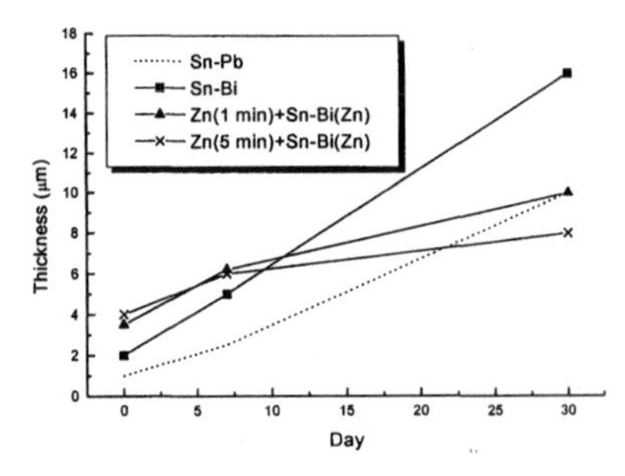

Fig. 2: Intermetallic layer thickness at the Sn solder alloy/ Cu interface.

Sn-Pb deposits show a not uniform intermetallic layer, fractured after aging. Sn-Bi deposits show a uniform layer, strongly growing with aging time. Sn-Bi deposited on Zn pretreated Cu shows a large intermetallic layer after reflowing, a possible consequence of the high diffusion coefficient of these alloys, but also the lowest increase of the layer with aging, as a consequence of the high deposit uniformity, as reported in figure 3 a-b.
The intermetallic thickness of as soldered Sn-Bi on Zn pretreated Cu is higher than that of as soldered Sn-Bi directly on Cu, but it does not affect greatly the mechanical properties, which depend mostly on the void presence at the interface and wettability problems as already shown. Of course, this result can explain the better peel strength of Sn-Pb with respect to all Sn-Bi layers, although the mechanical properties of the bulk Sn-Pb alloys resulted to be less than those of bulk Sn-Bi alloys. It could be possible to increase the peel strength of Sn-Bi on Zn pretreated Cu by decreasing the reflowing temperature. The decrease of the layer growth with Zinc additions gives a very interesting advantage to the so obtained layers.

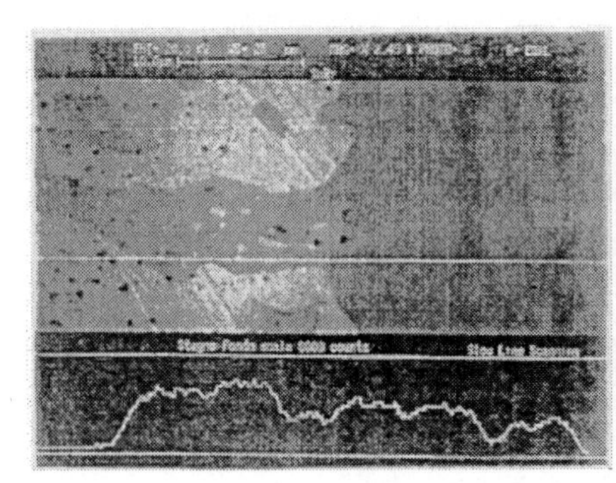

Fig. 3a: Sn-Bi electrodeposit on Cu sheet, with tin line profile, after soldering and 30 days aging.

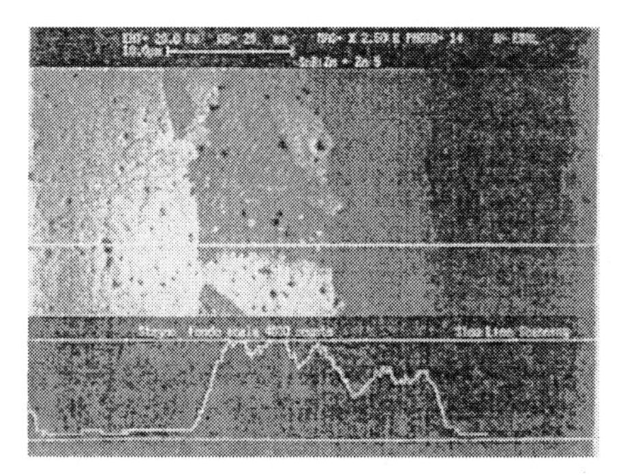

Fig. 3b: Sn-Bi electrodeposit on Cu sheet pretreated with Zn, with tin line profile, after soldering and 30 days aging.

A careful XPS-ESCA study was carried out on thin films of Sn-Bi eutectic alloy electrodeposited onto Cu foil, without and with Zn pretreatment (layer thickness < 0.1 μm), in particular examining the valence band region after various sputtering times. Fig. 4 shows the valence band region of a Sn-Bi/Zn/Cu layer sputtered for different times; the composition changes from outside, where Zn is maximum (Cu 1.5 % - Zn 76.5 % - Sn 22 % at), with Zn decrease and Cu increase (from Cu 66 % - Zn 20 % - Sn 14 % at to Cu 76 % - Zn 8 % - Sn 16 % at and Cu 81 % - Zn 5 % - Sn 14 % at). The states with lowest binding energy are normally attributed to 4s electrons of Cu, the following peak at about 3.6 eV to 3d electrons of Cu, the peak at about 11 eV to the 3d electrons of Zn. From the shoulder at low energy an influence of the Zn presence is seen with a density of states increase.

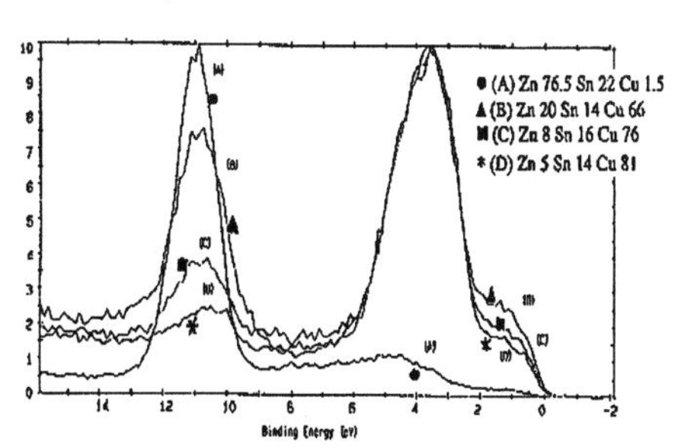

Fig. 4: XPS spectra of valence band for reflown Sn-Bi deposit on Zn pretreated Cu.

We also examined the behaviour of bulk metals: Zn, Cu and Sn (Fig. 5 shows the valence band region for bulk metals). The Cu 3d peak is shifted in the CuZnSn electrodeposited ED layer increasing from 2.5 to 3.6 eV, and also the Zn peak is shifted from 10 to 11 eV; this can be attributed to orbital hybridization.

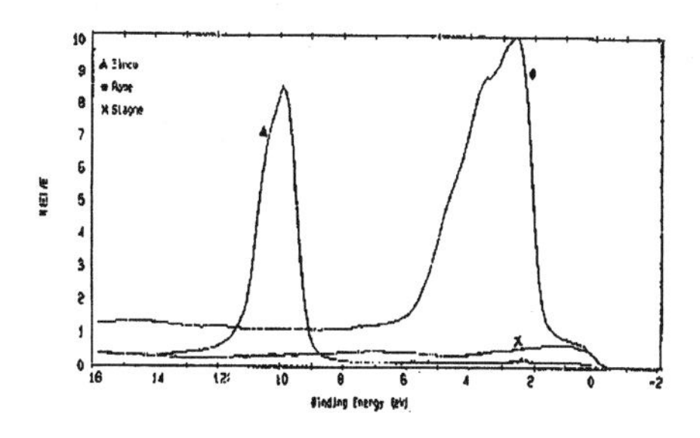

Fig. 5: Spectra of valence band region for bulk metals: Cu, Zn and Sn.

A comparison of the CuZnSn ED layer with bulk brass shows that the shift of the two peaks is quite different, a little more for the Cu 3d peak and much more for the Zn 3d peak, while the density of states at low energy is greatly increased for CuZnSn ED. ED Sn influence on the Cu density of states is much less, the Cu 3d peak is shifted only to 3.0-3.3 eV and the density of states near the Fermi energy is only slightly increased.
Comparison of CuZnSn ED layer with bulk brass and CuZn ED layer XPS spectra in fig. 6 shows the great increasing influence of Zn and Sn interaction on the density of states near the Fermi energy.
This is also supported from the comparison of ED CuZnSn, CuZn and CuSn layers, the behaviour of the ternary alloy is quite different from the two binaries at binding energy near to the Fermi level, with a higher number of states in this region.

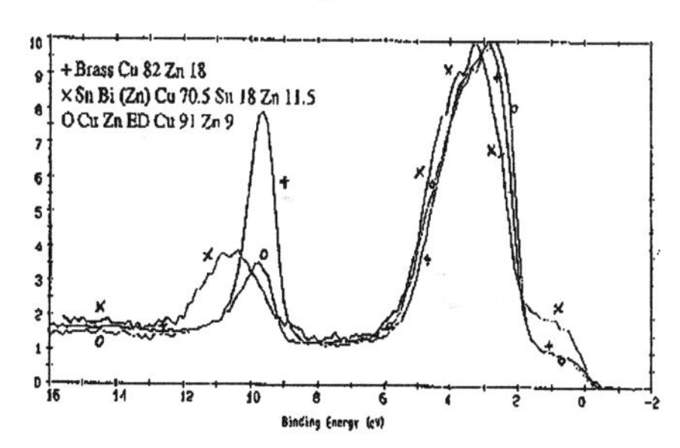

Fig. 6: XPS spectra of valence band for Sn-Bi on Zn pretreated Cu compared with CuZn.

2 - Zinc and Tin interaction with Benzotriazole

Zinc and Benzotriazole

Oxidation of Zinc pretreatment surface on copper can be prevented using organic coatings. Our studies indicate that benzotriazole forms a steady complex Zn(II)-BTA and as a consequence an adsorbed film on zinc surface is formed, like on pure copper in BTA solution.

A XPS-ESCA study was carried out to investigate the formation of a complex and electrochemical tests were done to measure corrosion inhibition of BTA adsorbed on zinc surface.

Experimentals

A solution of isopropylic alcool and 100 g/l of benzotriazole was prepared, using chemicals of reagent grade. Specimens of zinc electrodeposited on copper and of zinc bulk were pickled in diluted sulphuric acid, washed in distilled water and immersed in the solution for t = 1 min at T = 40 °C, rinsed with isopropylic alcool and dried in a nitrogen flux. ESCA-XPS measurements were carried out on PHI5600 with a monochromatic Al Kα X-ray source (1486.67 eV). The pass energy was 29.35 eV, sputtering rate 2 nm/min, the detector/sample angle was 45°.
Potentiodynamic runs were carried out with an EG&G 273A potentiostat-galvanostat computer controlled (scanning rate 0.5 mV/s) with SoftCorr™III software and an Ag/AgCl reference electrode was used. Corrosion tests were made in a solution containing:
Sodium Chloride 35 g/l
Boric Acid 20 g/l
pH = 4.5 with Hydrochloric Acid or Sodium Hydroxide.

Adsorbed Zn(II)-BTA film on zinc surface is confirmed by XPS data: on zinc sample, pretreated with BTA solution, considerable atomic concentration of Nitrogen on surface is always present. No detectable amount of Nitrogen was found on surface of zinc samples without BTA pretreatment. Nitrogen on zinc surface is a proof that adsorbed film has been formed. Table 4 reports XPS atomic concentrations obtained:

Table 4: XPS atomic concentrations on surface for zinc bulk without and with BTA solution pretreatment.

Element	**ZINC BULK** (% at.)	**ZINC BULK + BTA** (% at.)
C 1s	61.79	61.95
O 1s	26.65	16.61
Zn 2p3	11.56	8.32
N 1s	/	13.11

A careful study of photopeak and valence band region was carried out in order to verify the existence of a steady complex and the interaction between zinc and benzotriazole. Zinc 2p3 photopeak was considered and deconvolution of peak, as reported in figure 7, was made. Perfect agreement between measured peak and calculated convolution was obtained only considering three peaks (as in table fig. 7): one relative to pure Zn and one relative to Zn/Oxygen interaction; the third peak is attributed to Zn/Nitrogen interaction. Peak for Zn-H_2O is considered.

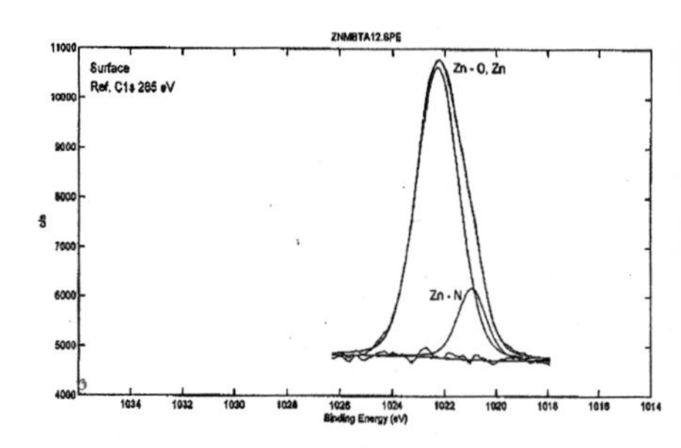

Peak	Binding energy (eV)	FWHM	Area (%)
Zn - N	1021.01	1.48	20.22
Zn - O, Zn	1022.29	1.86	79.78
ChiSquared = 1.201			

Fig. 7: Zinc 2p3 photopeak on surface and deconvolution for massive zinc pretreated with BTA solution. Table reports peaks data.

Also deconvolution in two peaks of Nitrogen 1s photopeak on surface for massive zinc pretreated with BTA was obtained: one peak was attributed to BTA Nitrogen and the other was attributed to Nitrogen/Zinc interaction, as reported in figure 8a.
Figure 8b shows a comparison between Nitrogen 1s photopeaks of copper treated with Entek 56 (commercial BTA solution) and of massive zinc treated with benzotriazole solution. Overlapping of the two peaks confirm that interaction BTA-copper is comparable to interaction BTA-zinc.

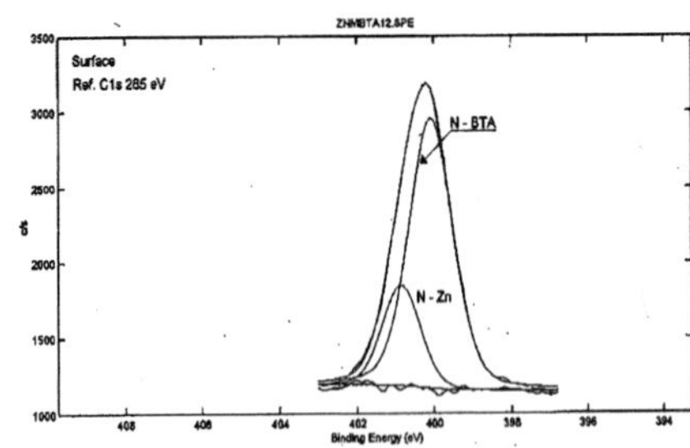

Peak	Binding energy (eV)	FWHM	Area (%)
N - BTA	400.08	1.34	76.76
N - Zn	400.86	1.16	23.24
ChiSquared = 1.288			

Fig. 8a: Deconvolution of Nitrogen 1s photopeak on surface for massive zinc pretreated with BTA and peaks data.

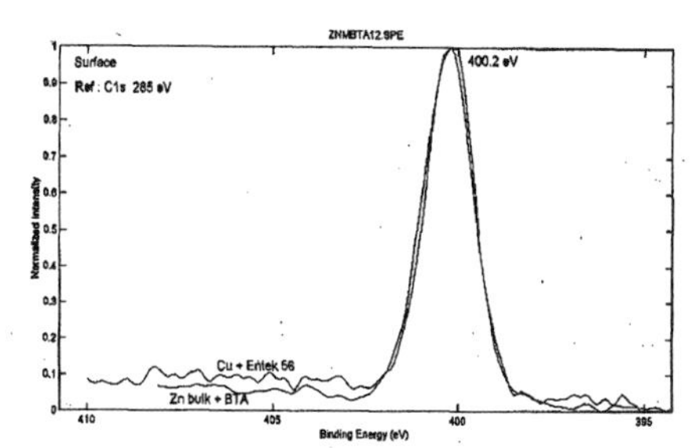

Fig. 8b: Nitrogen 1s photopeak for copper treated with Entek 56 and for massive zinc pretreated with BTA.

Behaviour of Auger LMM peak of zinc is altered by the presence of BTA: Auger peak for massive zinc is similar to Auger peak of zinc oxidized and considerably differs from pure Zinc Auger peak, as it can be seen in figure 9.

A study of valence band region was carried out and results are reported in figure 10. Peak attributed to 3d electrons of Zinc is shifted at about 9 eV by the pretreatment with BTA solution and the density of states near the Fermi energy is much less for Zinc treated than that of pure Zinc.

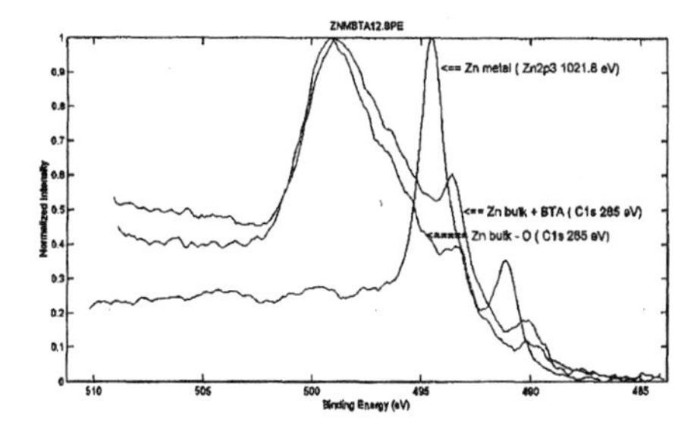

Fig. 9: Auger LMM peak for pure zinc, oxidized zinc and zinc with BTA treatment.

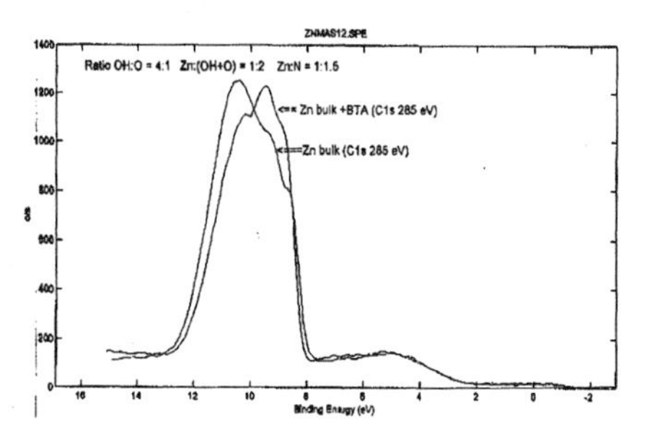

Fig. 10: Valence band spectra on the surface for massive zinc without and with BTA pretreatment.

In fig. 10 are also reported the calculated ratios of atomic concentrations of Oxygen present as oxide (ZnO) and as hydroxide ($Zn(OH)_2$ + H_2O Adsorbed). Ratio between atomic concentration of Zn, present as oxide and hydroxide, and atomic concentration of Oxygen, present as oxide and hydroxide and ratio between atomic concentration of Zn, which interacts with Nitrogen, and atomic concentration of Nitrogen, which interacts with Zinc, are shown.

Corrosion inhibition of zinc by surface adsorbed film of Zn(II)-BTA was investigated with electrochemical methods. Potentiodynamic curves were run on specimens of massive and electrodeposited zinc without and with BTA pretreatment and corrosion current densities were calculated measuring the polarization resistance. In table 5 the corrosion current are reported and figures 11 a-b show the potentiodynamic curves.

Table 5: Corrosion currents for massive zinc and ECD zinc without and with BTA treatment.

SAMPLE	CORROSION CURRENT
Zn	316.7 μA/cm^2
Zn + BTA	191.7 μA/cm^2
Cu + Zn ECD	175 μA/cm^2
Cu + Zn ECD + BTA	82.2 μA/cm^2

With BTA pretreatment of zinc, the corrosion current is half that of the zinc without pretreatment, showing that passivation occurs as a consequence of the Zn(II) complexation with benzotriazole. Also corrosion potential is affected by BTA treatment and an increase is observed for samples treated with benzotriazole: much more for massive zinc respect to ECD specimens.

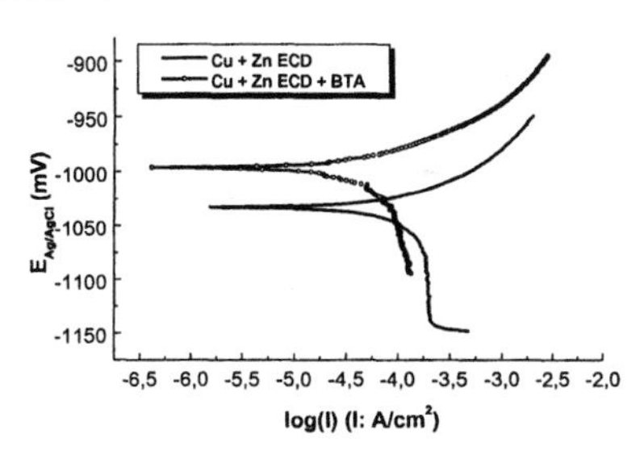

Fig. 11a: Potentiodynamic curves for ECD zinc without and with BTA.

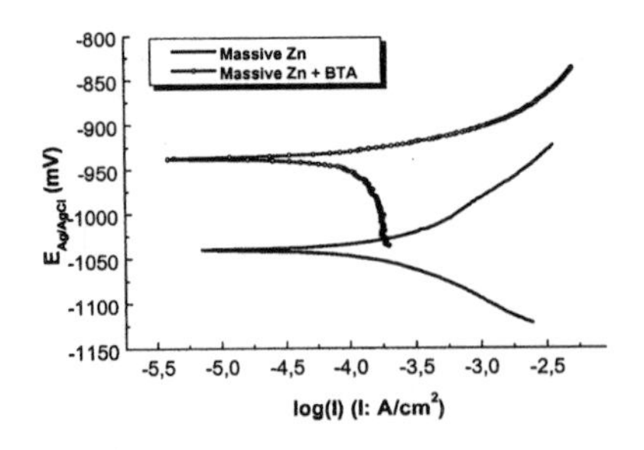

Fig. 11b: Potentiodynamic curves for massive zinc without and with BTA.

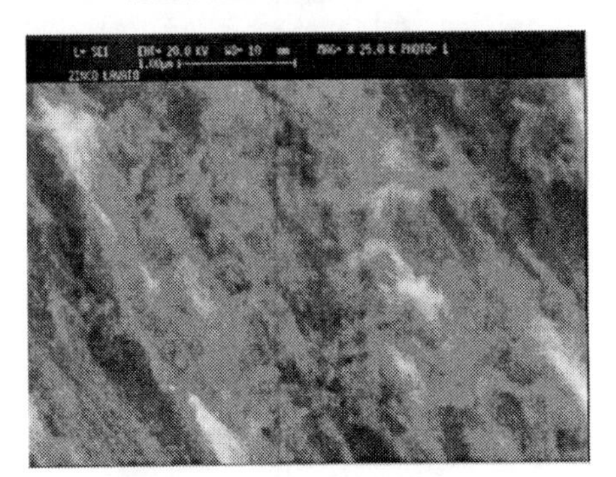

Fig. 12a: Massive zinc surface after pickling in sulphuric acid.

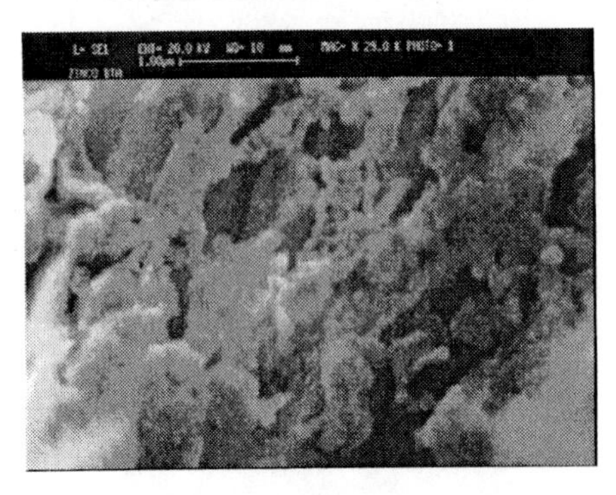

Fig. 12b: Massive zinc surface after cleaning and pretreatment with BTA solution.

Figures 12 a-b show how benzotriazole forms peculiar structure on zinc surface. Same structures can be seen on copper surface or on brass surface after treatment with BTA (Fig. 13).

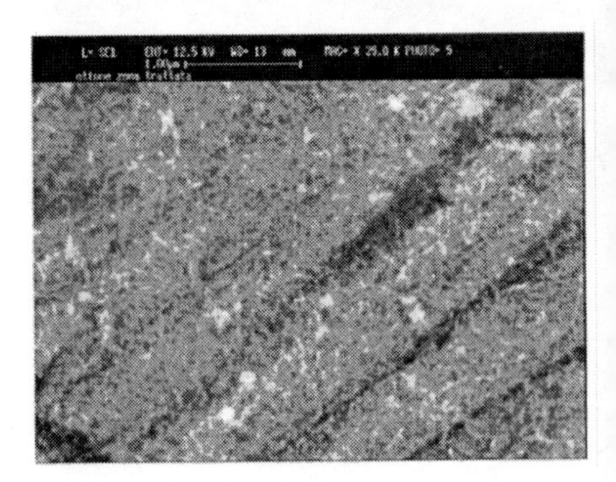

Fig 13: Brass surface after pretreatment with BTA.

Tin and Benzotriazole

In order to inhibit corrosion of tin and tin-zinc alloy, possibility of interaction between tin and benzotriazole was studied.

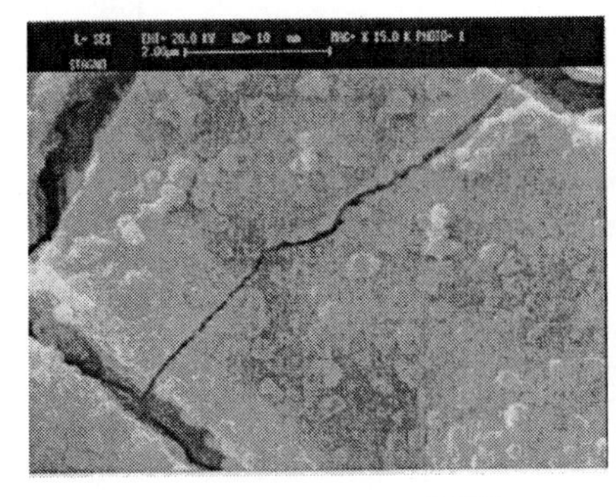

Fig. 14: Tin surface after treatment with BTA.

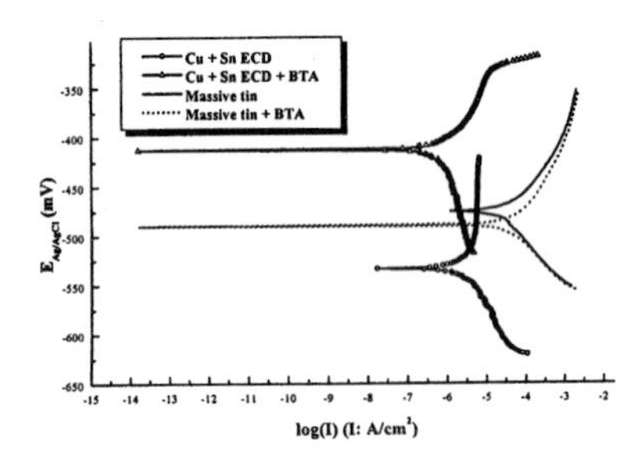

Fig. 15: Potentiodynamic curves for tin samples.

From XPS analyses on massive tin, no amount of nitrogen on surface was detected and calculated corrosion currents don't differ between samples treated with BTA or not treated. Only an irregular variation of

corrosion potential is shown in figure 15.
BTA seems not to be able to form a steady adsorbed film on tin surface, as shown in figure 14. Studies on tin-zinc alloy deposits show that BTA treatment is able to increase the wettability of the surface, measured with a wetting balance (with eutectic molten Sn-Pb alloy).

3 - Conclusions

The addition of zinc at the interface Cu/Sn alloy presents several interesting features: a decrease of the Cu and Sn interdiffusion and of the intermetallics formation at the interface; an increase of the wettability of the Cu surface by the tin alloy; an increase of the mechanical properties of the joint, in particular of the peel strength. This behaviour can be related to the different interactions in the metallic couples: Zn/Cu, Sn/Cu and Sn/Zn.
When Sn alloy are soldered to Cu, Sn interaction with Cu prevails. In fact, Sn has a negative enthalpy for forming an alloy with Cu; Zn behaves similarly, whilst Bi and Pb have positive alloy formation enthalpies (J/mole), as estimated from phase diagram behaviour and with the semiempirical method developed by Miedema. Thus, a Cu atom interacts preferentially with Sn and Zn, in the order; Sn and Zn show a positive interaction enthalpy and the atoms of the two metals prefer to cluster. A very thin layer onto copper gives an interface where Sn diffusion is not stopped, maintaining wettability, but is decreased, for the repulsive effect between Sn and Zn. Zn presence at the interface can even increase wettability, for its protective action towards oxidation.
Zinc can be protected from corrosion by a surface adsorbed film of benzotriazole. In fact BTA is able to form, like with pure copper, a steady complex Zn(II)-BTA. As a consequence, passivation and reduced corrosion of zinc result. No interaction exists between tin and BTA, but benzotriazole on tin-zinc alloy increases wettability.

References

1- P.L. Cavallotti, G. Zangari, V. Sirtori, "Electrodeposition of Sn-Bi(Zn) Solder Alloy for PCB Manufacture", ISHM Meeting, Venezia 1997
2- P.L. Cavallotti, G. Zangari, V. Sirtori, UK Pat.Appl.9608660.8 (26/4/96)
3- P.L. Cavallotti, G. Zangari, V. Sirtori, UK Pat.Appl.9608665.7 (26/4/96)
4- Y.N. Sadana, R.N. Gedye, S. Ali, Surf.Coat.Tech., 27, 151-166 (1986)
5- A.K. Suri, S. Banerjee, "Materials Science and Technology" Vol. 8 Ch. 1, VCH, N.Y., 1996
6- Wilson, US Pat. 5,039,576 (13/8/1991)

AUTHOR INDEX

SUBJECT INDEX